D1122377

PLUMBER'S HANDBOOK

Revised Edition

Howard C. Massey

Craftsman Book Company
6058 Corte del Cedro/ P.O. Box 6500 / Carlsbad, CA 92018

Acknowledgements

The author expresses his sincere thanks and appreciation

■ to Mr. A. T. Strother, Executive Director of the Plumbing Industry Program of Miami, Florida, for granting permission to use excerpts from his article "A History of Plumbing" which appeared in *The Florida Contractor Magazine*.

■ to the Florida Energy Committee and the Environmental Information Center of the Florida Conservation Foundation, Inc. for pertinent information necessary to authenticate the thermosiphon and pumped solar water heating system.

■ to American Standard for providing the plumbing fixture roughing-in measurements (as illustrated in Chapter 21).

■ to Josam Manufacturing Company for providing the plumbing fixture carriers (as illustrated in Chapter 21).

To my wife, Hilda, for her untiring assistance in editing and typing this manuscript.

Library of Congress Cataloging-in-Publication Data

Massey, Howard C.
   Plumber's handbook / by Howard C. Massey. -- Rev. ed.
      p.   c.m.
   Includes index.
   ISBN 1-57218-056-0
   1. Plumbing--Handbooks, manuals, etc.   I. Title.
TH6125.M37   1998
696'.1--dc21   98-18088
                                                        CIP

First edition ©1978 Craftsman Book Company
Second edition ©1985 Craftsman Book Company
Third edition ©1998 Craftsman Book Company

Contents

1

Plumbing and the Plumber

If you've chosen plumbing as your profession, you should find it one of the most challenging and satisfying of all construction trades. The many variations in design, layout, and installation methods present a challenge to any competent professional plumber. But notice that word *competent*. If you don't have a good knowledge of practical plumbing methods and of the minimum requirements of modern plumbing codes, you're going to be discouraged, frustrated, and confused.

Learning plumbing from a code book is a very difficult task. That's the reason for this manual. It's intended to help you grasp the important design and installation principles recognized as essential to doing professional-quality plumbing work. What you learn here should be applicable nearly anywhere in the U.S., regardless of the model code adopted by your jurisdiction. And if you're just learning the fundamentals of plumbing, you'll find this book much easier than reading and understanding the code.

Remember, however, that this book is not the plumbing code. All plumbers will have to refer to their local code from time to time. I'll emphasize the minor variations in model plumbing codes throughout this book, so you should easily recognize them as you read and compare sections of this book with your local code. *But the basic principles of sanitation and safety remain the same, regardless of the geographical location.*

The History of Plumbing

The art and science of plumbing came into being as mankind struggled against disease. The history of civilization is the history of plumbing. At the dawn of civilization, when two or three families gathered together to make a tribe, people drank from springs and streams. They made no provisions for the disposal of sewage and garbage. We can assume that when their site became fouled with kitchen refuse and human waste, they just moved on. If disease killed members of the tribe because they neglected the laws of sanitation, they didn't understand the cause and effect. They didn't know that lack of cleanliness breeds disease.

Archeologists, while digging in various parts of the world, have confirmed that even ancient civilizations developed plumbing systems for protecting health. At Nippur, in Babylon, archeologists uncovered an aqueduct made of glazed clay brick that dates back to 4,500 B.C. This aqueduct contained three lines of glazed clay pipe. Each section was 8 inches in diameter and 2 feet long, with a flanged mouth. Other excavations have revealed glazed clay pipe in jar patterns, concave and cone shapes and a sewage system complete with manholes.

On the island of Crete, some of the palaces of ancient kings were equipped with extensive water supply and drainage systems. The glazed clay pipe was found to be in perfect condition after 3,500 years.

Archeologists even discovered evidence of plumbing fixtures constructed of hard clay.

In ancient Greece, further advances were made in cleanliness. Greek aqueducts took pure water from mountain streams into cities. Sewers, which exist to this day, carried away waste to the surrounding rivers. They understood that bathing was a desirable habit. Greeks portrayed Hygeia, the goddess of health (from whose name we get the word "hygiene"), as supplying pure water to a serpent, the symbol of wisdom.

The ancient Egyptians also realized the value of sanitation. Moses was acquainted with the sanitary science of the Egyptians and used it in framing the code of laws found in the book of Leviticus.

The Romans in the time of Julius Caesar developed the principles of sanitation to a high art. Unlike the ancient Greeks and Egyptians, they were familiar with lead, which they imported from the British Isles. They called it *plumbum*. The word *plumbing* is derived from the Latin word for a worker in lead. The Romans used lead in many of the same ways we use it today.

Two thousand years ago the city of Rome had an adequate water supply and sewage disposal system. Water was piped from hills and mountains 50 miles distant from the city. To bring this water into Rome, great overhead aqueducts and underground tunnels were built of masonry. Branch lines carried water into the homes of the upper class for private bathrooms long before the development of the great public baths. Some baths in Pompeii had floors and walls of marble, with brass, bronze and silver fixtures.

From as far back as 600 B.C. Rome had an elaborate drainage system called the *Cloaca Maxima*. This main was 13 feet in diameter and was joined by many laterals. It was constructed from three concentric rows of enormous stones piled one on the top of another without cement or mortar. It still exists and is used today in the drainage system of *modern* Rome.

When Rome set out to conquer the world, they took their bathing habits with them. In what is now Great Britain, in the city of Bath, archeologists uncovered a Roman bath 110 feet long and 68 feet wide.

In the 12th century, trade guilds were first organized in England. The first apprenticeship laws were passed in 1562 during the reign of Queen Elizabeth. These laws required an apprenticeship of seven years and made apprenticeship in all crafts compulsory. It was not until 1814 that the compulsory clause was removed and apprenticeship was made voluntary. The first known master plumbers' association was organized in England and incorporated in the College of Heralds of London.

With the discovery of the New World, man, like his ancient ancestors, sought to escape the dark and dirty cities of Europe for a fresh campground.

Although America has become a symbol of high standards in plumbing and sanitation, progress in the early development of sanitation and plumbing was very slow. As the population of the early settlements increased, conditions deteriorated. Garbage and sewage dumped onto the ground and seepage from earth-pit privies polluted nearby wells.

Health conditions became so intolerable that eventually public sewers had to be installed underground and extended to each building. Although New York in 1782 installed the first sewer under the streets, Chicago is credited with having the first real city sewage system, constructed in 1855.

Plumbing as we know it today traces its roots back many centuries, but was not really perfected until the twentieth century. Many older Americans, reared without indoor plumbing, still remember the open well, the pitcher pump, the outhouse, and the Saturday night romp in the old wooden tub. The modern bathroom, city water, and the sewers of today are taken for granted. But don't forget that plumbers protect the health of our nation and the world.

Designing Drainage Systems

The private sanitary drainage system is the essential part of most plumbers' work. It includes all the pipes installed within the wall line of a building on private property for the purpose of receiving liquid waste or other waste substances (whether in suspension or in solution), and the pipes which convey this waste to a public sewer or a private, approved sewage disposal system.

You have to install the drainage system so it's not a health or safety hazard. Most municipal authorities have adopted codes to protect the public health. And since the majority of requests for clarification and resolution before any Board of Rules & Appeals center on the drainage system, it's vital that you understand this section of your code.

Although sanitary drainage and vent arrangements are the heart of the plumbing system, most experts agree that this is the most complex, misunderstood, and misinterpreted section of the code. Engineers, plumbers and plumbing officials frequently disagree about its intent and interpretation. That's why it's not surprising that most of the questions (and the isometric drawings) in the journeyman's and master's examinations are taken from this section of the code.

Although plumbing installation details may vary, the basic principles of sanitation and safety remain the same, wherever you work. You shouldn't have any trouble recognizing any minor changes from the basic rules I'll describe here as you read and compare this book with the code used where you work.

Isometric Drawings

Before you can understand drainage and vent systems, you must be familiar with isometric drawings. Isometric drawings are the way plumbing professionals communicate with each other. They're used by the plumbing contractor to estimate the cost of new work and to show the job foreman how to rough-in a particular job. Anyone who deals with plumbing work must be able to make and interpret isometric drawings. Although it may look difficult, with a little study you'll soon find that it's easy to read and make isometric illustrations.

You have only three basic angles to illustrate in a plumbing system: the horizontal pipe, the vertical pipe, and the 45-degree angle pipe. The only pieces of equipment necessary are a sharp No. 2 pencil and a 30-60-90 right triangle. If you follow the directions and practice the exercises in this chapter, you'll be able to produce your own isometric projections and understand those made by others.

Practicing Isometric Drawings

First, draw a circle with a dot in the exact center. Place the letter N at the top of the page to designate the direction of north. See Figure 2-1. In the top half of the drawing, the solid lines indicate the angles you'll need to illustrate the sanitary system. The horizontal broken line is just a constant to make the angles clear.

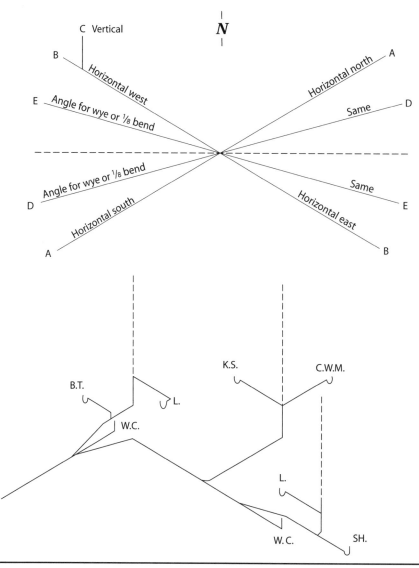

Figure 2-1
Isometric drawing illustrating the various angles

Square the short base of the triangle with the right edge of your paper and draw line A through the center dot. (This represents the north-south horizontal pipe.)

Again using the triangle, square the short base with the left edge of your paper and draw line B through the center dot. (This represents the east-west horizontal pipe.)

To draw line C (the vertical line), square the short base of the triangle with the lower edge of the paper and use the long base as a straightedge. Connect line C with the desired horizontal line. Line C represents the stacks in your isometric.

To find where to place line E, divide the area equally between the horizontal line B and the horizontal broken line. Then draw line E east and west through the center dot. (This shows the change in direction at the 45-degree fittings of either a wye or $^1/_8$ bend.) The same procedure will, of course, yield D north and south. The lower portion of Figure 2-1 shows a simple isometric drawing that includes all three basic angles you use to design rough plumbing for any building.

Take time to make sure you understand this process. You can't become proficient in the plumbing trade until you learn how to draw isometric layouts correctly.

Fittings Within an Isometric Drawing

The lines on isometric drawings represent pipe and fittings. Symbols show the location of the fixtures. The symbols you use are the same regardless of the type of pipe used.

Figures 2-2, 2-3, and 2-4 show typical isometric drawings and the no-hub pipe fittings they represent. I've numbered each fitting in these drawings to correspond with the drawing of the same fitting. Look at Figure 2-2. You'll see that the "horizontal twin tap sanitary tee" (also known as an "owl fitting") is fitting number 14 in the isometric drawing. This fitting per-

mits two similar fixtures to connect to the same waste and vent stack at the same level. In this case it connects two lavatories.

Fixture Abbreviations, Definitions and Illustrations

There are several common abbreviations used to identify various types of plumbing fixtures in isometric drawings and floor plans. For example, some writers will use the letter *L* to designate a lavatory. Others

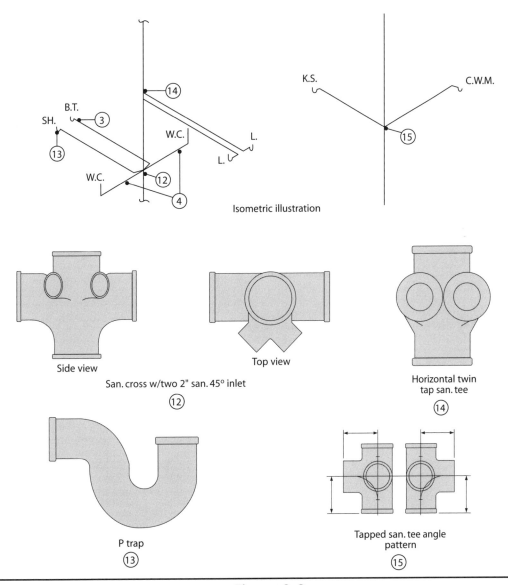

Figure 2-2
Fittings illustrated with isometric drawings

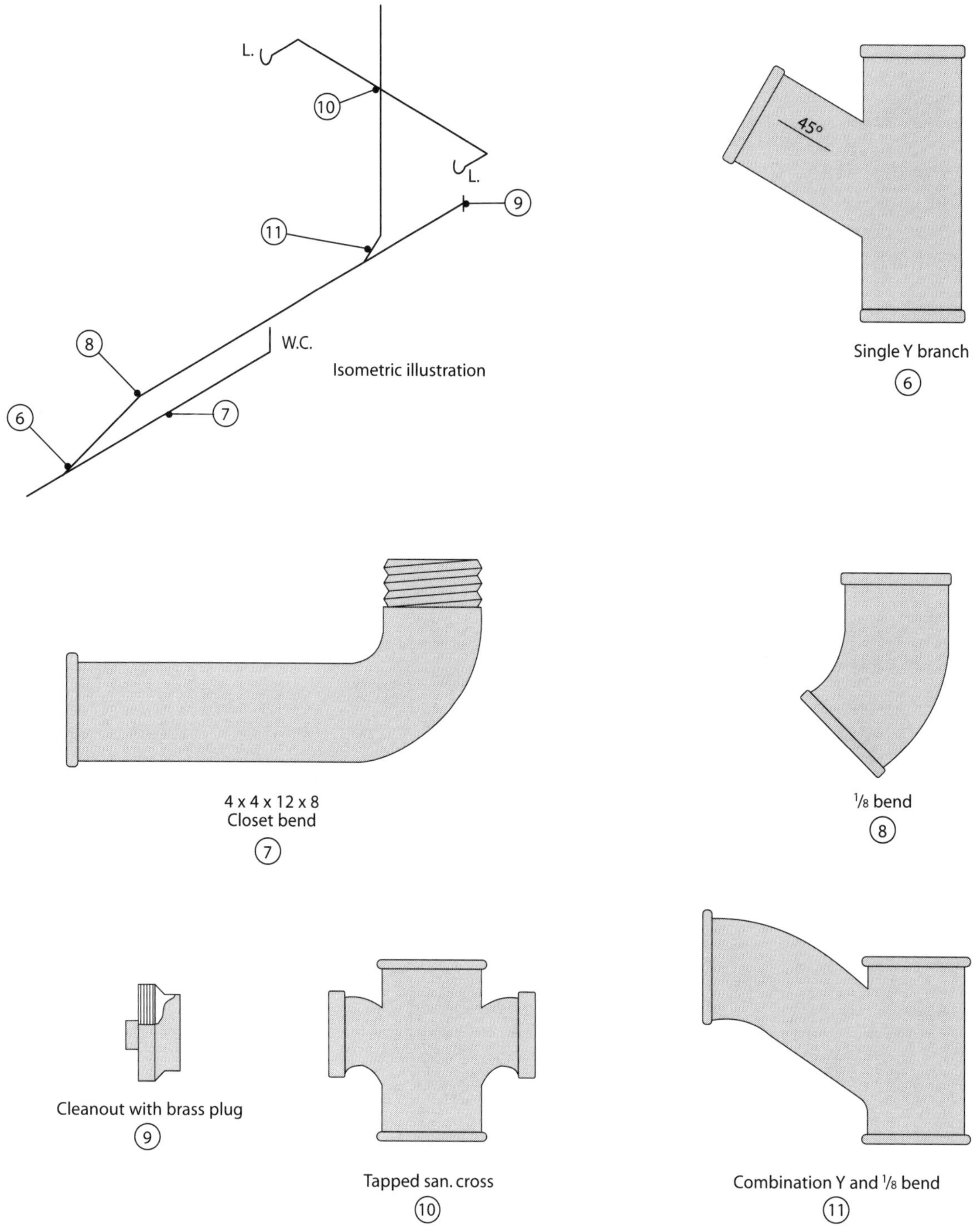
L.
L.
10
9
11
W.C.
Isometric illustration
8
6
7
45°
Single Y branch
6
4 x 4 x 12 x 8
Closet bend
7
1/8 bend
8
Cleanout with brass plug
9
Tapped san. cross
10
Combination Y and 1/8 bend
11

Figure 2-3
An isometric drawing and its fittings

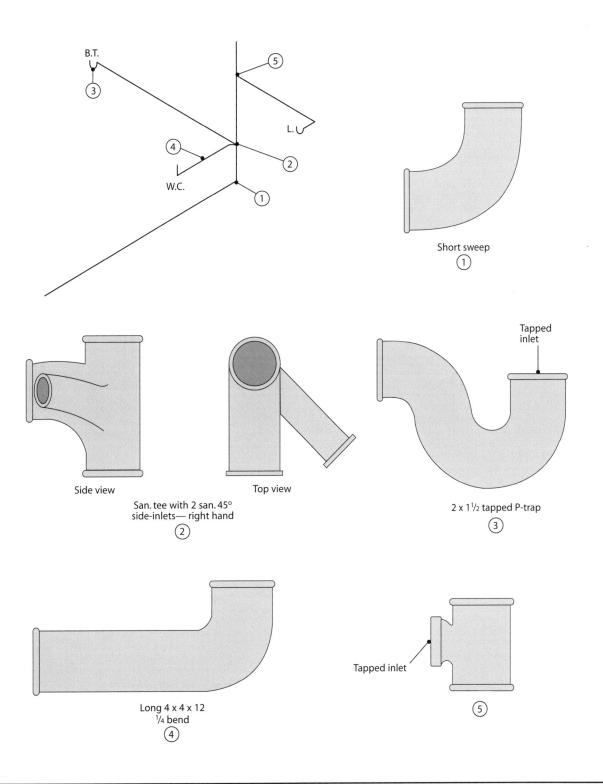

B.T.

③

④

W.C.

⑤

L. ∪

②

①

Short sweep
①

Side view

Top view

San. tee with 2 san. 45°
side-inlets— right hand
②

Tapped
inlet

2 x 1½ tapped P-trap
③

Long 4 x 4 x 12
¼ bend
④

Tapped inlet

⑤

Figure 2-4
Another isometric drawing and its fittings

may use *LAV*. I'll use the abbreviations in Figure 2-5 to identify plumbing fixtures in isometric drawings and floor plans in this book.

One of the first steps in becoming a plumber is learning to identify the basic piping arrangements as defined in your local code. Without this knowledge, you won't be able to design, lay out or install pipes and

Plumbing fixtures	Abbreviations
Bathtub	B.T.
Cleanout	CO
Clothes washing machine	C.W.M.
Kitchen sink	K.S.
Lavatory	L.
Shower	SH.
Water closet	W.C.
Vent through roof	V.T.R.

Figure 2-5
Typical plumbing fixture abbreviations

fittings. Figures 2-6, 2-7 and 2-8 show three sanitary isometric drawings which include all the major parts of basic drainage and vent systems. The illustrations identify the parts of the system with symbols, and each includes a legend showing what the symbols mean.

Figure 2-6 is a sanitary isometric drawing of a typical two-bath house, including a kitchen and utility room. It shows an installation on the flat connected to a public sewage system. Figure 2-7 is a typical one-bath house with a kitchen and utility room. The installation is on a stack system connected to a private sewage disposal unit (a septic tank). Installation on the flat requires less height than a stacked system. Figure 2-8 shows a typical battery of plumbing fixtures often found in a two-story public building.

I've made every effort to use simple, easily-understood language to clear up the complex and seemingly-contradictory wording of the code. The code may identify a particular section of pipe by several different terms with various definitions. The definitions we'll use for the parts of a drain, waste and vent system are on page 14. You'll find each of them illustrated in at least one of the isometric drawings.

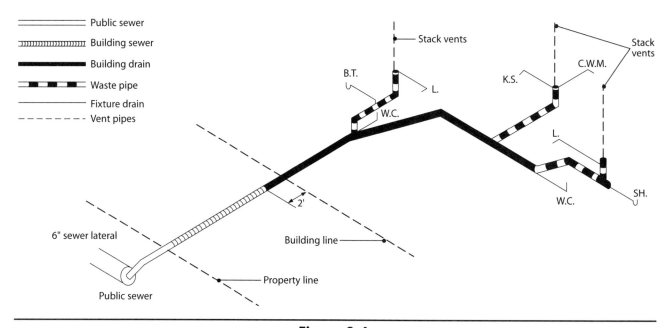

Figure 2-6
Isometric drawing of a two-bath house illustrated in graphic symbols

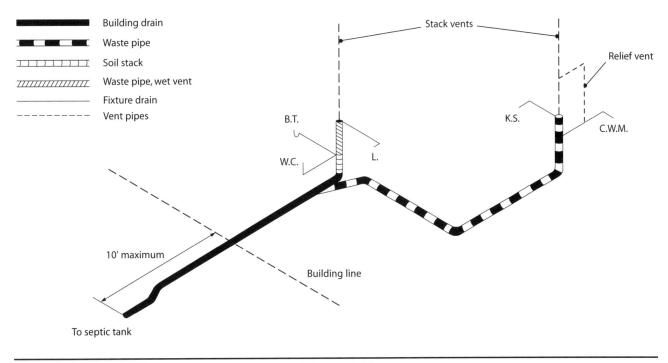

Figure 2-7
Isometric drawing of a one-bath house

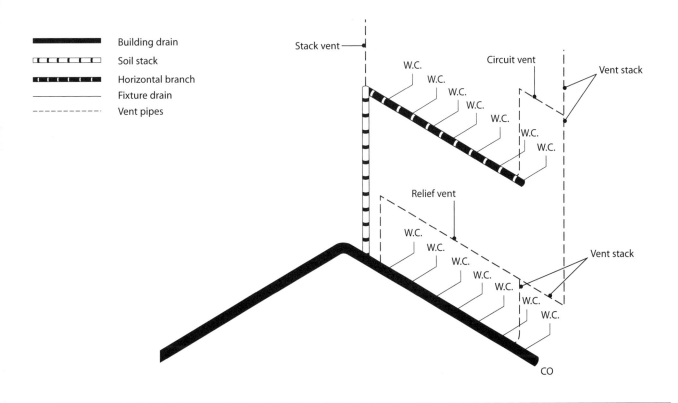

Figure 2-8
Isometric drawing of a two-story public building

Public Sewer

A public sewer may also be called a *municipal sewer*. This sewage collection system is located in a street, alley or a dedicated easement adjacent to each parcel of privately-owned property. Public sewers are common pipes installed, maintained and controlled by the local authorities, paid for through some form of taxation.

Most public sewers include a 6-inch sewer lateral from the main several inches past the property line of each lot. This makes it easy for individual property owners to connect to the system. When a homeowner or plumbing contractor pulls a permit for a sewer connection, they can check with the local municipal engineering department for the depth and location of the lateral.

Building Sewer

A building sewer is the part of the main horizontal drainage system that conveys sewage or other liquid waste from the building drain to the public sewer lateral. The building sewer begins at its connection to the 6-inch lateral a few inches within the property line. It terminates at its connection to the building drain 2 feet (more in some codes) from the outside building wall or line.

It's also known as a *private sewer*, since it's not controlled directly by the public authority. It's installed and maintained by the individual property owner.

Another name for the building sewer is a *sanitary sewer*, because it carries sewage that doesn't contain storm, surface, or ground water.

Building Drain

The building drain is the main horizontal collection system, exclusive of the waste and vent stacks and fixture drains. It's located within the wall line of a building. It carries all sewage and other liquid wastes to the building sewer, which begins 2 feet (more in some codes) outside the building wall or line.

It's also considered a *main*, since it acts as the principal artery to which other drainage branches of the sanitary system may be connected.

Fixture Drain

The fixture drain is the part of the drainage system that includes the pipe from the fixture trap to the vent serving that fixture. It may connect directly to a vertical vent stack above the floor or, in the case of a shower or bathtub, to the horizontal wet vent section beneath the floor. It's often referred to by plumbers as the *sink arm* or *lavatory arm*.

Waste Pipe

The waste pipe is the drainage pipe that carries liquid waste (but no fecal matter) from a fixture drain to the junction of any other drainpipe. In the code it's also called *liquid waste*, because it doesn't carry waste from water closets or bed pan washers. Another name is *wet vent*, since it carries liquid waste from plumbing fixtures (excluding water closets) to the building drain and serves as a vent for these fixtures as well.

Soil Stack

A soil stack is the vertical section of pipe that receives the discharge of water closets, with or without the discharge from other fixtures. Then it conveys this waste, usually to the building drain.

The *branch interval* performs the same function as the soil stack and becomes an integral part of the soil stack. The only difference is in its vertical height. It usually corresponds to a story height but it can never be less than 8 feet in length. Stacks also include any vertical pipe, including the waste and vent piping of a plumbing system.

Horizontal Branch

A horizontal branch is the part of a drainpipe that extends laterally from a soil or waste stack and receives the discharge from one or more fixture drains.

A more detailed and complete section on abbreviations and definitions is included in the Appendix at the end of the book.

How to Size the Drainage System

When you size individual pipes (vertical or horizontal) in a drainage system, there's one major determining factor: the maximum fixture unit load. But you also have to consider the types of fixtures used, the slope of the drainpipe, and the vertical length of drainpipes.

Your code book contains tables which list the various fixture load values. Use them to compute the total fixture load for any kind of plumbing system. In this book, I'll provide typical tables that we'll use in sizing the examples in this chapter. The first one is Figure 2-9, which gives the fixture units per fixture for both the *UPC* and the *SPC*. It includes the most common residential fixtures.

The residential or commercial fixture units may vary between codes, but in most cases that doesn't affect the pipe sizes. Let's consider just one variation. With a 1¹/₂ inch trap, the *UPC* counts 3 fixture units for a bathtub (with or without an overhead shower). Both the *SPC* and the *IPC* count 2 fixture units for this particular fixture.

Now look at Figure 2-10. Use Figure 2-9 to tabulate the total fixture unit load for each pipe size in the drawing, using both the *UPC* and the *SPC*. Of course, Figure 2-9 is just a sample. Your code book includes a complete table of fixture unit values and trap sizes for residential and commercial fixtures. Figure 2-11 shows my tabulation for the plan in Figure 2-10.

Special Fixtures

Figure 2-12, *Special fixtures* (identified in some codes as *Fixture unit equivalents* or *Fixtures not listed*), is for fixtures that aren't included in Figure 2-9. These are fixtures, equipment or appliances with an intermittent flow and an indirect connection to the drainage system, usually found in commercial buildings. Some of these special fixtures are drinking fountains, bottle coolers, milk or soft drink dispensers, ice making machines and coffee urns.

Figure 2-12 is based on maximum drain or trap size up to 3 inches and the fixture unit values in the *UPC* and *SPC*. Let's use it to size a special fixture to illustrate. A coffee urn comes from the manufacturer equipped with a ¹/₂- or ³/₄-inch drain from the drip

Fixture	Standard Plumbing Code		Uniform Plumbing Code	
	F.U. value as load factor	Minimum trap size (in)	F.U. value as load factor	Minimum trap size (in)
Bathtub (with or without overhead shower)	2	1¹/₂	3	1¹/₂
Shower stall, residential	2	2	2	2
Lavatory, residential	1	1¹/₄	1	1¹/₄
Water closet, private	4	3	3	3
Kitchen sink, residential	2	1¹/₂	2	1¹/₂
Clothes washing machine	3	2	3	2
Note Fixture units and trap sizes may vary from those listed above. Check local code requirements.				

Figure 2-9
Fixture units per fixture

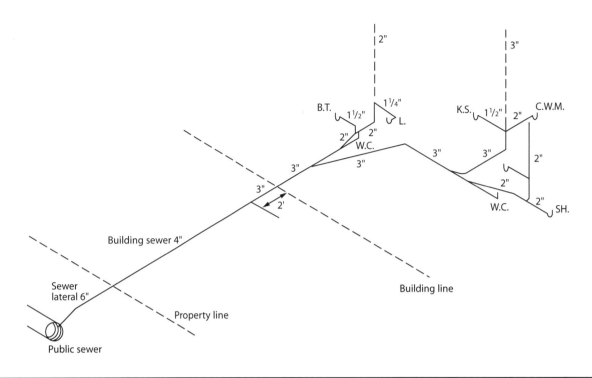

Figure 2-10
Plumbing drainage system illustrated and sized on the "flat"

Fixture type and number	Standard Plumbing Code Total F.U.	Uniform Plumbing Code Total F.U.
1 bathtub (with or without overhead shower)	2	3
1 shower	2	2
2 lavatories	2	2
2 water closets	8	6
1 kitchen sink	2	2
1 clothes washer	3	3
Total	**19**	**18**

Figure 2-11
Cumulative fixture unit load for Figure 2-10

Standard Plumbing Code		Uniform Plumbing Code	
Fixture drain or trap size (in)	Fixture unit value	Fixture drain or trap size (in)	Fixture unit value
1¼ and smaller	1	1¼	1
1½	2	1½	3
2	3	2	4
2½	4		
3	5	3	6

Note Fixture units vary considerably between codes using the same size traps as listed above. Check local code requirements.

Figure 2-12
Special fixtures

pan. It has a waste drain smaller than $1^1/_4$ inches. You can see in Figure 2-12 that it's rated as 1 fixture unit. You'll need the table from your local code when sizing any building drainage system.

Continuous and Intermittent Flow Devices

Some devices (pumps, sump ejectors, air conditioning equipment and similar devices) discharge into a drainage system with a continuous flow. The code allows 2 fixture units for each gallon per minute of flow. The manufacturer usually provides the flow rate information.

Figure 2-13 shows intermittent flow rates up to 50 gallons per minute (based on the *UPC*). You'll have to check with your local building code for flow rates exceeding 50 gallons per minute. Some codes don't address intermittent flow rates.

If you list the number and types of fixtures and refer to these three tables, you'll be able to calculate the maximum fixture unit load for any building. Your code also has additional tables, footnotes and subsections for sizing vertical drainage, horizontal drainage and vent pipes, based on fixture units and other conditions. Model codes vary greatly in the way they assign fixture units. You'll see what I mean when you compare the *SPC* in Figures 2-14 and 2-15 with the *UPC* in Figure 2-16.

Uniform Plumbing Code	
GPM	**Fixture unit value**
Up to $7^1/_2$	1 F.U.
8 to 15	2 F.U.
16 to 30	4 F.U.
31 to 50	6 F.U.

Flow rates exceeding 50 GPM are determined by the administrative authority. Check local code requirements.

Figure 2-13
Discharge capacity in GPM for intermittent flow only

Standard Plumbing Code			
Maximum number of fixture units that may be connected to any portion[1] of the building drain or the building sewer[2]			
Diameter of pipe (inches)	**Fall per foot**		
	$1/_8"$	$1/_4"$	$1/_2"$
2	—	21	26
$2^1/_2$	—	24	31
3	20[3]	27[3]	36[3]
4	180	216	250

[1]Not over 2 water closets or bathroom groups can connect to any building drain less than 3" in diameter
[2]The minimum size of a building sewer is 4" in diameter
[3]A maximum of 2 water closets
Note Pipe size, fall per foot and fixture units may vary between codes. Check local code requirements.

Figure 2-14
Building drains, horizontal branches and sewers

Standard Plumbing Code				
Maximum number of fixture units that may be connected to:				
Diameter of pipe[4] (inches)	**Any horizontal fixture branch[1,4]**	**One stack of 3 stories in height or 3 intervals**	**More than 3 stories in height**	
			Total for stack	**Total at 1 story or branch**
$1^1/_4$	1	2	2	1
$1^1/_2$	3	4	8	2
2	6	10	24	6
$2^1/_2$	12	20	42	9
3	20[2]	30[3]	60[3]	16[2]
4	160	240	500	90

[1]Does not include branches of the building drain
[2]Not over 2 water closets
[3]Not over 6 water closets
[4]A branch drain or stack can't be smaller than 3" when serving a water closet

Figure 2-15
Horizontal fixture branches and stacks

Uniform Plumbing Code						
Diameter of pipe (in)	**1¼**	**1½**	**2**	**2½**	**3**	**4**
Maximum units drainage piping[1]						
Vertical	1	2[2]	16[3]	32[3]	48[4]	256
Horizontal	1	1	8[3]	14[3]	35[4]	216
Maximum length drainage piping						
Vertical (ft)	45	65	85	148	212	300
Horizontal (ft)	—	—	—	—	—	—
Vent piping, horizontal and vertical, maximum units	1	8[3]	24	48	84	256
Maximum lengths (ft)	45	60	120	180	212	300

[1]Excluding fixture drain
[2]Except sinks, urinals or dishwashers
[3]Except 6-unit traps or water closets
[4]Maximum of 4 water closets or 6-unit traps permitted on any vertical stack. Maximum of 3 water closets or 6-unit traps allowed on any horizontal branch drain or drain pipe.
[5]Based on ¼ inch fall per foot. For pipe fall at ⅛ inch per foot, multiply horizontal fixture units by factor 0.8 and add to allowed fixture units.

Note The diameter of an individual vent pipe should not be smaller than 1¼ inches. Individual vent piping should never be smaller than one-half the diameter of the drain pipe to which it is connected. Fixture unit load values for drainage and vent piping shall be taken from Figures 2-9 and 2-11, this book. Up to one-third of the total allowed length of any vent may be installed horizontally. When the entire length of a vent pipe is increased by one pipe size, the maximum length limitations listed in this figure do not apply.

Figure 2-16
Maximum fixture unit loading and maximum length of drainage and vent piping

Figure 2-14 (*Building drains, horizontal branches and sewers*) lists the pipe size, fall (slope) per foot and maximum fixture units for each horizontal pipe. The most generally-accepted fall per foot for horizontal pipe is ¼ inch, though ⅛ inch is also usually acceptable. Sometimes you can use falls of ¹⁄₁₆ or ½ inch per foot for difficult installations involving special conditions, but you should get prior approval from your administrative authority. (The *UPC* has no separate building drain or sewer table.)

You'll need Figure 2-15 to compute the fixture units, pipe size and permitted length of soil and waste stacks for multistory buildings.

Figure 2-16 is all-inclusive. It lists the pipe size, ¼ inch fall per foot (⅛ inch is also acceptable, see footnote 5), and maximum fixture units for horizontal and vertical drainage pipe for single or multistory buildings.

Use Figures 2-14 through 2-16 *only* for the examples in this book — not on your jobs. They include only the pipe sizes illustrated in Figure 2-12. Your code book will include larger sizes.

Figures 2-17 and 2-18 are identical layouts for a five-story building. Figure 2-17 follows the *SPC* while Figure 2-18 follows the *UPC*. As we size them, you can compare the maximum fixture units under the two codes.

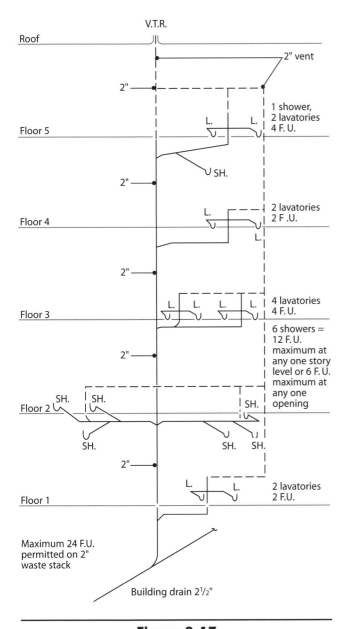

Figure 2-17
*Plans showing maximum 24 F. U.
on 2" waste stack (SPC)*

Figure 2-18
*Plans showing 2½" waste stack
having 24 F. U. (UPC)*

Standard Plumbing Code

Use Figure 2-15 for this example. Look at line three, a 2-inch waste pipe. The maximum number of fixture units for a 2-inch stack is 24 (column 4). Column 5 shows that total fixture units at any one floor opening carrying waste into this 2-inch stack can't exceed 6. Remember, this is a *waste stack*, not a *soil stack*. This

means, according to our definition, that this waste pipe, because of its size, conveys *only liquid waste not containing fecal matter.*

Now turn back to Figure 2-9 (*SPC*) and note the fixture units given for each type of plumbing fixture. Figure 2-15, line 3, shows that you can install 12 showers (total 24 F.U.) on this 2-inch stack at *different* floor levels. Or you can install, at any *one* story level, three showers (total 6 F.U.) on a branch drain connected to this 2-inch stack. You can install 24 residential lavatories (total 24 F.U.) on this 2-inch stack at *different* floor levels, or six lavatories (total 6 F.U.) on a branch drain connected to this 2-inch stack at any *one* story level.

Now size the building drain on the basis of a $1/4$-inch fall per foot. In Figure 2-14, first column, find your 2-inch pipe diameter. Line one shows it can carry up to 21 fixture units. Since you need 24 fixture units, a 2-inch horizontal waste pipe won't be large enough. However, line two shows that the next larger size, $2^1/2$ inches, will accommodate the 24 fixture units for this building.

The showers and lavatories have the same fixture unit values in both codes. You may wonder why there's this difference in pipe sizes since the layout and fixture unit values are identical. The major difference between codes is in *how* the fixture units and other restrictions are applied to a plumbing system.

Uniform Plumbing Code

Now let's size the waste stack and building drain in Figure 2-18 under the *UPC*. Using Figure 2-16, look at the *Vertical* line in column 3. We have a 2-inch vertical drainage pipe (waste stack). Its maximum load is 16 fixture units with some restrictions (see the footnotes). Notice that a 2-inch pipe is too small. So move on to column 4 (a $2^1/2$ inch pipe). It can carry up to 32 fixture units. The vertical stack size for this layout would have to be $2^1/2$ inches.

To size the building drain in Figure 2-18, look again at Figure 2-16. The fixture units and pipe sizes for horizontal pipe are based on a slope of $1/4$ inch per foot (see footnote 5). You see that the *Horizontal* line shows the number of fixture units you can use with the various pipe sizes listed at the top of each column. To accommodate 24 fixture units you must use a 3-inch building drain, as shown in column 5.

The layouts shown in Figure 2-17 (*SPC*) and Figure 2-18 (*UPC*) illustrate that different pipe sizes may be required when you install 24 fixture units in a five-story building. (Incidentally, this isometric layout is acceptable to most codes.)

More Comparisons of the Plumbing Codes

Now let's find the cumulative fixture unit load for the two-bath house illustrated in Figure 2-10. In the table in Figure 2-11, you see that the total fixture units in the *SPC* is 19. In the *UPC* for this same building, it's 18. To compare and size the drainage pipes included in Figures 2-14, 2-15 and 2-16, you'll also note that the fixture unit difference between these two codes has *no effect on the pipe sizing in Figure 2-10.*

Figure 2-10 shows the building sewer is sized at 4 inches, even though the calculation table for Figure 2-10 shows *19* fixture units for the *SPC* and *18* fixture units for the *UPC*. This may seem confusing, as Figure 2-14 (*SPC*) shows that a 3-inch pipe would be adequate at a $1/4$-inch fall per foot. It can carry *27* fixture units. *Nevertheless, the size of the building sewer is controlled by footnote 2.* This clearly states that by their standards, the minimum size of a building sewer is 4 inches.

Now let's check the other code. In the *UPC*, the building sewer size is controlled by a subsection called *Size of Building Drain.* It says: "The minimum size of any building sewer shall be determined on the basis of the total number of fixture units drained by such sewer. No building sewer shall be smaller than the building drain."

As you look at Figure 2-16 (*UPC*) you see that at $1/4$-inch fall per foot, a 3-inch horizontal pipe is large enough to carry up to 35 fixture units! Since we have only 18 fixture units, the sewer by this code can be 3 inches.

It seems that our tabulation is complete. But there may be certain restrictions, limitations and exceptions imposed on a drainage system. *Any such exceptions will always supersede the established pipe sizes and fixture units in any code drainage table.* Before you size the drainage pipes for any building, be sure you're familiar with the footnotes (usually located at the bottom of each table). Also check any other relevant subsections scattered throughout the code book.

Some Unique Code Agreements and Disagreements

Now let's check code comparisons for other common areas of the drainage system. More often than not they'll agree. One code may address a subject,

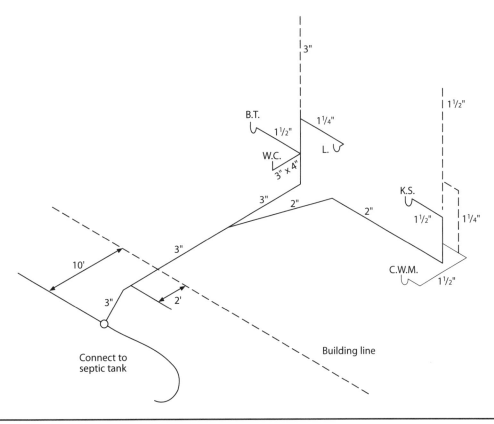

Figure 2-19
Plumbing drainage system illustrated and sized on the "stack"

while the other may not address it at all, or will do so in a different way. It may take your best effort to understand the drainage system and the vent system (Chapter 3 in this book). Considered the most complex sections of the code, they're often misunderstood and misinterpreted.

▌ Codes generally agree that the following are *prohibited*: (1) fixtures with waste openings larger than the waste pipe to which they connect and (2) fixtures that convey greasy waste on a 2-inch *waste stack* that vents lower fixtures.

▌ A *waste stack* is referred to as a *soil stack* when water closets or similar fixtures are installed. In such instances, the size is increased to 3 inches and is governed by footnotes below Figures 2-15 and 2-16.

▌ The *SPC* permits using heel or side inlet quarter bends under certain conditions. The *UPC* calls it a prohibited fitting.

▌ The *UPC* states that no waste connection can be made to a closet bend or stub of a water closet or similar fixture. The *SPC* doesn't address this.

▌ The building sewer shown in Figure 2-19 is connected to a septic tank. Though this is a building sewer, it's not classified as such by some codes because its developed length doesn't exceed 10 feet. Therefore, the part of the sewer pipe exceeding the 2-foot limit (more, in some codes) beyond a building exterior wall may be considered part of the building drain and sized at 3 inches.

▌ Take a look at Figure 2-20, an addition with a bathroom at the rear of an existing building. (Figure 2-20 can also illustrate similar requirements for an accessory building on the same lot.) The plot plan shows the sewer from the new addition running around the outside of the existing building and connecting to the existing sewer in the front yard. The new sewer pipe is sized at 3 inches, when some codes would

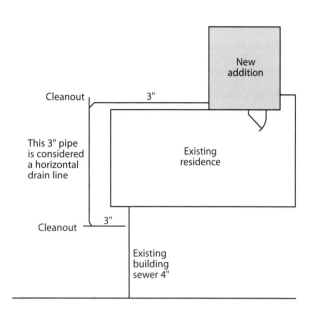

Figure 2-20
Sewer installation for addition

normally require 4 inches. Why? The code that usually requires a 4-inch pipe has made an exception in this case. It's considered to be a horizontal drain so you can size it according to Figure 2-14 (*SPC*) or Figure 2-16 (*UPC*).

Codes permit the same exception for accessory residential buildings located on the same lot with an existing building that share a single building sewer. (For similar commercial buildings you usually have to get special approval from your administrative authority.)

▪ The building drain in Figure 2-10 is sized at 3 inches. For all buildings with one or two water closets, the minimum-size building drain is 3 inches. If this building had a third bathroom, some codes require increasing the building drain to 4 inches at the junction of flow from all three water closets. The *UPC* permits up to three water closets on a 3-inch building drain. See Figure 2-14, footnote 1, and Figure 2-16, footnote 4.

Again referring to Figure 2-10, the waste pipe (also called a *wet vent* in this design) is sized at 2 inches. Why? The code states that the minimum-size vent (wet or dry) to serve a water closet is 2 inches.

Also in Figure 2-10, consider the pipe leading to the kitchen sink and clothes washing machine. Figures 2-14 and 2-16, at $1/4$-inch fall per foot, show that a 2-inch pipe is enough to convey the fixture units of these two fixtures. Why the 3-inch pipe? Again, footnotes and subsections in the code modify the plumbing possibilities.

▪ Most codes agree that (1) you need a minimum $2^1/2$-inch pipe if you're installing a waste on a cross installation with a pump discharge fixture (a clothes washing machine, for instance), and (2) other than accessory buildings, each building requires at least one minimum-size vent stack of not less than 3 or 4 inches extending through the roof.

Each bathroom has a 2-inch wet vent. Most codes require a 3-inch pipe up through the sanitary tap cross. You can satisfy the code requirements most economically by making this vent stack the main vent for this building.

▪ Figure 2-19 shows another plumbing design in which the bathroom vent is used as the main or minimum-size vent stack of 3 or 4 inches for the building. The kitchen sink and clothes washing machine are installed on a 2-inch waste pipe in this instance. Why is this legal? Again, most codes agree that a 2-inch pipe can be used if the fixture connections to the waste stack are at different levels, and if a relief vent is used on the clothes washing machine fixture drain.

Sizing Drainage Pipes, Multistory Building

To illustrate how to size the drainage in a multistory building, we'll use the drainage figures in this book. Of course, larger pipe sizes are listed in your code. We'll calculate pipe sizes from Figures 2-14 and 2-16. Again, we'll compare the *Standard* and *Uniform Plumbing Codes*' sizing methods. The load is accumulated at the base of each stack. The sizes shown in Figure 2-21 are based on a fall of $1/4$ inch per foot.

The only difference in the two codes for sizing the building drain is noted in Stack C. On a 2-inch drain at $1/4$-inch slope per foot, the *SPC* allows up to 21 fixture

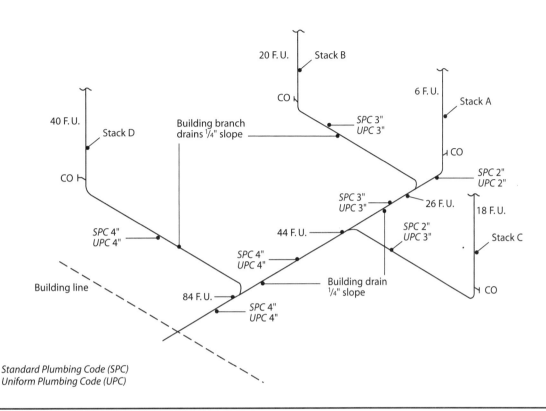

Figure 2-21
Sizing the building drain and branch drains

units, while the *UPC* allows only 8. To accommodate the 18 fixture units illustrated in Figure 2-21, the *UPC* requires a 3-inch pipe. Always check your local code.

To size the vertical drainage pipes in a multistory building, start with the top drainage fixture units and work down to the building drain. Accumulate the total fixture loads at the base of the stack. That total determines the size and length of the entire vertical stack. The vertical waste or soil stack must be the same size throughout its length. Keep in mind that a larger horizontal branch drain can't empty into a smaller vertical waste or soil stack.

The fixture load for sizing the stacks in Figure 2-22 is limited to those listed in the drainage portion of Figures 2-15 and 2-16. Again, we'll compare the sizing requirements from the *Standard* and *Uniform Plumbing Codes*. Look at Figure 2-22. For Stack A, with 128 fixture units, both codes require a 4-inch stack. For Stack B, with 59 fixture units, you see that the *SPC* requires only a 3-inch stack while the *UPC* requires a 4-inch stack.

Figure 2-23 may need some clarification. Note that stacks A and B extend to the upper floors in this multistory building, and that Section C reflects a horizontal fixture branch drain connecting several fixtures to the building drain. We'll use Figures 2-14 and 2-16 to size the building branch drain and building drain, using the cumulative fixture unit values at the base of each stack.

Stacks A and B connect to the building drain or building branch drain. Since several fixtures connect to the drainpipe, Section C is *not* a building drain. Using the *SPC* (Figure 2-14), it must be sized as a 3-inch horizontal fixture branch. But if you're using the *UPC* (Figure 2-16), this pipe isn't considered special. You'd size it at 3 inches as a horizontal drainage pipe subject to the total fixture units and types of fixtures.

Look now at Figure 2-24. A soil or waste stack can't be smaller than the largest horizontal branch pipe connected to it. There's one exception: "A 3" × 4" water closet bend shall not be considered a reduction in pipe size."

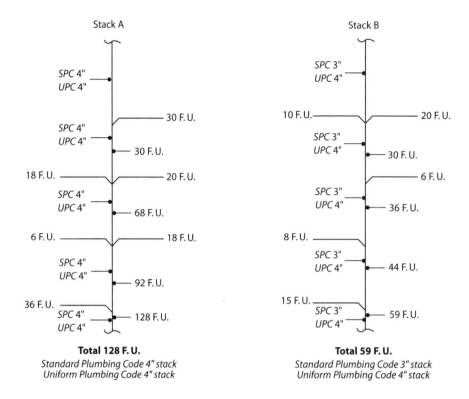

The difference in the two codes is the number of drainage fixture units allowed on a vertical drainage stack. For example: The *Standard Plumbing Code* permits up to 60 F. U. on a 3" stack and up to 500 F. U. on a 4" stack. The *Uniform Plumbing Code* permits up to 48 F. U. on a 3" stack and up to 256 F. U. on a 4" stack.

Figure 2-22
Sizing vertical stacks

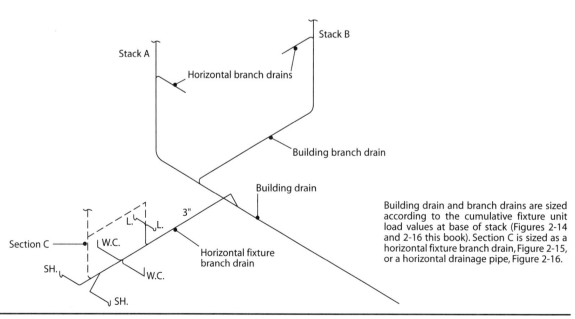

Building drain and branch drains are sized according to the cumulative fixture unit load values at base of stack (Figures 2-14 and 2-16 this book). Section C is sized as a horizontal fixture branch drain, Figure 2-15, or a horizontal drainage pipe, Figure 2-16.

Figure 2-23
Sizing a horizontal fixture branch drain

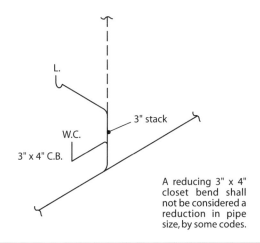

Figure 2-24
Reducing closet bend illustrated

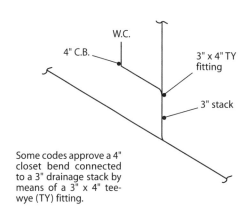

Figure 2-25
4" closet bend connecting into 3" stack illustrated

Some codes allow a 4-inch horizontal fixture branch drain to connect to a 3-inch stack or horizontal building drain under certain conditions. Let's look at a couple of these, as illustrated in Figures 2-25 and 2-26.

Future Fixtures

Provisions may be required in some buildings for the installation of future fixtures. When this occurs, consider the number and types of fixtures when sizing drain and vent pipes. Properly-sized pipes must then terminate with properly-plugged pipe or pipes. See Figure 2-27.

Vertical Offset in Drainage Pipes

You can consider as straight any vertical stack with an offset of 45 degrees or less. Size it as a straight vertical stack. However, a relief vent is required (1) in buildings having 10 or more branch intervals (10 or more stories high), or (2) if a horizontal branch drain connects to the stack within 2 feet above or below the offset. See Figure 2-28.

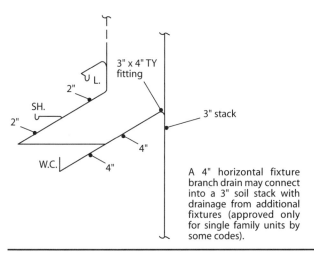

Figure 2-26
4" drain connecting into 3" stack illustrated

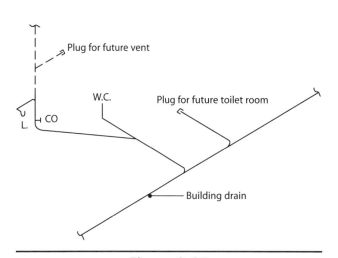

Figure 2-27
Drainage and vent pipe plugged for future installation

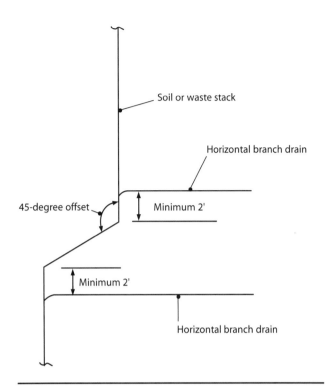

Figure 2-28
Example of 45-degree offset in vertical stack

Horizontal Offset in Vertical Drainage Stack

Here's how to size vertical stacks having an offset of more than 45 degrees (see Figure 2-29). The vertical portion above the offset may be sized as a regular stack. Compare Figures 2-15 and 2-16 for sizing guidelines. Size the horizontal portion as a building drain using Figures 2-14 and 2-16. The vertical portion below the offset can't be smaller than the horizontal pipe. Horizontal branch drains shouldn't be connected to the vertical stack within 2 feet below or above the offset unless they're properly vented. See Chapter 3 for venting details.

Horizontal Branch Drains

Most codes require that horizontal branch drains in multistory buildings have a minimum separation of 8 feet. Figure 2-30 illustrates this.

Suds Pressure Zones

Following the close of World War II, soap and chemical industries began to develop and market many new products. Synthetic detergents provide plentiful

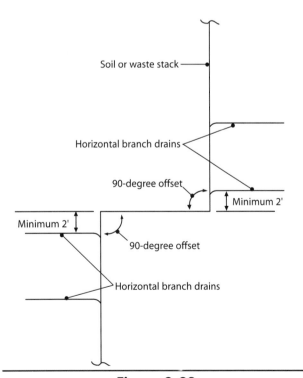

Figure 2-29
Example of horizontal offset in vertical stack

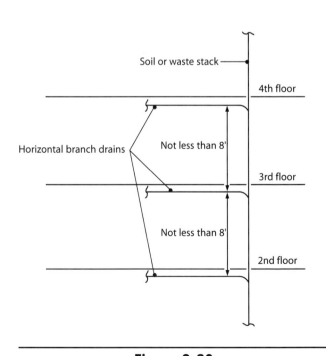

Figure 2-30
Example of minimum 8' separation of horizontal branch drains

quantities of long-lasting suds in both hard and soft water. The down side is that the increased use of these products has increased the volume of suds in household wastes — and their attendant problems.

Multistory residential buildings always get the worst of it. When upper floor fixtures and appliances discharge wastes containing detergents, they mix with the suds-producing ingredients as they rush down the inner wall of the stack. These suds settle into the lower sections of the drainage pipes. The air that accompanies wastes is forced down into the suds in the lower sections. This mixture compresses and forces the suds to move through any available path of relief. Improperly-designed drainage and vent piping may serve as relief paths, perhaps causing suds to bubble up into lower floor fixtures.

Different codes address this particular problem with varying degrees of concern, or perhaps not at all. Where these conditions exist, check with local authorities for any special design that may be required.

Some codes consider bathtubs, laundries, clothes washing machines, kitchen sinks and dishwashers as suds-producing fixtures and appliances. Figure 2-31 shows the likely location of suds pressure zones when large numbers of these fixtures are used.

Fixtures which produce sudsy detergents will normally discharge at an upper level into a soil or waste stack. In tall residential buildings which serve fixtures at lower levels, avoid connecting the drainage and vent piping for the lower fixtures to any suds pressure zone. Figure 2-32 illustrates one way to install waste pipe from suds-producing fixtures so it's acceptable to most codes.

The *UPC* doesn't require special drainage designs for single-family residences and buildings that don't exceed three stories. But it outlines certain restrictions in residential buildings more than three stories high. It prohibits suds-producing piping from connecting into a drainage pipe within 8 feet of any vertical-to-horizontal change of direction. Check local code requirements.

In designing a drainage system with suds-producing zones, there are two ways to determine the distance for pipe connections to horizontal and vertical lines. See Figures 2-31 and 2-32.

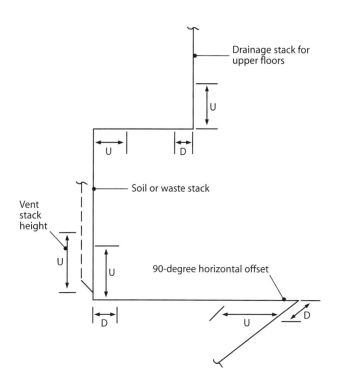

Suds pressure zones are represented by letters "D" = down and "U" = up. The length of pressure zones is determined by the size of drainage pipe. See Figure 2-34. Suds pressure relief vent pipe sizes are determined by drainage pipe size. See Figure 2-33.

Figure 2-31
Suds pressure zones (NSPC)

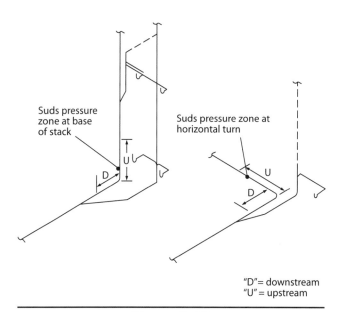

"D" = downstream
"U" = upstream

Figure 2-32
Suds producing zones (NSPC)

The *UPC* has a flat 8-foot separation regardless of the pipe size used. Other codes accept separation connections in pipe diameters for upstream and downstream suds-producing zones. For example: 40 pipe diameters *upstream* and 10 pipe diameters *downstream*. It works like this: A 4-inch vertical stack *upstream* would equal 160 inches, or 13.3 feet (4" × 40 ÷ 12 = 13.3 feet). A 4-inch vertical pipe downstream would equal 3.3 feet (4" × 10 ÷ 12 = 3.33 feet).

Figures 2-33 and 2-34 should help you avoid the necessity of individual pipe size calculations when designing drainage systems affected by suds. Though most codes accept these calculations, it's always wise to check. You can see that Figure 2-33 shows relief venting, while Figure 2-34 shows drainage pipes upstream and downstream.

Drainage Systems Below Sewer Level

Where a drainage system or a part of a drainage system can't discharge to the building sewer or public sewer by gravity flow, you're required to discharge the wastes into an approved sump. Automatic pumping equipment then will lift the contents and discharge it into the building gravity drainage system. Be sure the sump and pumping equipment are of adequate capacity for the volume and types of liquids to be conveyed.

Remember, only waste that requires lifting can discharge into a sump. Size, install and vent the drainage system for this waste like a standard gravity system. Other drains must discharge by gravity into the building gravity drainage system.

Sewage Ejectors

Sumps and receiving tanks must be in accessible locations for inspection, repairs and cleaning. Constructed of concrete, metal or other approved materials, they must also be tightly covered and vented. Be sure the concrete sump walls and bottom are reinforced and designed to acceptable standards. Metal sumps or tanks must be thick enough to serve their intended purpose and be protected from external corrosion.

For public use, most codes require a duplex pumping system. This permits the pumps to discharge waste alternately. Also, in case of repair, one pump can remain in service.

You must install a check valve and gate valve in the discharge line between the pump and the building gravity system. In the case of a horizontal building gravity drain, the sump discharge line must connect from the top through a wye branch fitting. The connection has to be at least 10 feet from the base of any soil or waste stack.

Drainage pipe size	Relief vent size
1½"	2"
2"	2"
2½"	2"
3"	2"
4"	3"

Note A suds relief vent, relieving the nonpressure zone, shall be provided at each suds pressure zone where such connections are installed.

Figure 2-33
Suds pressure relief vents (NSPC)

Drainage pipe size	"U" upstream	"D" downstream
1½"	5'0"	1'6"
2"	7'0"	1'6"
2½"	8'0"	2'0"
3"	10'0"	2'6"
4"	13'0"	3'6"

Figure 2-34
Suds pressure zones for various size drainage pipes, upstream and downstream (NSPC)

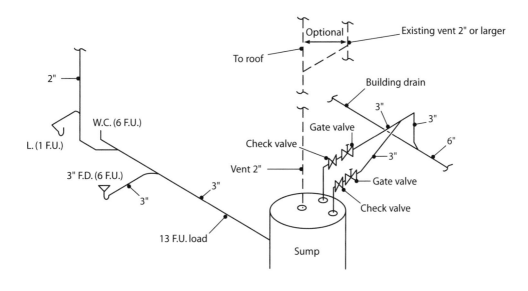

Figure 2-35
Sewage ejector "public use" sump

Always install the gate valve on the discharge side of the check valve. It must be the full-way type with corrosion-resistant working parts. It's important not to overload the building drainage system. The *UPC* allows 2 fixture units for each gallon per minute of flow. The discharge piping must be at least 3 inches in diameter. See Figure 2-35 for piping details.

Sumps receiving clear water waste from floor drains, air conditioning condensate drains and the like need not have a cover or be vented.

Review Questions (answers are on page 275)

1. What are the two parts of a private sanitary drainage system?

2. What is the aim of municipal codes in relation to drainage systems?

3. On which section of the code do the Boards of Rules and Appeals have the most requests for clarification and resolution?

4. From what section of the code are most of the questions and isometric drawings on the journeyman and master's examinations taken?

5. What purpose do isometric drawings serve?

6. List two reasons why it's important for a plumbing contractor to know how to make and interpret isometric drawings.

7. What are the three basic pipe angles you need to illustrate in an isometric drawing for a plumbing system?

8. What do the lines on an isometric drawing represent?

9. What is the purpose of a horizontal twin tap sanitary tee?

10. How are plumbing fixtures identified in isometric drawings and floor plans?

11. What is another term for a public sewer?

12. What are the two other terms used for a building sewer?

13. Why is a building drain also known as a main?

14. Plumbers often refer to a fixture drain by what other terms?

15. What kind of waste is carried by a waste pipe?

16. What other term is used to identify a waste pipe?

17. What is a soil stack?

18. What function does a branch interval serve?

19. What is a horizontal branch?

20. What's the main factor you use to determine pipe size within a drainage system?

21. Besides maximum fixture unit value, what three additional factors must you take into consideration when sizing drainage piping?

22. For what purpose would you use the tables in your code book that list the various fixture load values?

23. How are special fixtures connected to the drainage system?

24. Name two devices that are considered special fixtures.

25. Name two continuous and intermittent flow devices that you can connect to a drainage system.

26. How many fixture units does the *Uniform Plumbing Code* allow for each gallon per minute of flow from continuous flow devices such as sump ejectors?

27. What is the major difference between codes?

28. What is the generally-accepted fall per foot for horizontal pipe?

29. What will restrictions, limitations and exceptions always supersede in any code book?

30. How do most codes regard fixtures with waste openings larger than the waste pipe to which they need to connect?

31. What does the code call a stack that receives the discharge from a water closet?

32. What's the minimum size vent required by the code to serve a water closet?

33. What's the code-accepted minimum size for a main vent stack in a building?

34. When sizing drainage piping in a multistory building, at what point do you accumulate the fixture unit load?

35. What procedure must you follow in sizing vertical drainage pipes in a multistory building?

36. How does the size of a vertical waste/soil stack vary from one end to the other?

37. What's the one exception to the rule that a soil or waste stack can't be smaller than the largest horizontal branch pipe to which it connects?

38. What are the main considerations when sizing drain and vent pipes for future fixtures?

39. How do you define a vertical stack with an offset of 45 degrees or less when you're sizing it?

40. How do you size the horizontal portion of a vertical stack that has an offset greater than 45 degrees?

41. In a vertical stack, what's the minimum distance for an offset above or below the horizontal branch?

42. What's the minimum required separation of horizontal branch drains in a multistory building?

43. Since World War II, how have synthetic detergents affected the characteristics of household waste?

44. Name two suds-producing fixtures or appliances.

45. What problems can occur when suds-producing fixtures and appliances discharge into an improperly-designed drainage and vent system?

46. When it's impossible for waste to drain by gravity into the building drainage system, where must it be discharged?

47. Why should sumps and receiving tanks be accessibly located?

48. Why do most codes require that sumps for public use be equipped with a duplex plumbing system?

49. What two devices are required in the discharge line between the pump and the gravity system?

50. According to most codes, what's the minimum acceptable size for a sump discharge pipe?

3

Designing Vent Systems

Vent pipes are vital to the successful functioning of plumbing fixtures and the sanitary drainage system. Although the English inventor Joseph Braman first developed the water closet in 1778, water closets couldn't be installed for common use in buildings until a way was found to protect the fixture trap seals. Siphoning and back pressure in building drainage systems destroyed the trap seal and allowed objectionable odors and sewer gases to escape into the building.

A theory about protecting the traps of plumbing fixtures by installing vents was proposed and tested in the late 1800s. Further field testing helped to determine the distances from traps and the sizes of vent pipes needed to serve various plumbing fixtures. Finally, in 1875, it was found that by extending all vent pipes to the atmosphere above the building's roof, any odor from the system would escape from the building.

This breakthrough in the development and design of the sanitary drainage system eliminated resistance to indoor plumbing systems. Plumbing installations in buildings soon became routine design features. By the late 1800s, many cities began establishing separate plumbing codes to protect the health of people in densely populated areas. Plumbing practices varied considerably from municipality to municipality, and the early code requirements in each area reflected these variations.

Eventually, the U.S. Department of Commerce launched a comprehensive effort to standardize the plumbing code. Scientific experiments conducted by the National Bureau of Standards formed the basis for many of the new plumbing requirements.

States authorized the establishment of examining boards made up of qualified plumbers who, along with other agencies, were requested to write plumbing regulations and amend these regulations when necessary to keep abreast of new materials and installation methods. Although plumbing practices still vary from area to area, the variations today are usually minor.

Types of Vents

A *vent system* consists of all the vent pipes of a building. It may include one or more pipes installed to provide a free flow of air to and from a drainage system. This prevents back pressure or siphonage from breaking the water trap seals serving the fixtures.

Here are the types of vents you'll need to be familiar with to install drain, waste and vent systems:

Battery Venting — *Battery venting* uses a branch or circuit vent to vent a group of two or more similar adjacent fixtures that discharge into a common horizontal waste or soil branch. Some of these fixtures may be the above-floor type shown in Figure 3-1, and others may be the floor-outlet type shown in Figure 3-2.

Branch Vent — A *branch vent* connects one or more individual vents to a vent stack (Figure 3-1).

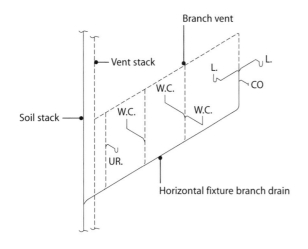

Figure 3-1
Branch vent

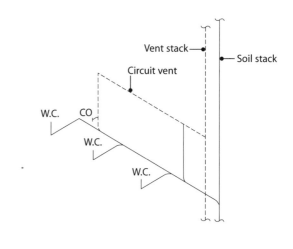

Figure 3-2
Circuit vent

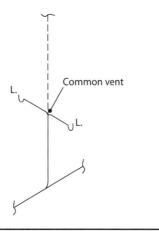

Figure 3-3
Common vent

Circuit Vent — A *circuit vent* functions much like a *branch vent*. Circuit vents serve two or more fixtures and rise vertically from between the last two fixture traps located on a horizontal branch drain. They are connected to the vent stack as shown in Figure 3-2.

Common Vent — A *common vent* is a vertical vent that serves two fixture branches installed at the same level. In Figure 3-3 these are two lavatories.

Continuous Vent — A *continuous vent* is the vertical portion that's a continuation of the drain to which it's connected (see Figure 3-4). It's also known as a *stack vent*.

Dry Vent — A *dry vent* is that portion of a vent system that receives no sewage discharge (Figure 3-5).

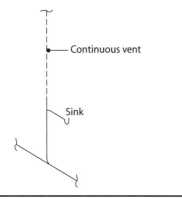

Figure 3-4
Continuous vent

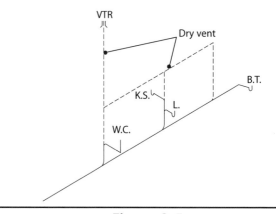

Figure 3-5
Dry vent

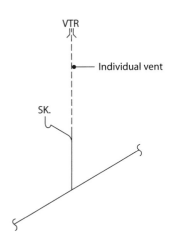

Figure 3-6
Individual vent

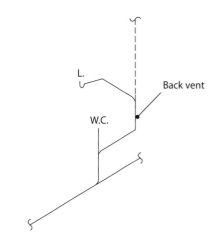

Figure 3-7
Back vent

Individual Vent — An *individual vent* (Figure 3-6) is a pipe installed to vent a single fixture trap. It may connect to the existing vent system above the fixture served or terminate through the building roof into the open air. It's also known as a *back vent* (see Figure 3-7).

Loop Vent — A *loop vent* is similar to a circuit vent except that a loop vent loops back and connects to the stack vent, as shown in Figure 3-8, and not to the vent stack.

Main Vent — The *main vent* is the principal pipe of a venting system to which vent branches may be connected (Figure 3-9). The main vent must connect at the base of a soil or waste stack below the lowest horizontal branch.

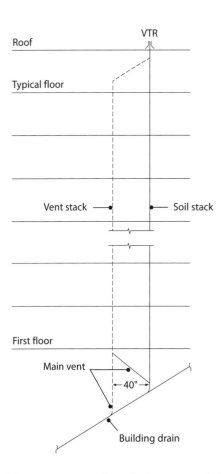

Note Main vent to connect at base of soil or waste stack. *Standard Plumbing Code* will accept main vent to connect to building drain, but requires a maximum 40" separation between the two stacks.

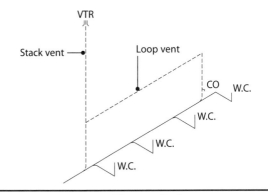

Figure 3-8
Loop vent

Figure 3-9
Main vent

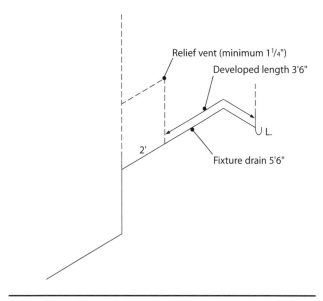

Figure 3-10
Single fixture relief vent (SPC)

Relief Vent — The *relief vent* is sometimes referred to in the trade as a "re-vent." Its primary function is to provide a route for the circulation of air between the drainage system and the vent of a plumbing system. Relief vents are shown in Figures 3-10 and 3-11.

Side Vent — A *side vent* is a vent that connects to a horizontal drain pipe through a fitting at an angle no greater than 45 degrees to the vertical (see Figure 3-12).

Stack Vent — A *stack vent* is nothing more than the extension of a soil or waste stack (dry section) up and through the roof of a building (Figure 3-13).

Stack Venting — *Stack venting*, as shown in Figure 3-14, is a method of venting fixtures through the soil or waste stack.

Vent Header — A *vent header* is a single pipe that receives the connection of two or more vent pipes and then connects to the main vent stack or extends to the atmosphere separately at one point (Figure 3-15).

Vent Stack — A *vent stack* is the vertical portion of a vent pipe. Its primary purpose is to provide circulation of air to and from all parts of a drainage system. Vent stacks are shown in Figures 3-1, 3-2, 3-9 and 3-15.

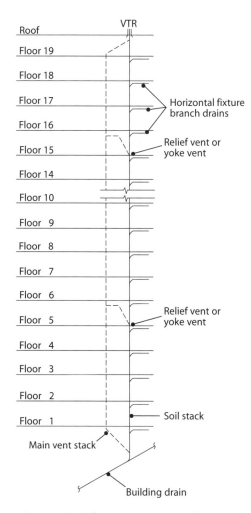

Note The diameter of relief vents must not be smaller than the diameter of the main vent they connect to. Relief vents must be located on each 5th floor.

Figure 3-11
Relief vents

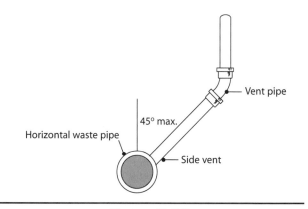

Figure 3-12
Side vent

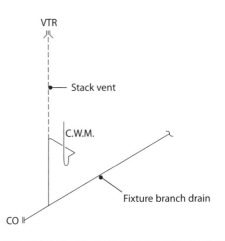

Figure 3-13
Stack vent

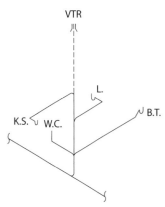

Note Each fixture drain connects independently to the stack. The tub and/or shower and water closet connect to the stack at the same level.

Figure 3-14
Stack venting

Wet Vent — A *wet vent* is a waste pipe that vents and conveys waste from fixtures other than water closets. Figures 3-16 through 3-21 show various *wet vent* systems.

Yoke Vent — A *yoke vent* is a pipe connecting upward from a soil or waste stack to a vent stack for the purpose of preventing pressure changes in the stack. A yoke vent is shown in Figure 3-22.

Venting procedures that are not listed here or defined specifically in most codes will be discussed and illustrated later in this chapter.

Problems Created by Inadequate Venting

In rural areas where inspections aren't required, or in municipalities where inspections are lax, inadequate sizing and arrangements of vent pipes can cause problems. Following are some of the problems related to poor venting:

- Plumbing fixtures may drain slowly, as if a partial stoppage exists.

- Water closets may need several flushes to remove the contents from the bowl.

- Back pressure (known as *positive* pressure) within the drainage pipes may force sewer gases up and through the liquid seals and into the building.

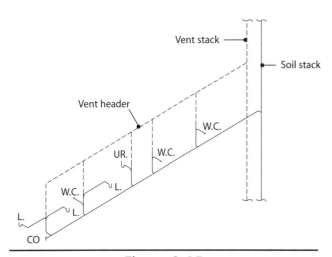

Figure 3-15
Vent header

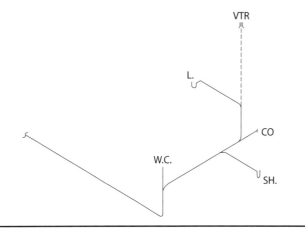

Figure 3-16
Stacked wet vented system

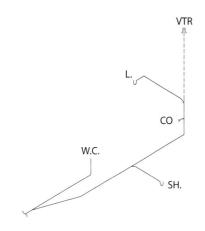

Figure 3-17
Flat wet vented system

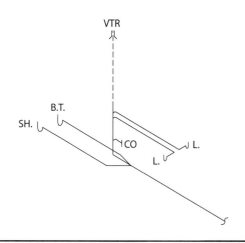

Figure 3-18
Wet vented shower and bathtub

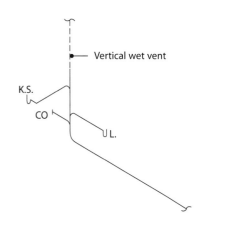

Figure 3-19
Lavatory wet venting

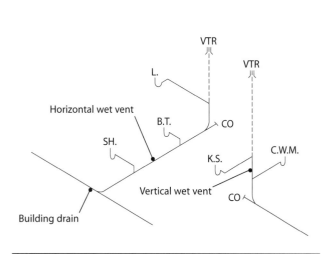

Figure 3-20
Horizontal and vertical wet venting

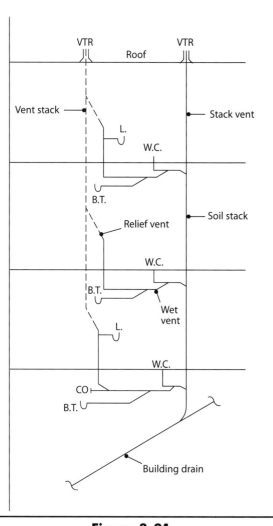

Figure 3-21
*Alternate method of wet venting single bathrooms
in multistory buildings*

▮ Plumbing fixtures located a greater distance from a vent pipe than is permitted by code may, when the contents are released, siphon the liquid trap seal. This siphoning action is known as *negative* pressure.

▮ In very cold climates, as the warm moist air flows up and out of the vent pipes and makes contact with the cold atmosphere, there's a danger of frost forming on the inside of the pipes. This can restrict the free flow of air within the drainage system.

The plumbing codes have regulations for preventing these problems. For example, in regions where there's a danger of frost forming inside a vent pipe, the codes have the following regulations:

The *Standard Plumbing Code* requires that a vent extension through a roof be at least 3 inches in diameter. When it's necessary to increase the vent size, the change in diameter must be made at least 12 inches *inside* the building.

The *Uniform Plumbing Code* mandates that vent terminals be at least 2 inches in diameter but never smaller than the required vent pipe. The diameter increase has to be made at least 12 inches *inside* the building, below the roof, and must terminate at least 10 inches above the roof. See Figure 3-23.

Correctly sizing and arranging the vents as well as understanding how the parts work together is fundamental to installing an adequate vent system.

How to Size the Vent System

You size and arrange vent pipes to relieve pressure that builds up as plumbing fixtures discharge water into the sanitary drainage system. This free flow of air within the system keeps the back pressure or siphoning action from destroying fixture trap seals.

The sizing of a vent pipe, like sizing its cousin, the soil and waste pipe, depends on the maximum fixture unit load, its developed length, the type of plumbing fixtures, and the diameter of the soil or waste stack it serves. You can use Figure 3-24 to compute the size and permitted length of vent stacks for multistory buildings. We've included it for you to use to follow

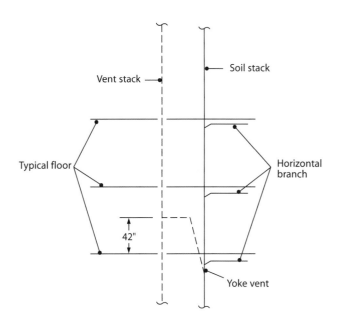

Figure 3-22
Yoke vent

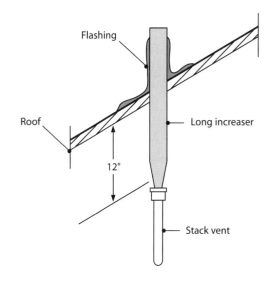

Figure 3-23
Cold-climate vent terminals

Diameter of soil or waste stack (in)	Maximum fixture units	Size and max. length of vent piping (ft)				
		1¼"	1½"	2"	2½"	3"
1¼	2	30	—	—	—	—
1½	8	50	150	—	—	—
1½	10	30	100	—	—	—
2	12	30	75	200	—	—
2	20	26	50	150	—	—
2½	42	—	30	100	300	—
3	10	—	30	100	200	600
3	30	—	—	60	200	500
3	60	—	—	50	80	400

Figure 3-24
Size and length of vent pipes (SPC)

along with our examples. However, it's just a partial table. For larger sizes and lengths, you'll have to consult the code.

How do you interpret Figure 3-24? Look at lines 4 and 5, both involving a 2-inch waste stack. In line 4, the 2-inch waste stack serves 12 fixture units. If the height of the building doesn't exceed 30 feet, a 1¼-inch vent stack *may* be used. If the building exceeds 30 feet, but is less than 75 feet, you can use a 1½-inch vent.

Finally, for a building height of more than 75 feet but less than 200 feet, you *must* use a 2-inch vent stack. These are the minimum sizes permitted by the code. It's OK to upgrade a plumbing system by using larger vent pipes — as long as the vent pipe doesn't exceed the diameter of the soil or waste stack it serves.

Here's something you'll want to take notice of in Figure 3-24. Even though the 2-inch waste pipe on line 5 is the same as the one on line 4, the height of the vent pipe must decrease as the maximum fixture unit load increases. A larger load requires more air flow, and air is restricted more and more as a pipe grows longer.

Always check your code for any footnotes or subsections that place restrictions or limitations on a vent pipe or vent system. These footnotes or subsections always supersede the established tables in the code.

Basic Sizing Principles

Here are the basic principles for sizing venting, as summarized in most codes:

▮ For each building with a single building sewer receiving the discharge of a water closet, there must be at least one vent stack (of not less than 3 or 4 inches) extending through and above the building roof.

▮ If there's an accessory building or buildings located on the same lot, with one common building sewer, use the size tables in the code (Figure 3-24) to find the minimum size vent stack or stacks to serve each accessory building. There's one exception to this basic rule: Should a water closet be located in the accessory building, the vent stack must be no smaller than 2 inches.

▮ No vent, wet or dry, for a water closet can be less than 2 inches in diameter.

▮ The diameter of the vent stack can't exceed the diameter of the soil or waste stack to which it connects.

▮ The diameter of an individual vent stack can't be less than 1¼ inches nor less than one-half the diameter of the drain to which it's connected. In other words, if a drain is 3 inches, the minimum size vent is 1½ inches. If a drain is 4 inches, the minimum size vent is 2 inches.

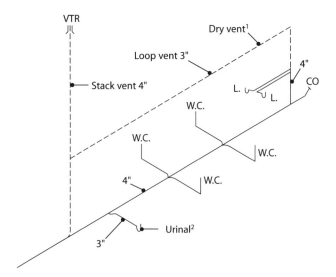

Notes

1) The pipe of the dry vent section of a circuit or loop vent may have a diameter of one pipe size less than the diameter of the pipe of the horizontal soil or waste pipe it serves. This may vary in some codes.

2) Floor mounted urinal traps or similar fixtures installed downstream from a water closet must be 3 inches in diameter where permitted by code, or be re-vented.

Figure 3-25
Sizing a circuit or loop vent

▌ The pipe of the dry vent section of a loop or circuit vent can be only one pipe size smaller than the diameter of the horizontal soil or waste drain it serves. For example, if the horizontal drain is 3 inches in diameter, the minimum dry vent section of pipe should be 2 inches.

▌ When the vent stack or stack vent size doesn't control the sizing of a loop or circuit vent (as in Figure 3-25), most codes will accept one of the following sizing methods:

1) The dry vent section of a loop or circuit vent may be one pipe size smaller than the horizontal soil or waste drain it serves.

2) This vent section is acceptable if the diameter is one-half the size of the horizontal soil or waste drain it serves. *When sizing loop or circuit vents, always check code requirements.*

▌ Any plumbing fixture that exceeds the maximum code-established distance from the vent pipe must have a minimum size relief vent of $1^1/4$ inches. (Refer to Figure 3-10.)

▌ The following rules illustrate the relationship of the horizontal and vertical vent pipe to the total permitted pipe length:

1) The maximum length of any vent pipe is determined by the diameter, as shown in Figure 3-24.

2) When the vent diameter is increased one pipe size for its entire length, the maximum length limitation does not apply (*Uniform Plumbing Code*).

3) The horizontal section of a vent pipe (see A in Figure 3-26) must not exceed one-third of the total permitted length of vent pipe as illustrated by the vertical section B.

4) Some codes will allow the horizontal section (Figure 3-26, A) to be installed up to 20 feet in length.

▌ A vent pipe must rise vertically to a point at least 6 inches above the flood level rim of the fixture it serves before it can be offset horizontally and connect to any other vent. See Figure 3-27.

▌ Vent stacks may sometimes have a dual role. In some installations they both supply and remove the air from a drainage system, and serve as cleanouts for inserting a cleaning cable. An installation must meet these requirements in order to qualify for this dual role:

1) It's a one-story building with not more than one 90-degree change in direction in the drainage system.

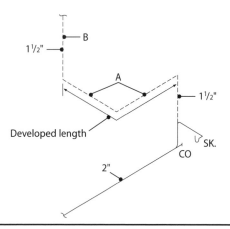

Figure 3-26
Maximum horizontal length of vent pipe

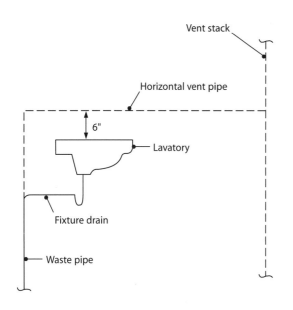

Figure 3-27
*Minimum height of vent pipe above
floor-level rim of fixture*

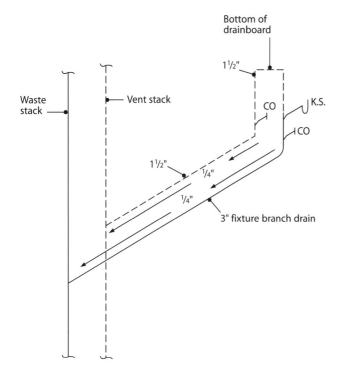

Figure 3-28
Venting an island sink, example 1

2) The vent stack is vertical throughout (without any offsets) and extends up through the roof.

3) The vent stack is the same size as the waste pipe it serves.

4) You don't reduce the vent stack beyond these minimums:

> 4 inch can't reduce to less than 3 inch
> 3 inch can't reduce to less than 2 inch
> 2 inch can't reduce to less than $1^1/_2$ inch.

Special Venting Systems

The following installations require special venting systems.

Venting Island Sinks

Island sinks, and other fixtures located away from walls and partitions, are becoming more and more common in construction. As you might guess, you have to use special venting arrangements for them. Some codes don't address island fixtures, but most will accept one of the designs illustrated in the following examples.

Example 1

The multistory building shown in Figure 3-28 has a 3-inch branch drain extending from the waste or soil stack to the island fixture. The 3-inch pipe extends up to the sink outlet. A $1^1/_2$-inch vent section extends up to the bottom of the drainboard and then loops back down below the floor and connects to the vent stack. It's sloped to drain dry. For this installation, the cleanouts must be full size and accessible.

Example 2

Figure 3-29 shows a special arrangement for a single-family home. The building drain is installed to pass beneath the island sink location in the kitchen. There are two 3-inch accessible cleanouts installed above the finished floor and a 3-inch sanitary tee installed to receive the fixture drain. In an installation like this, the dry vent portion is generally sized at $1^1/_2$ inches.

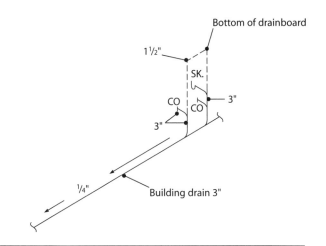

Figure 3-29
Venting an island sink, example 2

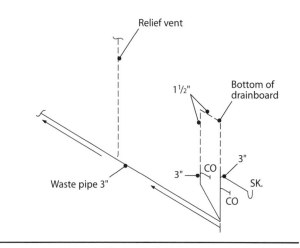

Figure 3-30
Venting an island sink, example 3

Example 3

The island sink in Figure 3-30 is located at the end of a horizontal waste pipe. In such cases, the vent portion may connect through a wye at the base of the vertical waste pipe. This is called a return vent. With this installation you must install a relief vent downstream as near to the fixture as possible. Most codes accept pipe sizes and cleanouts as illustrated in this example.

Example 4

Figure 3-31 shows an arrangement that's clearly acceptable under the regulations of the *Uniform Plumbing Code*, and probably most other codes as well. To vent this island sink according to code requirements, follow the guidelines shown in this example:

1) The island sink and similar equipment traps must be roughed-in above the floor. The vent is extended at least as high as the underside of the drainboard, and then turned downward to connect to the horizontal sink waste pipe immediately downstream from the vertical sink waste pipe.

2) You connect the return vent to the horizontal waste pipe through a wye-branch fitting. Provide a foot vent at the base of the vertical fixture vent by using a wye-branch fitting roughed-in below the floor that extends to the nearest partition. This vent should then extend separately through the roof or connect to another vent if one is available.

3) Make sure the fittings below the floor are the drainage type and that the vent has a minimum slope of $1/4$ inch per foot back toward the drain pipe. The return bend under the drainboard must be a one-piece fitting or it should be assembled using a 45-degree, a 90-degree and then another 45-degree elbow (in that order).

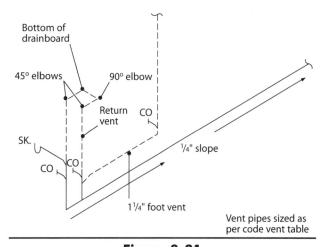

Figure 3-31
Venting an island sink, example 4

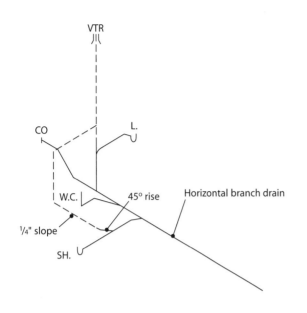

Figure 3-32
*Re-venting a minor fixture located downstream
from a water closet*

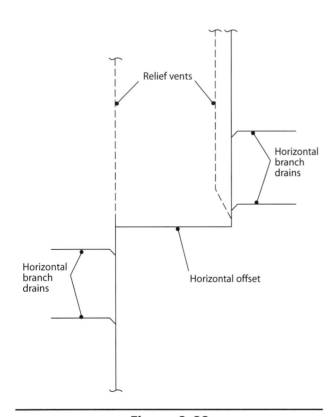

Figure 3-33
Venting horizontal offset in a vertical stack

4) You can use the code vent table in your code book to size the pipes.

5) The island sink waste pipe upstream from the return vent can't be used for any other fixture. You'll also need to install an accessible cleanout in the vertical portion of the foot vent.

Re-venting a Minor Fixture

When a floor-outlet type minor fixture, such as a shower or bathtub, is located downstream from a major fixture (a water closet, for example), you must re-vent the minor fixture. This is necessary to prevent siphonage of the minor fixture trap. Figure 3-32 illustrates the correct venting procedure.

Venting a Horizontal Offset in a Vertical Stack

Figure 3-33 illustrates how to vent a horizontal offset in a vertical stack. You must provide a relief vent at the top of the lower section and a vent at the base of the upper section, above the offset. Use the vent tables in your code book to size the pipes.

Wet Vents

Some codes permit this special method of venting in single-family residences and apartment units. It provides adequate protection for fixture trap seals located along the vent (see Figure 3-34). Since it's a single piping system, it's economical and can serve to vent several adjoining plumbing fixtures located on the same floor level.

A wet vent may be vertical or horizontal (refer back to Figure 3-20). As a wet vent, it can convey waste only from fixtures with low unit ratings. This excludes water closets and similar fixtures. Because a wet vent serves a dual purpose, the plumbing code has placed a number of restrictions on its use. The restrictions, pipe sizes and maximum capacities for wet vents are as follows:

▌ A horizontal wet vent can't exceed 15 feet and may only receive the discharge from fixture branches.

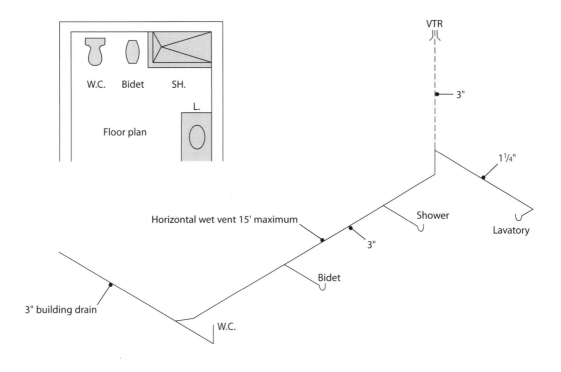

Figure 3-34
Horizontal wet vent

- A 2-inch horizontal wet vent can convey up to four fixture units but can't be used for urinals, pressure fixtures or for sinks, with or without garbage disposers.

- A 2^1/$_2$-inch horizontal wet vent can convey up to 10 fixture units. Water closets and fixtures requiring a waste opening greater than 2 inches can't use a wet vent.

- A 3-inch horizontal wet vent can convey up to 16 fixture units. Water closets or other fixtures requiring a waste opening greater than 3 inches aren't permitted.

- A 4-inch horizontal wet vent can convey up to 32 fixture units. Water closets or other fixtures requiring a waste opening greater than 4 inches aren't permitted.

Remember to always check local codes for their specific requirements regarding wet vent installations.

Venting Fixtures at Different Levels

You can use a common vent when two fixtures are connected to a stack at different levels. The vertical drain (wet vent) must be one pipe diameter larger than the highest fixture drain, but never smaller than the lower fixture drain. Figure 3-35 shows two examples of venting fixtures at different levels.

Sump Vents

When there's a sub-building drain in a basement that conveys body waste to a sump or retaining tank, you must install a local vent for the tank. A sump or retaining tank must have a gas- and air-tight cover. A local vent will permit the air within the tank to escape as the sewage enters. When the sewage is ejected, the vent will permit air to re-enter the tank. This keeps the sub-building drainage system from becoming air-locked and useless.

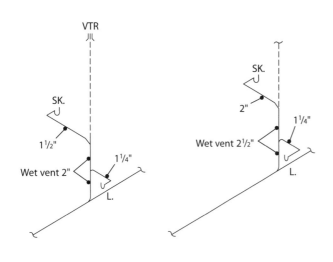

Figure 3-35
Wet venting fixtures at different levels

The following venting procedures are accepted by most plumbing codes:

▌ You must use a vent at least 1¹/₂ inches in diameter for a sump receiving body waste.

▌ The sump vent may extend independently up and through the building roof.

▌ The sump vent may be connected to an existing vent system of the same size (1¹/₂ inches) or to a larger size vent pipe.

Sumps receiving clear water waste from buildings, subsoil drains, floor drains, air conditioning condensate drains, etc., don't need a cover or a vent.

Vertical Combination Waste and Vent

You'll usually use this special method for receiving the waste from and venting specific plumbing fixtures installed in high-rise buildings. You can actually use it to advantage in any multistory building where plumbing fixtures are located directly over each other on different floor levels. Because it's an installation using a single vertical pipe riser, it's both economical and practical.

Use the table in Figure 3-36 for computing the size and permitted length of combination waste and vent stacks. This is only a portion of the complete table you'll find in the *Standard Plumbing Code.*

The code limits the size of the stack, the number of fixture units it may receive, the type of plumbing fixtures, and the total length of the combination waste and vent. Figure 3-37 illustrates the workings of a combination waste and vent pipe. In this illustration, Figure 3-37A shows a four-story building with eight kitchen sinks (without garbage disposers or dishwashers), back to back, with a total of 16 fixture units. Figure 3-37B indicates a six-story building with six kitchen sinks (without garbage disposers or dishwashers), one on each floor, for a total of 12 fixture units. Again, the number of total fixture units and the maximum length (height) of the pipe determine the diameter of the stack.

If you check Figure 3-36, you'll find that the combination waste and vent in Figure 3-37A should be 3 inches to accommodate the 16 fixture units. Since a four-story building is approximately 40 feet high, the combination waste and vent pipe remains within the 50-foot limit shown in the table.

In Figure 3-37B, it's important to note that the diameter of the combination waste and vent pipe must be 4 inches, even though the 12 fixture units are fewer than the 16 in Figure 3-37A. Since a six-story building exceeds the maximum length permitted by the code for a 3-inch combination waste and vent pipe, you have to use the 4-inch pipe. *The length of the pipe always takes precedence over the number of fixture units involved.*

Codes prohibit the discharge of water closets or urinals of all types into a combination waste and vent stack.

Diameter of stack (in)	Fixture units on stack	Maximum length (ft)
2 (no kitchen sink)	4	30
3	24	50
4	50	100
5	75	200

Figure 3-36
Vertical combination waste and vent (SPC)

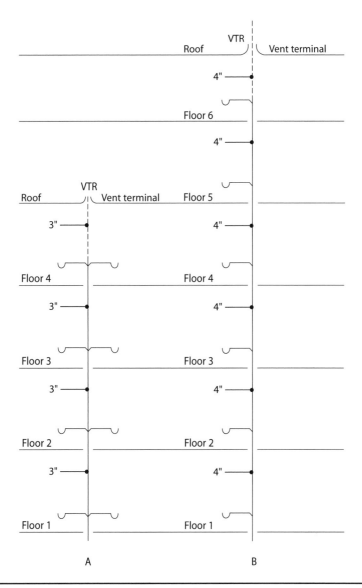

Figure 3-37
Vertical combination waste and vent (SPC)

The *Uniform Plumbing Code* does not accept the type of vertical combination waste and vent system as illustrated in Figure 3-37. To conform to the *Uniform Plumbing Code,* you must meet the following additional provisions:

▎ Vertical wet venting is limited to vertical waste piping receiving wastes from fixtures with low unit ratings.

▎ The ratings of each fixture can be no more than two fixture units.

▎ Only four such fixtures can be used.

▎ The maximum number of fixture units on this vertical wet vent system is eight.

▎ All wet vented fixtures must be within the same story.

▎ No vertical wet vent can exceed 6 feet in developed length.

When in doubt about which provisions apply in your area, check with your local code. Some codes may not accept this type of installation at all.

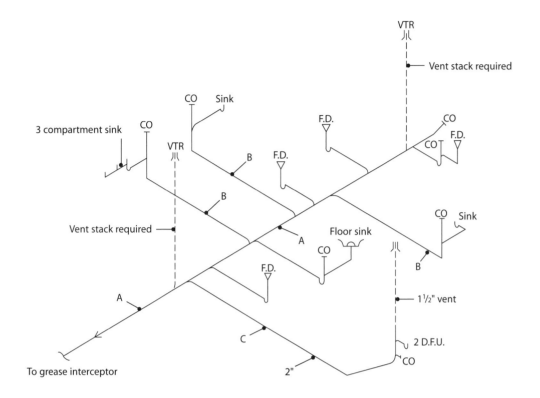

Notes

A Minimum size of main waste pipe (A) is 6", for a total of 96 F.U. For systems that exceed 96 F.U. the main waste pipe must be 2 pipe sizes larger than the required sizes as listed in Figures 2-15 and 2-16 for building drains and horizontal branch drains.

B Branch drains (B) for horizontal combination waste and vent system, as illustrated, must be 2 pipe sizes larger than the required sizes as listed in Figures 2-15 and 2-16 for building drains and horizontal drains.

C Plumbing fixtures with up to 3 D.F.U. may connect to drainage system by conventional method. Size according to Figures 2-15 and 2-16.

Figure 3-38
Horizontal combination waste and vent system

Horizontal Combination Waste and Vent System

Certain plumbing fixtures (usually found in restaurants or certain other commercial establishments, such as supermarkets or large warehouses with floor drainage) are permitted to use this special system for receiving waste and for venting. It's a one-pipe method you can use where special plumbing fixtures and regular plumbing fixtures aren't adjacent to walls or partitions.

The system is limited to sinks, dishwashers, floor sinks, indirect waste receptors, floor drains or similar applications. The trap of the fixture isn't always individually vented for this type installation. See Figure 3-38.

There's usually code agreement on the basic principles of this type of installation. However, don't install a horizontal combination waste and vent system until you've had your plans and specifications approved by your local administrative authority.

Special Requirements for Horizontal Combination Waste and Vent Systems

The plumbing code has placed a number of restrictions on the use of this type of system. They are:

▐ Because of possible venting problems, appurtenances (certain types of equipment) delivering large quantities of water or sewage shouldn't discharge into a horizontal combination waste and vent system.

- The horizontal waste pipe and each fixture trap must be at least two pipe sizes larger than the code requires for regular systems.

- The vertical waste pipe must be two pipe sizes larger than the fixture outlet.

- There must be a cleanout installed in the top of the connecting waste tee.

- You must size the floor sink and waste pipe from the floor sink to the trap for the total fixture units. If installed underground, they must not be less than 2 inches.

- Be sure to provide a vent at the upstream end of the waste pipe as well as downstream from all fixtures in the system.

- Design the system to assure that the vertical distance from fixture or drain outlet to the trap (weir) doesn't exceed 24 inches.

- In large installations where you can't avoid long runs, install relief vents at intervals of not more than 100 feet.

- Some codes require that you have a separate vent, installed in an approved manner, for a branch drain exceeding 15 feet.

- You should size vents in accordance with code requirements, but be sure you extend the same size pipe as the waste pipe to a point 6 inches above the flood level rim of the highest fixture before you make any reduction in size.

Vent Terminals

The following requirements apply to the termination of vent pipes:

- Terminate vent pipe extensions at least 6 inches above the roof. This ensures that gases and odors in all parts of the drainage system will discharge well above the roof surface.

- Where vent pipes penetrate the building roof, you must install approved flashing to make the joint watertight.

- You may not install a vent terminal for a sanitary system within 10 feet of any door, window that opens, or ventilating opening, unless it extends at least 3 feet above (or 2 feet in some codes) the top of that opening. See Figure 3-39.

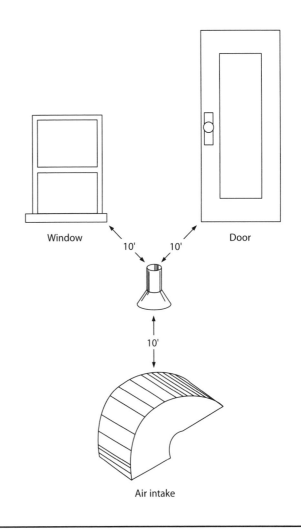

Figure 3-39
Code requirements for vent terminals

- A vent pipe terminal must be at least 10 feet from any lot line, or any mechanical air intake opening for air conditioning units.

- Horizontal vent terminals extending through a wall should turn upward and be effectively screened.

- Where danger of frost closure exists, any vent extension through a roof must be a minimum of 3 inches in diameter (or 2 inches in some codes). If it's necessary to increase the size of the vent terminal, you must make the change in diameter inside the building, at least 12 inches below the roof, and terminate the vent pipe at least 10 inches (6 inches in some codes) above the roof. (Refer back to Figure 3-23.)

- You must terminate a vent 6 inches above any fire wall.

- Do not allow vent terminals to be used to support flags, TV aerials, clotheslines or other similar items.

- Never terminate a vent under the overhang of any building.

- If a roof is used as a sun deck, solarium or similar function, the vent must extend a minimum of 7 feet (5 feet in some codes) above the roof deck.

- A vent pipe installed outdoors must be at least 10 feet above the surrounding ground, located at least 3 feet (10 feet in some codes) from any lot line, and be securely supported. This type of installation is generally used in venting a sewage system located in a trailer park. (For more information regarding trailer park installations, see Chapter 10.)

- Some codes limit the height at which you can use certain materials, such as cast iron, copper or screw pipe or plastic piping, for vent extensions. When conditions call for an extension, always check your code for approved materials.

Aggregate Cross-sectional Area of Pipe

The code is careful to insist that you vent a building drainage system by means of one or more pipes. The rule of thumb is that the aggregate cross-sectional area can't be less than that of the largest required building sewer.

How does that play out for the plumber? You can use the following formula to find the cross-sectional area of a building sewer. You'll probably find this formula is familiar to you.

The square of the diameter of a circle multiplied by .7854 equals the area of the circle.

Nominal pipe size (in)	Cross-sectional area (sq. in.)
1½"	(1.5 x 1.5 x .7854) = 1.7671
2"	(2 x 2 x .7854) = 3.1416
2½"	(2.5 x 2.5 x .7854) = 4.9087
3"	(3 x 3 x .7854) = 7.0686
4"	(4 x 4 x .7854) = 12.5664

Figure 3-40
Aggregate cross-sectional area of pipes

For example, if a building sewer is 4 inches, the cross-sectional area of the sewer equals $4 \times 4 \times .7854 = 12.566$ square inches. So, the total cross-sectional area of the vent pipes can't be smaller than the cross-sectional area of the sewer, or 12.5664 square inches.

To find the total aggregate cross-sectional area of pipe, you calculate the total cross-sectional area of all the vent pipes and then add them together. For instance, if you have a vent system consisting of two 2-inch vent pipes and one 3-inch vent pipe, your calculations would look like this:

Vent # 1	$2 \times 2 \times .7854 =$	*3.1416 sq. inches*
Vent # 2	$2 \times 2 \times .7854 =$	*3.1416 sq. inches*
Vent # 3	$3 \times 3 \times .7854 =$	*+ 7.0686 sq. inches*
		13.3518 sq. inches

As you can see, the total cross-sectional area of these three vents (13.3518 square inches) is not less than the building sewer (12.566 square inches), so they're adequate for this particular example.

Figure 3-40 shows the cross-sectional areas of various size pipes up to 4 inches. You can use this for calculating vent sizing. Use the formula above to calculate larger pipe sizes.

Review Questions (answers are on page 277)

1. What are the two causes of fixture trap seal loss that vent systems protect against in normal fixture use?

2. What is the function of a branch vent?

3. What is battery venting?

4. How is a common vent used in a plumbing system?

5. What is a continuous vent?

6. What other term is used to describe a continuous vent?

7. What is the term used to describe a vent that does not receive any sewage discharge?

8. How many fixture traps does an individual vent serve?

9. What is another term used to describe an individual vent?

10. How does a loop vent differ in function from a circuit vent?

11. How is a main vent defined in the code?

12. What is the primary function of a relief vent?

13. What is the trade name for a relief vent?

14. What is the definition of a side vent?

15. What is the definition of a stack vent?

16. What is the function of a vent header?

17. What is the primary purpose of a vent stack?

18. What two purposes does a wet vent serve?

19. What purpose does a yoke vent serve?

20. Name two problems that may occur if vent pipes are not sized and arranged properly.

21. What is another term used to describe back pressure?

22. What can occur if you have negative pressure in a fixture drain?

23. According to the *Standard Plumbing Code,* what's the minimum diameter required for a vent extension through the roof in a frost-prone climate?

24. According to the *Uniform Plumbing Code,* how far above the roof must a roof vent in a cold climate terminate?

25. What does the free flow of air within the sanitary drainage system prevent?

26. How must the height of a vent pipe change as the maximum fixture unit load increases?

27. What is the vent stack requirement for a building with a single building sewer?

28. If a water closet is located in an accessory building, what's the minimum size vent accepted by code?

29. What is the minimum size "dry" vent allowed by code when venting a water closet?

30. What is the smallest individual vent stack size permitted by code?

31. What is the minimum size vent that can be used for a 3-inch drain pipe?

32. Since loop or circuit vents are rarely used, what should you always do when laying out a sizing plan that includes such vents?

33. What factor determines the maximum length of any vent pipe?

34. What distance separation is required between a horizontal vent pipe and the flood level rim of the fixture served?

35. What is the term used to describe a kitchen sink that's located away from walls or partitions?

36. When a shower (or any minor fixture) is located downstream from a water closet, what precaution must be taken?

37. How many vents must you use to vent a horizontal offset in a vertical stack and what are they?

38. From what type of fixtures can a wet vent be used to convey waste?

39. What length restrictions do most codes place on horizontal wet vents?

40. Name two fixtures that can't convey waste through a 2-inch horizontal wet vent.

41. How many fixture units will some codes allow on a 3-inch horizontal wet vent?

42. What do you call the vertical drain between two fixtures connected to a stack at different levels?

43. What kind of vent is required by code for a sump that receives body waste?

44. Why is a local vent needed on a sump?

45. What is the minimum size vent required for a sump receiving body waste?

46. What are the venting requirements for sumps receiving clear water waste?

47. In what type building do some codes permit the use of vertical combination waste and wet vent piping?

48. What fixtures are prohibited from discharging into a combination waste and vent stack?

49. In what type of establishments are horizontal combination waste and vent systems usually installed?

50. Name three fixtures that may connect to a horizontal combination waste and vent system.

51. What's a major requirement before you can install a horizontal combination waste and vent system?

52. Why do codes prohibit the connection of appurtenances delivering large quantities of water to a horizontal combination waste and vent system?

53. What's the minimum size requirement for floor sink waste pipes installed underground in a horizontal combination waste and vent system?

54. How high above the roof must a vent terminal extend?

55. What's the minimum distance from a door that you may install a terminal for a sanitary vent system?

56. What does the code require to be installed on a horizontal vent that extends through a wall and turns upward?

57. Under what area should you never terminate a vent pipe?

58. At what height above ground level must a vent pipe terminate when it's installed outdoors?

59. What's the "rule of thumb" for determining the minimum aggregate cross-sectional area required for the vents used in venting a building drainage system?

60. What is the formula you use to find the cross-sectional area of a pipe?

Fixture Traps

By the mid 1850s, plumbing fixtures had been installed in a number of homes in New York City. For these early installations, the plumbers installed a handmade trap in the drain of individual fixtures. Of course, the intent was to prevent the entrance of odors and sewer gases. In those days, fixture traps often lost their water seals due to siphonage and back pressure. That led to two public health problems. First, rats could travel freely from building to building. And second, offensive odors from decomposing sewage in drainage systems permeated the buildings.

All efforts to use check valves and other specially-designed traps to prevent fixture trap seal loss proved ineffective. The principle of venting fixture drains was simply not known at that time. But the design and regulation of plumbing systems progressed rapidly after about 1900.

Building Traps

Health officials required building traps installed on each sanitary or combined building sewer. These proved to be generally effective. See Figure 4-1. They also provided a *secondary* safeguard to keep rats, vermin, sewer gases and odors out of a building. (The individual fixture traps provided the *primary* safeguard.)

That requirement for building traps was a big advance at the time. But since the development of modern collection, drainage and venting systems, most model codes don't require (and in fact actually *prohibit*) building traps.

In areas where sewer gases are extremely corrosive, there may even be a risk of explosions in the public sewer system, especially in larger cities. Occasionally you'll hear news reports of manhole covers blowing off, causing considerable damage. In those areas, local authorities may still require a building trap installed in the building drain line.

So building traps haven't entirely vanished. In the rare situations where they're required, install them as shown in Figure 4-1. It's important to install the cleanouts correctly. Because the building trap is a collection point, it invites stoppages. As always, refer to your local code before you plan the installation of a building trap.

Fixture Traps

The code requires that plumbing fixtures connect directly to the sanitary drainage system and include a water seal trap. Each fixture must be separately trapped, unless they have integral traps built into the fixture body (like a water closet). Integral traps have to meet appropriate manufacturing standards.

Traps are designed and constructed to provide a liquid seal. And they have to provide this protection without materially affecting the flow of sewage or other waste liquids.

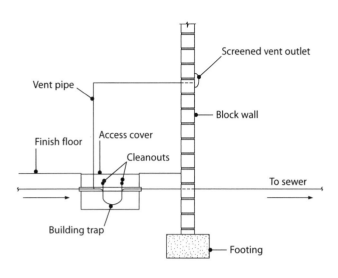

Figure 4-1
*Building trap detail — isometric
(not always code-approved)*

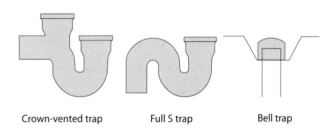

Figure 4-2
*These traps are **prohibited** by code*

Because of the trap's unique importance in protecting public health, the code places many restrictions and limitations on its use. Some of these are listed below. We'll discuss other requirements separately in more detail later.

Here are some of the special trap requirements:

▌ Fixture traps must be self-cleaning, with the exception of interceptor traps.

▌ No trap outlet may be larger than the fixture drain to which it's connected. For example, you can't connect a $1^1/_2$-inch trap to a $1^1/_4$-inch drain by using a reducer.

▌ You can't use any trap which depends on the action of movable parts to retain its seal.

▌ You can't use any trap with a single barrier or partition.

▌ Most model codes prohibit bell traps, crown-vented traps, pot traps, running traps, $^3/_4$ S traps, full S traps, drum traps, and traps with slip-joint nuts and washers on the discharge side of the trap above the water seal. See Figure 4-2.

▌ Each fixture trap must have a water seal that's between 2 and 4 inches deep. See Figure 4-3. The only exception is for interceptor traps, which all require deeper seals.

▌ Fixture traps installed below concrete floors on fill (or otherwise concealed) can't have trap cleanouts.

▌ All traps must be installed level in relation to their water seals. This prevents negative action and self-siphonage. A pipe that's not level also creates a bad joint for the fixture tailpiece. See Figure 4-4.

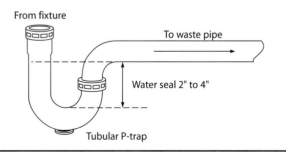

Figure 4-3
Minimum and maximum trap water seal

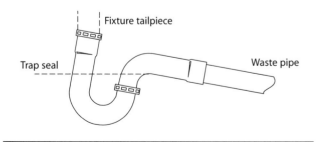

Figure 4-4
Improperly installed trap

■ Each plumbing fixture must be separately trapped by a water seal trap with the following exceptions: water closets, urinals or similar fixtures that have integral traps. No fixture can be double trapped. See Figure 4-5.

■ Two- or three-compartment sinks or laundry tubs of equal depth may connect to a single trap with a continuous waste. See Figure 4-6.

■ Two or three lavatories adjacent to each other may connect to a single trap if (a) the waste outlets don't exceed 30 inches, center to center, and (b) the trap is centrally located between the three lavatories. See Figure 4-7.

■ Restaurant, commercial and industrial sinks served by a single trap can't receive the discharge from a food waste disposal unit. Each food waste disposal unit must be separately trapped. See Figure 4-8.

■ A residential food waste disposal unit may be trapped separately from any other fixture or fixture compartment. In fact, it's *required* by some local codes, just like the commercial sink in Figure 4-8.

■ According to some codes, a residential food waste disposal unit may discharge through the continuous waste of a sink. When that's allowed, you must use a directional tee or other approved method served by a single trap. See Figure 4-9.

■ Some codes permit a domestic clothes washer to use the same trap serving a laundry tray, if it's adjacent to the laundry tray. See Figure 4-10.

■ Codes prohibit the connection of clothes washers or laundry trays to a trap serving a kitchen sink.

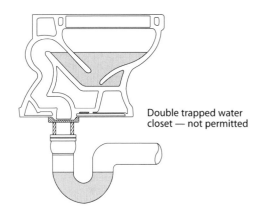

Figure 4-5
No fixture can be double trapped

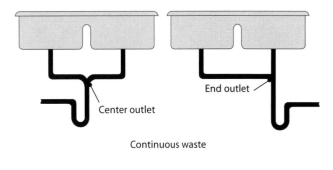

Figure 4-6
Two-compartment sink with one trap, approved

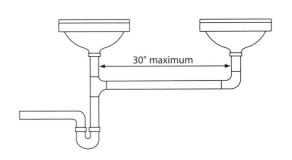

Figure 4-7
Two lavatories with one trap, approved

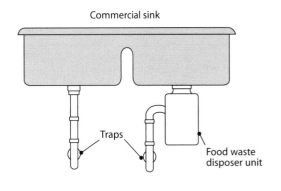

Figure 4-8
*Commercial food waste disposer unit
must be separately trapped*

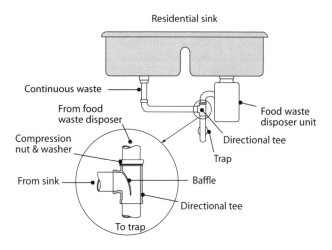

Figure 4-9
Typical residential food waste disposer installation using directional tee

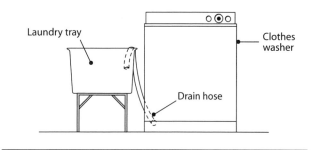

Figure 4-10
Clothes washer and laundry tray using same trap

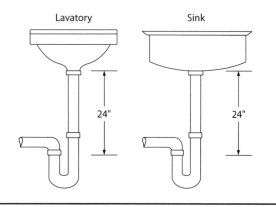

Figure 4-11
Maximum vertical drop from fixture waste outlet to the trap water seal

▪ Materials for concealed fixture traps (bathtubs, showers, floor drains and similar fixtures) must be cast brass, cast iron, lead, ABS plastic, PVC plastic, or other approved materials. Cleanouts are not allowed on concealed fixture traps.

▪ Except for fixtures with integral traps, exposed or accessible traps must be of cast iron, cast brass, lead, 17 gauge tubular brass, or copper. Some codes will accept 20 gauge tubular traps. In a plastic system, you must use ABS or PVC traps. Exposed fixture traps may be equipped with cleanouts.

▪ Traps for chemical, acid or corrosive wastes must be constructed of borosilicate glass, high silicon cast iron, lead pipe with walls at least $1/8$ inch thick, or other approved materials.

▪ There's a maximum vertical drop from a fixture waste outlet to the trap water seal. For kitchen sinks, lavatories, showers, bathtubs, and laundry tray fixtures, the allowable vertical drop is 24 inches. This further prevents self-siphonage of the fixture trap water seal. The shorter the distance between these two points, the more efficient the fixture trap will be. See Figure 4-11.

The *Standard Plumbing Code* has one two-part exception: (a) The vertical length of a washing machine standpipe can't exceed 48 inches. (b) The vertical standpipe inlet can't be less than 34 inches above the finished floor.

▪ The vertical drop of the pipe serving floor-connected fixtures with integral traps (water closets, floor mounted urinals and similar fixtures) can't exceed 24 inches. See Figure 4-12.

▪ Floor drains are considered fixtures, so they're governed by the 24-inch tailpiece (pipe) limit shown in Figure 4-11. But since it's often difficult to meet that limit, most codes approve the installation method shown in Figure 4-13. It shows a floor drain installed in a room where the horizontal building drain is 4 feet below the finished floor. The vertical rise from a horizontal drainage line serving a floor drain can't exceed 6 feet. The horizontal waste line can't be less than 3 feet from the vertical rise. Anything less than 3 feet would form an S trap, which is prohibited by code.

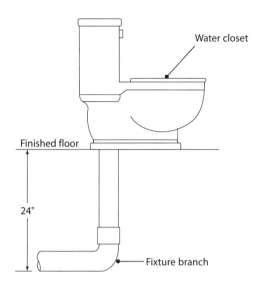

Figure 4-12
Maximum vertical drop from floor-mounted fixtures with integral trap

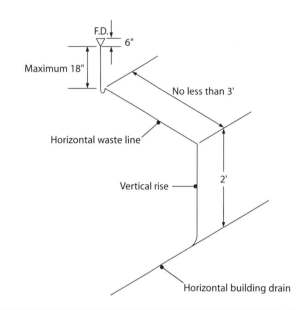

Figure 4-13
Special floor drain installation illustrating maximum vertical drop

Fixture Trap Sizes

One of the main reasons for code-established fixture trap sizes is to ensure they'll drain the fixture rapidly. Never use sizes smaller than those listed in Figure 4-14 (from the *Uniform Plumbing Code*). You'll find a more complete listing of trap sizes in your code book.

Horizontal Distance of Fixture Trap from the Vent

In the 1920s, the U.S. Department of Commerce made a substantial effort to establish standards for many plumbing requirements. Most codes now reflect those standards. But unfortunately, model codes differ greatly in several requirements, including the distance of traps from vents and fixture drain sizes. Refer to the table adopted by your local authority for this particular section of the code.

Figure 4-15 shows the discrepancies. The first table shows the requirements in the *Standard Plumbing Code* and the *International Plumbing Code*. The second table is from the *Uniform Plumbing Code*. Both

Plumbing fixture type	Trap size (in)
Bathtub (with or without overhead shower)	1½
Bidet	1¼
Clothes washing machine standpipe	2
Dental lavatory	1¼
Dishwasher, domestic	1½
Drinking fountain	1¼
Food waste disposer, commercial	2
Food waste disposer, domestic	1½
Kitchen sink, domestic	1½
Lavatory, domestic	1¼
Lavatory, public	1½
Laundry tray (1 or 2 compartment)	1½
Shower stall or drain	2
Sink, commercial	1½

From the UPC® with permission of the IAPMO ©1997

Figure 4-14
Minimum size of nonintegral traps

Standard Plumbing Code (1994) and International Plumbing Code (1997)		
Size of fixture drain (in)	Size of trap (in)	Distance from trap
1¼	1¼	3'6"
1½	1/2	5'0"
2	2	6'0"
3	3	10'0"
4	4	12'0"

Uniform Plumbing Code (1997)		
Size of fixture drain (in)	Size of trap (in)	Distance from trap
1¼	1¼	2'6"
1½	1½	3'6"
2	2	5'0"
3	3	6'0"
4 and larger	4	10'0"

Figure 4-15
Horizontal distance of fixture trap from vent at slope of ¼" per foot

codes require the fixture drain slope to be ¼ inch per foot. But each has a different distance for the developed length of the fixture drain to the vent.

Let's look at one example. For a 1¼-inch trap, the *Standard Plumbing Code* sets a limit from trap to vent of 3½ feet. The *Uniform Plumbing Code* sets a limit from trap to vent of 2½ feet. Does it make sense to have these differences in two of our major model codes? I don't think so. If a 1¼-inch trap will drain a

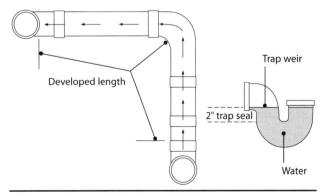

Figure 4-16
Developed length of fixture drain

fixture properly at 3½ feet in Alabama, it'll do the same thing in California. This is only one of many areas of the plumbing code where standardizing would be helpful to the trade. We can only hope that plumbing code writers will strive for greater standardization in the near future.

Here are the general principles for a horizontal run between fixture trap and vent (accepted by most codes):

▌ The closer the trap to the vent on a minimum slope, the better.

▌ Every fixture trap must be protected against siphonage and back pressure and have a vent piping system that permits the free flow of air under normal usage.

▌ Measure the developed length of a fixture drain from the crown weir of a fixture trap to the vent pipe. The measurement, including offsets and turns, must be within the prescribed limits set forth in your code book. See Figure 4-16.

▌ In some cases, because of the fixture location in a bathroom, kitchen or utility room, you'll have to exceed the allowable length of the fixture drain. When this happens, you must install a relief vent (re-vent), as illustrated in Figure 4-17.

▌ The traps of floor drains, depending on drain pipe size, are limited by their distance from an individual vent or a vented drainage line. See Figure 4-18.

Unvented Lavatories

When a horizontal waste branch exceeds the code-accepted length, some codes permit the installation of up to three lavatories or one kitchen sink without re-venting, if:

1) The horizontal waste branch is at least 2 inches in diameter throughout its length.

2) The fixture waste outlets connect into the side (not top) of the branch drain.

3) The branch drain slopes no more than ¼ inch per foot.

Before making this type installation, check your local code. See Figure 4-19.

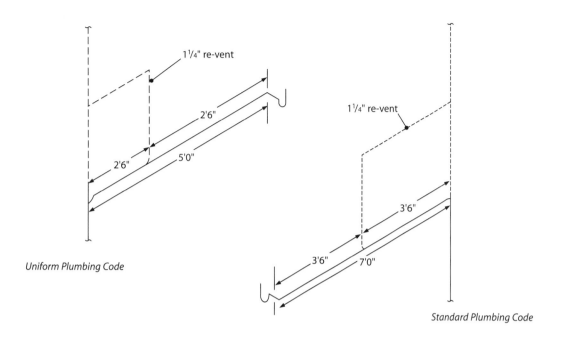

Figure 4-17

Fixture drain re-vent, 1¼-inch fixture drain
(fixture arm)

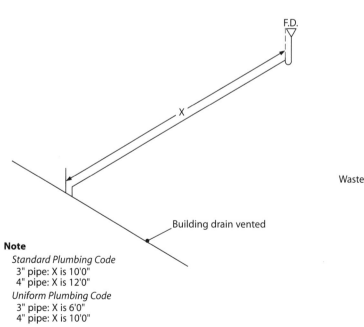

Figure 4-18

Maximum distance of floor drain
from building drain

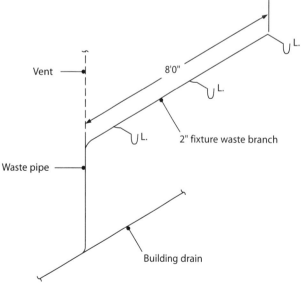

Figure 4-19

Three lavatories installed on one horizontal waste
branch (accepted by some codes)

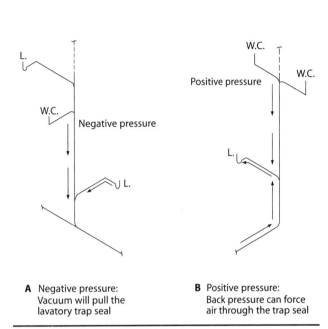

A Negative pressure:
Vacuum will pull the
lavatory trap seal

B Positive pressure:
Back pressure can force
air through the trap seal

Figure 4-20
*Major fixtures (water closets) can't be installed
above minor fixtures (lavatories)*

Broken Trap Seals

There are five ways most likely to break the trap seals: trap siphonage, back pressure, wind effect, evaporation and capillary attraction. Improperly installed plumbing systems can cause trap seal loss by siphonage and back pressure. The other problems aren't related to the installation, and they're far less likely to occur.

Figure 4-20 shows two examples of trap siphonage due to improper design. Figure 4-21 shows a broken trap seal as a result of improper installation.

Parts of a Fixture Trap

You're probably already familiar with the parts of the two most commonly-used traps in the plumbing trade. If not, Figure 4-22 should help you with the parts of a one-piece trap. Figure 4-23 will help you with the parts of a two-piece tubular trap. These simple things could be important on a plumbing examination.

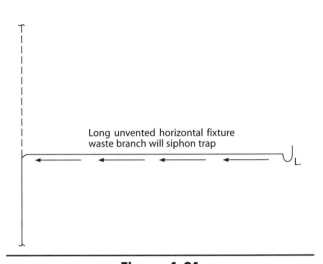

Figure 4-21
This type of installation is not permitted by code

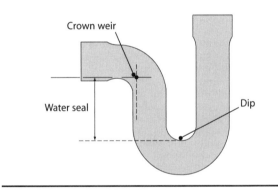

Figure 4-22
One-piece trap

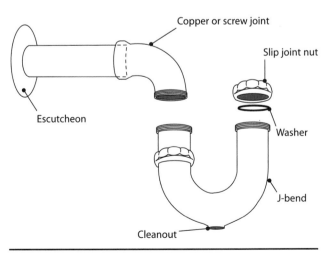

Figure 4-23
Two-piece tubular trap

Review Questions (answers are on page 280)

1. What is the purpose of a fixture trap?

2. Under what rare instances are building traps still required today?

3. When plumbing fixtures are connected directly to the drainage system, what must they be equipped with?

4. What must not be affected in the protection of a liquid seal?

5. What is the only kind of trap that doesn't have to be self-cleaning?

6. What rule governs the size of a trap outlet compared to its connecting fixture drain?

7. When can you use a trap that depends on the action of movable parts to retain its seal?

8. Name two traps prohibited by most model codes.

9. What is the minimum depth of fixture trap seals?

10. What is the maximum depth of fixture trap seals?

11. What kind of trap is exempted from the normally-required depth of a trap water seal?

12. Where should you locate the cleanouts when you install fixture traps below a concrete floor on fill?

13. When installing a fixture trap, what do you consider when determining its correct level?

14. Why may a water closet not have a separate trap?

15. When can two or three lavatories adjacent to each other use a single trap?

16. Where must a single trap be located when three lavatories are connected to it?

17. According to code, when may a fixture be double trapped?

18. When may a food waste disposal unit in a restaurant discharge through a pot sink trap?

19. When, according to some codes, may a food waste disposal unit discharge through a continuous waste of a sink served by a single trap?

20. Under what circumstances can a domestic clothes washer use the same trap that serves a laundry tray?

21. Name two materials that the code approves for concealed fixture traps.

22. What does the code prohibit concealed fixture traps from having?

23. When a tubular trap is used, what minimum gauge must it be?

24. Name two acceptable materials commonly used for accessible fixture traps.

25. What code-approved materials may be used for chemical, acid or corrosive wastes?

26. If lead pipe is used to convey chemical, acid or corrosive wastes, what is the required wall thickness?

27. What is the maximum vertical drop from a shower outlet to the trap water seal?

28. What is the maximum vertical drop of a pipe that serves floor-connected fixtures which have integral traps?

29. What is the maximum vertical drop of floor drains to the trap water seal?

30. What is one of the main reasons that fixture trap sizes were established by code?

31. What must every fixture trap be protected against?

32. When figuring the developed length of a fixture drain, what measurement, as well as the distance from the crown weir to the vent pipe, must be included?

33. In some instances, because of the fixture location, the fixture drain may exceed the limits set by code. When this occurs, what must be installed?

34. What are two adverse reactions to a fixture trap when a plumbing system is improperly installed?

5

Cleanouts

Years ago, cleanouts weren't required on drainage piping. If you had an obstruction, you'd have to cut a hole in the blocked drainage pipe and insert a cleaning cable to remove it. Then you'd patch the hole with a cement mixture or other impervious material. Sooner or later your patch job would likely deteriorate and allow raw sewage to seep out. Then you have a health hazard for anybody in the area.

The major plumbing maintenance problem is, in fact, clogged drains. The most common cause of clogged drains is a foreign object or some substance that's not intended for a drainage pipe. Here are the most common causes of stoppages:

1) Foreign objects lodged in the drainage pipe: These can be pencils, toys, baby diapers, paper tissues, toothbrushes, sanitary napkins, over-rim bowl deodorants, or anything else you can imagine. I've even found apples!

2) Accumulation of hair or other matter: There's no avoiding the problem of normal body hair accumulation. But there's no reason to make it worse. Pets should never be bathed in bathtubs or sinks of any kind. Avoid the use of bath oils. And don't use a clothes washer without a built-in lint catcher.

3) Deposits of grease or other substances: Grease, cooking oils, butter, gravy and coffee grounds can clog any kitchen drain. These materials should go out with the household garbage, not dumped into the kitchen drain, even if you have a garbage disposer.

All drainage piping is subject to stoppages, no matter how well designed the system. Accessible cleanouts will save you valuable time and the owner unnecessary expense. We're fortunate that cleanouts are now an absolute requirement for the drainage system. Today's model codes specify locations, distance between cleanouts, size, and other requirements.

The Importance of Accessible Cleanouts

Consider the relevant code when you design and install cleanouts for the building drain. It's essential that you provide complete access to all parts of the drainage system. Let's explore code requirements for this.

Location of Cleanouts

1) You must have a cleanout or test tee where a building sewer connects to the public sewer lateral at the property line. A test tee at this location serves a dual purpose. It serves as a cleanout and access for a test plug to perform a water test on a building sewer. See Figure 5-1. Some authorities require that a cleanout be extended up to the finished grade (Figure 5-2), while others don't. If you extend the cleanout to grade in an area with foot traffic, you'll have to countersink the cleanout head (Figure 5-3) to protect people from tripping and the cleanout from damage.

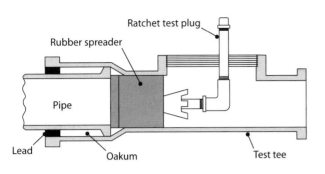

Figure 5-1
*Test tee used for water testing building sewer
and as a cleanout*

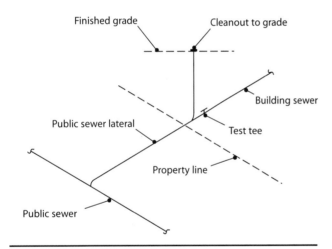

Figure 5-2
Cleanout extended up to finished grade

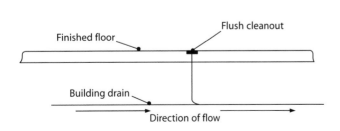

Figure 5-3
Countersunk cleanout

2) In most cases you'll have to locate a full-size cleanout outside the building at the junction of the building drain and the building sewer, usually within 2 feet (more in some codes) of the building line. Some administrative authorities don't require this cleanout if there are other cleanouts upstream accessible to a 75-foot sewer cable. If you put a cleanout at this location, it must be a two-way fitting that permits upstream as well as downstream rodding. See Figure 5-4. This fitting may or may not have to be brought to grade, depending on local code requirements. And again, in an area that carries foot traffic, you'll have to countersink it (Figure 5-3).

3) Cleanouts are required on building sewers 4 inches and larger (Figure 5-5). Separation distances vary according to pipe size and the code that applies. Always check your local code requirements. For example, with a 4-inch pipe, the model codes have three different requirements:

▌ *Uniform* and *International Plumbing Codes*: 100 feet

▌ *Standard Plumbing Code*: 80 feet

▌ *National Plumbing Code*: 75 feet

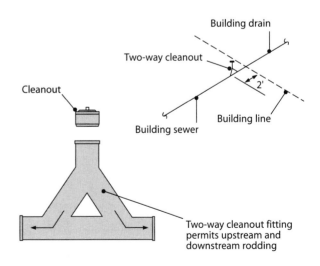

Figure 5-4
Two-way cleanout (shown isometrically)

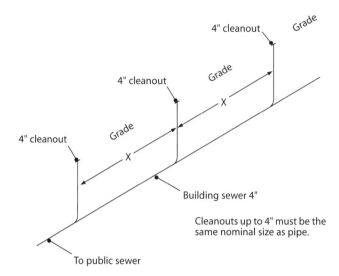

Cleanouts up to 4" must be the same nominal size as pipe.

Note Separation distances "X" vary according to code used. See text.

Figure 5-5
Cleanouts required

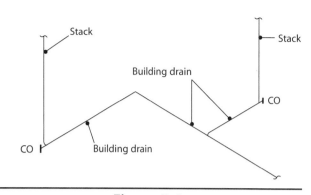

Figure 5-6
A cleanout must be provided at change of direction greater than 45 degrees

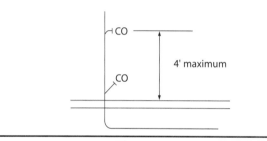

Figure 5-7
Provide a cleanout at the base of all required stacks (either location is acceptable)

4) Accessible cleanouts are required on all horizontal drainage piping. Check your local code for the separation distances. For each change of direction greater than 45 degrees in a building drain, you must provide a cleanout. See Figure 5-6.

5) You can put cleanouts in horizontal drainage lines beneath concrete flooring if they're accessible and flush with the finished floor. Cleanouts in walkways, hallways, and rooms must have countersunk plugs to prevent tripping or accidental injury. See Figure 5-3.

6) You can install a cleanout in the base of an exposed or concealed vertical stack not more than 4 feet above the finished floor, as shown in Figure 5-7. If you can't extend the cleanout to an accessible outside location, use a cleanout tee in the vertical stack. The cleanout plug must be accessible. If it's concealed, the cleanout must have a removable cover plate or access door to permit rodding. The raised head on the cleanout plug may be tapped to accept the long screw that holds it in place (Figure 5-8).

7) When you extend a cleanout from the base of a stack to the outside of a building, it creates a dead end. You can't locate the cleanout more than 5 feet from the building wall. See Figure 5-9.

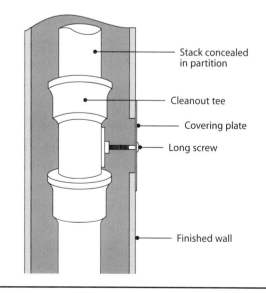

Figure 5-8
Cleanout on vertical stack

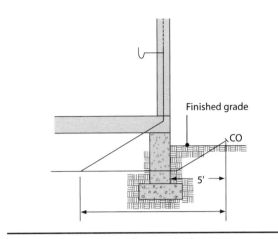

Figure 5-9
Dead end created with cleanout extension

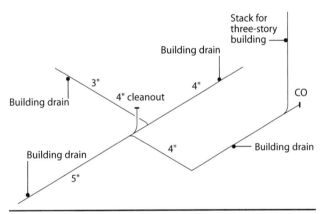

Figure 5-10
Where cleanouts are required

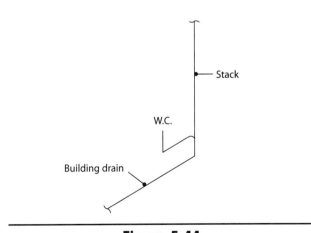

Figure 5-11
Water closet outlet is considered an equivalent cleanout by most codes

8) Every code requires a cleanout at the upstream end of the pipe for each horizontal drainage line.

9) For additional cleanout requirements and cleanout locations, see Figure 5-10. The stack needs a cleanout at its base because the change of direction is greater than 45 degrees and it's higher than one story. Where building drain pipes join together, requiring the drain size to increase in size, use a cleanout of the same size, brought to grade or finished floor level.

Cleanout Exceptions

1) The code generally considers a water closet as a substitute for a regular cleanout. The water closet fixture drain need not have a cleanout. See Figure 5-11.

2) An exposed P trap connected to the drainage pipe with a slip joint or ground joint connections doesn't require a cleanout at the base of the stack.

3) P traps that accept the discharge from residential clothes washers, floor drains, shower drains or tub drains with removable strainers don't require cleanouts.

4) In one-story buildings, you can use the roof stack terminal as a cleanout if it's permitted by your code. This is a controversial section of the code, so always check first. Figure 5-12 shows the situation. A two-way cleanout in the building drain permits rodding upstream to the base of stack A, so you don't need a cleanout in its base. But stack B *will* require a cleanout at its base. Although it's properly sized, it's not vertical throughout, as required by code. Stack C has the same diameter (2 inches) as the waste pipe it serves and is vertical up to and through the roof. So you don't need a cleanout in its base.

5) Rain leaders require a cleanout (Figure 5-13).

Cleanout Clearances

Every cleanout installed on 3-inch or larger pipe needs at least an 18-inch clearance to allow room for rodding. For cleanouts smaller than 3 inches, leave a clearance of at least 12 inches for rodding. See Figure 5-14.

Direction of Flow

The cleanouts you install must allow someone to insert a sewer cable in the direction of flow to clear blockages. Use two-way cleanouts to permit upstream and downstream rodding whenever practical. See Figures 5-3 and 5-4.

Prohibited Uses for Cleanouts

Never use cleanout openings to install another fixture or a floor drain unless you get written permission from your local authority. And if you do change the use of a cleanout, you must provide another cleanout of equal access and capacity.

Cleanout Material and Design

The required thickness for cleanout ferrule bodies is the same as that required for pipe and fittings of like material. The cleanout plug must extend at least $^1/_4$ inch above the hub. Any cleanout plug for new installations must be made of heavy brass or plastic at least $^1/_8$ inch thick, with a raised nut or a recessed socket for removal. The ferrule and cleanout plug must have ANSI standard tapered pipe threads. When you have to replace a damaged cleanout plug, you can use a heavy replacement lead or nylon plug. See Figure 5-15.

Cleanout Sizes

Cleanouts must be the same nominal size as the pipe in which they're installed, up to 4 inches. See Figure 5-16. You can also use a 4-inch cleanout for building drains 4 inches and larger. For building sewers 8 inches and larger, most codes require manholes.

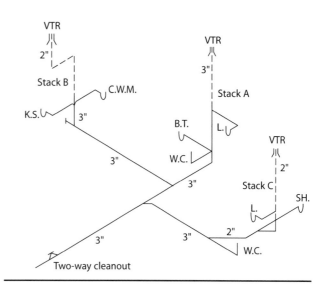

Figure 5-12
Additional cleanout locations

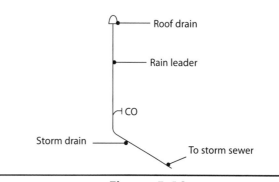

Figure 5-13
Cleanout required for rain leader

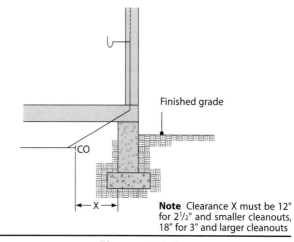

Note Clearance X must be 12" for 2½" and smaller cleanouts, 18" for 3" and larger cleanouts

Figure 5-14
Cleanout clearances

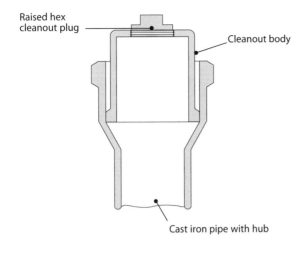

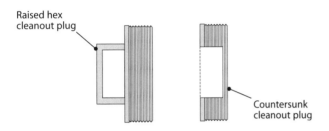

Figure 5-15
Cleanout design

Standard Plumbing Code, 1996		Uniform Plumbing Code, 1997	
Nominal pipe size (in)	Nominal cleanout size (in)	Nominal pipe size (in)	Nominal cleanout size (in)
1¼	1¼	1½	1½
1½	1½	2	1½
2	2	2½	2½
3	3	3	2½
4 and larger	4	4 and larger	3½

Note Some codes require a 6″ cleanout for an 8″ and larger building drain. The *Standard Plumbing Code* requires manholes for exterior drainage pipes 8″ and larger. Some codes require manholes only when pipes are larger than 10″. Always check your code for verification.

Figure 5-16
Cleanout sizes

You must locate a manhole with an approved cover at every change in direction, pipe size, alignment, grade or elevation. Straight runs must have manholes no more than 400 feet apart (*Standard Plumbing Code*) or 300 feet apart (*Uniform Plumbing Code*). Approved manhole covers must be used. See Figure 5-17.

Exterior Cleanouts

When you install drainage lines under blacktop or other paved surfaces (in commercial locations, for example), you may have to terminate a standard type cleanout in an area that carries vehicular traffic. If you do, install an approved cleanout box. That will protect the drainage pipe and cleanout from surface loads. A cleanout box is generally made of concrete or metal, fitted with an extra heavy removable cover for rodding. See Figure 5-18.

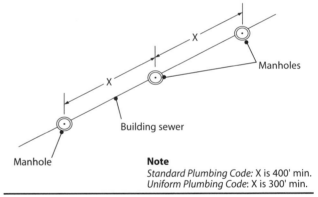

Note
Standard Plumbing Code: X is 400' min.
Uniform Plumbing Code: X is 300' min.

Figure 5-17
Manhole spacing

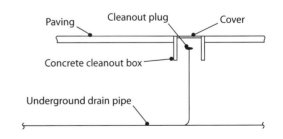

Figure 5-18
Cleanout and piping installed under paved area

Review Questions (answers are on page 281)

1. What procedures did plumbers use to clear stoppages before cleanouts became an essential part of the drainage system?

2. Name three important requirements for cleanouts regulated in today's model codes.

3. What is the most common plumbing maintenance problem?

4. What is the most common cause of clogged drains?

5. Name two things commonly found in the kitchen that are likely culprits in causing a clogged drain.

6. What dual purpose does a cleanout (or a cleanout tee) provide when installed where a building sewer connects to the public sewer lateral?

7. What is the minimum size of a cleanout installed in a 4-inch pipe at the junction of a building drain and a building sewer?

8. What is the name of the fitting that permits upstream as well as downstream rodding?

9. When a cleanout is extended to grade in an area subject to frequent traffic, what type cleanout head should be used?

10. According to the *Uniform Plumbing Code*, what is the maximum separation distance between 4-inch cleanouts?

11. What must be installed near the base of each vertical waste or soil stack?

12. When a dead end is created by a cleanout, what is the maximum distance it can extend outside the building wall?

13. What plumbing fixture is sometimes considered a substitute for a cleanout?

14. What is not needed in a P trap into which floor drains with removable strainers discharge?

15. When is the only time a roof stack terminal in a one-story building may be used as a cleanout?

16. What does the code require rain leaders be equipped with when they connect to a horizontal storm drain?

17. What clearance is required by code for a 2-inch cleanout?

18. In what direction should cleanouts be installed?

19. What are two prohibitions for the use of cleanout openings?

20. What must cleanout plugs be equipped with?

21. What is the smallest size cleanout acceptable for a 6-inch building drain, according to the *Uniform Building Code*?

22. When may a building sewer be installed without manholes?

23. According to the *Standard Plumbing Code*, what is the maximum distance between manholes on a sewer installed on a straight run?

24. What does the code require when a standard-type cleanout terminates in an area where there's vehicular traffic?

Interceptors, Special Traps and Neutralizing Tanks

Grease, oil, flammable waste, sand, plaster, lint, hair, glass and acids are objectionable — even harmful — to building drainage systems, public sewers, sewage treatment plants or septic tanks. That's why the code requires these materials to be intercepted, separated or neutralized before they enter any private or public drainage system.

Commercial buildings primarily use *interceptors* or *separator traps* to accumulate and recover objectionable substances in liquid waste before they can enter the drainage system. When you install interceptor traps, be sure to locate them where they're readily accessible. They don't do any good unless there's a way to maintain the interceptors and remove the accumulated matter.

Interceptors, *separators* and *neutralizing tanks* are available in many types and sizes to fit their particular functions. Make sure you have approved detailed drawings, specifications and locations clearly shown on your working blueprints before you begin installation. And keep in mind that waste that doesn't require intercepting, separation or neutralizing can't discharge through them.

Grease Interceptors

Grease interceptors are required in these commercial buildings:

- restaurants
- hotel kitchens
- cafeterias
- bars
- clubs
- supermarkets
- meat processing plants

Grease interceptors aren't generally required in single-family residences, private living quarters, apartment buildings or establishments that sell only take-out food. There are two types of grease interceptor installations, called simply *inside* and *outside* installations.

Inside Installations

Most codes don't provide established sizing methods for this type of installation. Instead they either recommend design criteria or leave it to the discretion of the local health department or plumbing officials.

Depending on code approval, architects will probably choose an inside interceptor for small restaurants or other businesses generating small amounts of grease. Grease interceptors must have a grease retention capacity of 2 pounds for each GPM of flow. Figure 6-1 gives you the most common code sizing criteria for small establishments. Grease interceptors for small restaurants require at least 1.5 hours retention time.

Number of fixtures connected	Required rate of flow per minutes (gal)	Grease retention capacity (lb)
1	20	40
2	25	50
3	35	70
4	50	100

Figure 6-1
Sizing criteria for inside grease interceptor

Figure 6-2 shows a typical factory-built cast iron unit. This kind of interceptor is available in sizes ranging from 40 pounds to the maximum 100 pounds permitted by code. Figure 6-3 shows typical installations for three common inside interceptors — two floor-mounted and one below the floor.

Small grease interceptors like these always need an approved flow control fitting located where it's visible and accessible. The flow can't exceed the rated capacity of the interceptor. Flow control fittings with adjustable or removable parts are prohibited. You can see the flow control fitting in all three installations in Figure 6-3.

Each fixture discharging into a grease interceptor (Figure 6-3 A) must be individually trapped and vented in an approved manner. But notice there's isn't a trap in Figure 6-3 B. You can omit a fixture trap for a single fixture if it meets the following conditions:

▮ The horizontal distance between the fixture outlet and the grease interceptor is no more than 4 feet (5 feet in the *International Plumbing Code*).

▮ The vertical fixture tailpiece or drain isn't more than 30 inches.

Locating an Inside Grease Interceptor

▮ The interceptor must be easily accessible for inspection, cleaning and removal of grease at all times.

▮ Service personnel must have access without using ladders or moving heavy objects.

▮ Never install the interceptor in a part of the building where food is handled.

▮ Locate the interceptor as close as possible to the fixture it serves.

In some parts of the country you can buy water-cooled interceptors which speed up the coagulation of the grease. While this improves the efficiency of the interceptor, it's not always the safest system. The jacket contains a circulating potable water supply for cooling purposes. If it fractures or corrodes, there's the

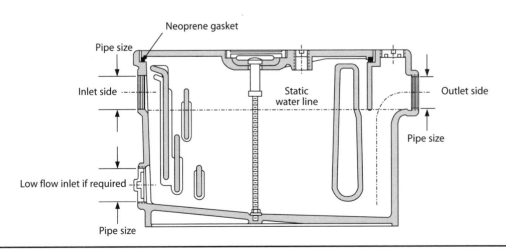

Figure 6-2
Typical inside grease interceptor

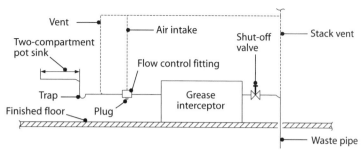

A Floor-mounted interceptor on same floor as sink

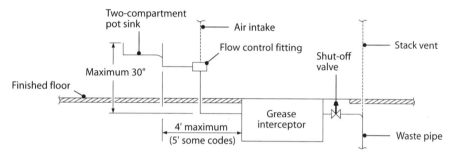

Notes Maximum drop from sink outlet to grease interceptor inlet is 30". Maximum horizontal distance from sink outlet to grease interceptor inlet (without trapping sink) is 4' (5' *IPC*).

B Recessed interceptor, flush with finished floor

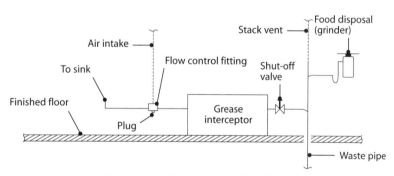

C Floor-mounted interceptor with food disposer

Figure 6-3
Typical installations of inside grease interceptors

potential for a potable water cross-connection. Codes almost always prohibit the use of such interceptors. Don't waste time installing a water-cooled unit unless you're certain that it has local plumbing official approval.

Codes generally prohibit food waste disposals (grinders) and commercial dishwashers from discharging through a grease interceptor. See Figure 6-3 C. Local plumbing officials may make an exception

for dishwashers if they're satisfied that the temperature of the waste water and/or the action of detergents won't affect the operation of the interceptor.

Outside Installations

Most codes require sizing grease interceptors for fully-equipped commercial restaurants according to seating capacity (meals served) and hours of operation

Liquid capacities of grease interceptors for eat and/or drink establishments (gal)			
Up to 50 persons	51 to 100 persons	101 to 150 persons	151 to 200* persons
750	1,500	2,250	3,000

*For restaurants with more than 200 fixed seats, multiply the additional seats by 15 gallons.

Figure 6-4

Liquid capacities of grease interceptors for restaurants

(i.e., 8, 16 or 24 hours). The formula should also provide for a retention period for cooling and separation of grease. For example, a fully-equipped restaurant open for business 8 to 24 hours requires a retention time of 2.5 hours. A single-service kitchen open for the same period requires a retention time of only 1.5 hours.

Most codes don't provide established sizing methods for commercial grease interceptors but leave it to the discretion of the health department or the local plumbing official. They usually require sizing commercial grease interceptors by the number of fixed seats or meals served at peak hours. However, each code has a different formula. Identical restaurants in different geographical locations often require interceptors of different sizes.

We'll look at two model codes — a typical local code and the *Uniform Plumbing Code*. The local code has a sizing table for restaurants with seating capacities up to 200 persons (Figure 6-4 in this book). As a rule-of-thumb guide for restaurants with seating capacities larger than 200, multiply the *additional seats* by a waste flow rate of 15 gallons. Neither the hours open nor the type of equipment used is a factor.

The *Uniform Plumbing Code* has a more complicated formula. It covers both single-serve and fully-equipped kitchens and generally requires a much larger grease interceptor. See Figure 6-5. Here's the formula it uses:

Number of meals per peak hour × waste flow rate × retention time × storage factor = interceptor size (liquid capacity)

Now let's compare the two code formulas. Our sample restaurant serves 200 meals per peak hour, with two peak hours each day. The local code would require a grease interceptor that could accommodate a liquid capacity of 3,000 gallons (Figure 6-4).

Number of meals per peak hour[1]	×	Waste flow rate[2]	×	Retention time[3]	×	Storage factor[4]	=	Interceptor size (liquid capacity)
1. Meals served at peak hour								
2. Waste flow rate								
A. With dishwashing machine						6 gallon flow		
B. Without dishwashing machine						5 gallon flow		
C. Single-service kitchen						2 gallon flow		
D. Food waste disposer						1 gallon flow		
3. Retention time								
A. Commercial kitchen waste, dishwasher						2.5 hours		
B. Single-service kitchen, single serving						1.5 hours		
4. Storage factors								
Fully-equipped commercial kitchen						8 hours (1 peak hour)		
						16 hours (2 peak hours)		
						24 hours (3 peak hours)		
Single-service kitchen						1.5 peak hours		
Self-service laundries						1.5 hours		

Figure 6-5

Sizing criteria for outside grease interceptor from the Uniform Plumbing Code

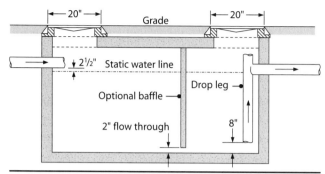

Figure 6-6
Outside grease interceptor detail

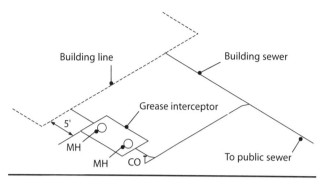

Figure 6-7
Outside grease interceptor location

The *Uniform Plumbing Code* (Figure 6-5) would calculate like this:

200 meals × 6 gallons flow rate × 2.5 hours retention time × 2 peak hours (16 hours) = 6,000 gallons capacity

That's exactly twice as large as required by the local code. And you've probably already figured out why. The *UPC* formula *does* consider the number of peak hours the restaurant is open. Each eight-hour period includes one peak hour. There are also other factors that contribute to its sizing: whether there's a dishwashing machine and/or a food waste disposer, and whether the retention time is 1.5 or 2.5 hours.

As a plumber you probably won't be asked to install outside grease interceptors. You'll only connect the piping to the inlet and outlet openings provided. But you still need to know how to size, design and locate grease interceptors. Many questions on the plumbing exams come from this section of your code.

Grease Interceptor Design (Outside)

Your local code governs the materials that can be used for outside grease interceptors. The most commonly-accepted material is concrete, either precast or poured-in-place. In some areas, the code may also approve steel, fiberglass-reinforced polyester or polyethylene. The structural design criteria for outside grease interceptors are fairly uniform across the country. Figure 6-6 shows the principal requirements.

▪ The inlet invert must discharge at least 2¹/₂ inches above the liquid level line.

▪ The outlet tee must extend to within 8 inches of the bottom of the tank.

▪ For maintenance, provide two minimum 20-inch diameter cleanout manholes (one over the inlet and one over the outlet tee) brought to grade. Manhole covers must be gas-tight.

▪ For cleaning purposes, very large grease interceptors require a minimum of one manhole per 10 feet of length. (Example: a grease interceptor 21 feet long would require three manholes, one over the inlet tee, one over the outlet tee, and one in the middle.)

▪ Grease interceptors must be designed and installed so that they won't become air bound. (Grease interceptors don't require a local vent, as adequate air can pass through the vents of the fixture they serve.)

▪ Most codes require grease interceptors to have at least two compartments with fittings designed for grease retention. See Figure 6-6.

▪ In areas subject to vehicular traffic, grease interceptors must have adequate reinforcement and cover.

▪ Waste discharge piping from an interceptor may be connected to the existing building sewer. See Figure 6-7.

Grease Interceptor Location (Outside)

Observe the following code restrictions when installing an outside grease interceptor on a building site.

▪ Locate the grease interceptor where there's easy access for inspection, cleaning and removal of intercepted grease.

■ An interceptor may serve only one establishment. There's one exception, however. Some codes will accept one or more centrally-located grease interceptors in large shopping centers. It may serve several grease-generating businesses, provided the shopping center management assumes, in writing, full maintenance responsibility.

■ The code prohibits installation of an outside grease interceptor closer than 5 feet to a building foundation or private property line.

Greasy Waste Systems

Greasy waste lines are designed and installed as a separate drainage system. Waste discharge from fixtures and equipment in establishments that generate grease may connect to the building sewer after passing through a grease interceptor. This includes scullery sinks, pot and pan sinks, dishwashing machines, soup kettles, garbage can washers, and floor drains that may receive kitchen spills. Floor drains and floor sinks that receive waste from certain fixtures and appliances in or near the kitchen (except commercial food disposers) must also connect to the greasy waste system.

Toilets, urinals and other similar fixtures must never waste through a grease interceptor.

There are two types of code-approved greasy waste systems. These are the *conventional greasy waste system* and the *combination waste and vent system*. The two are quite different, so we'll take a closer look at each.

Conventional Greasy Waste System

As the term implies, the conventional system is the most common and the one you'll most often work with. Both small and large restaurants require similar design and installation methods.

The layout and installation of a greasy waste system is no different from that of a building sanitary drainage system. To size the pipes and locate the vents, floor drains and fixtures, just follow the requirements outlined for DWV systems in your local code. An acceptable conventional greasy waste system for a small restaurant with an outside interceptor is shown in Figure 6-8.

Combination Waste and Vent System

The combination waste and vent system is much less common. Codes permit its use only where structural conditions prevent the installation of a conventional system. The combination system provides the horizontal wet venting of a series of traps with a common waste and vent pipe. At best, it's a custom-designed system for locations where you can't provide venting in the usual manner.

Some codes will approve the use of a combination system where conventional venting isn't practical:

■ Extensive floor drainage

■ Group shower drains

■ Floor sinks in supermarkets

■ Demonstration or work tables in school buildings

■ Similar applications where fixtures aren't located adjacent to walls or partitions

Some codes recommend that you don't connect grease-producing restaurant kitchen equipment to a combination waste and vent system because it's not self-scouring. Other codes don't object to this usage. Check your local code for specific requirements.

All codes require oversized combination system piping. Drainage pipe sizes are at least two pipe sizes larger than those required for a conventional system. This balances the system because the flow line is low enough in the waste pipe to allow adequate air movement in the upper portion. This prevents the loss of trap seals or the possibility of an air-lock condition.

To retain the grease for proper disposal, some codes require installation of an inside grease interceptor as close as possible to any grease-generating fixture or equipment. Figure 6-9 shows a combination system with an inside grease interceptor. Figure 6-10 details how to install three of the fixtures from Figure 6-9.

Commercial kitchen equipment schedule
1. Hand sink
2. Garbage disposer
3. Milk dispenser
4. Ice maker
5. Automatic glass cleaner
6. Bottle cooler
7. Three-compartment sink
8. Pot sink
9. Coffee maker

F.D.   = floor drain
F.S.   = floor sink
I.W.   = indirect waste
S.D.   = sanitary drain
G.W.D. = greasy waste drain

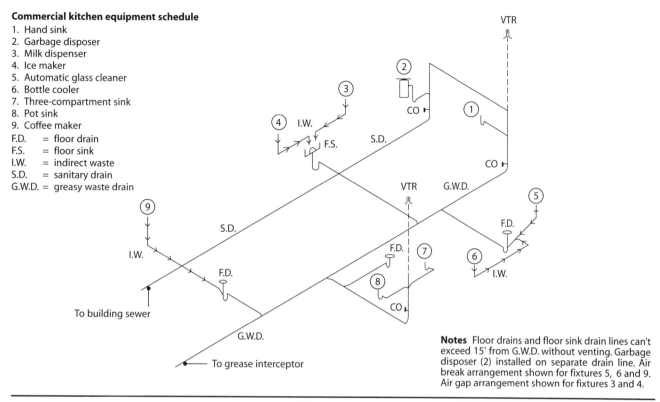

Notes Floor drains and floor sink drain lines can't exceed 15' from G.W.D. without venting. Garbage disposer (2) installed on separate drain line. Air break arrangement shown for fixtures 5, 6 and 9. Air gap arrangement shown for fixtures 3 and 4.

Figure 6-8
Conventional greasy waste collection system

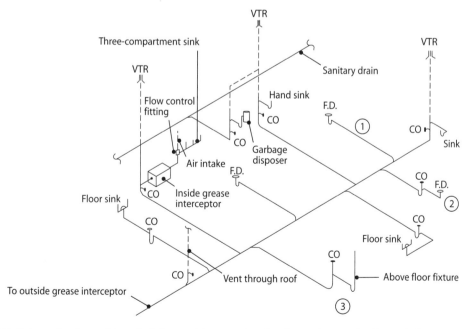

Notes Main waste pipe and branch drains must be two pipe sizes larger than required by code for conventional systems. Garbage disposer connects to the building sanitary drain. Inside grease interceptors *optional* for this illustration, as an outside grease interceptor is provided. Figure 6-10 illustrates an enlarged design for fixtures 1, 2 and 3 of the above drawing.

Figure 6-9
Greasy combination waste and vent system

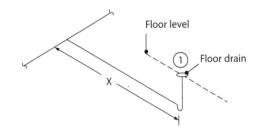

Distance X shall not exceed 15' without venting.

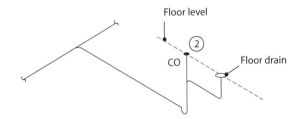

Floor drain used as indirect waste receptor with
cleanout at floor level.

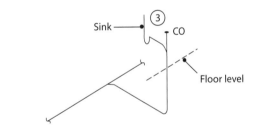

Normal roughing for above-floor fixture in a combination
waste and vent system.

Figure 6-10
*Examples of how to install fixtures 1, 2 and 3
as illustrated in Figure 6-9*

Most codes accept these requirements for a greasy
combination waste and vent system:

▪ Each waste pipe and each trap must be at least
two pipe sizes larger than the sizes required for a
conventional system. (Pipe sizes recognized for
this purpose are 2, 2½, 3, 3½, 4, 4½, 5 and 6
inches, and so on.) For example, a 2-inch pipe
would increase to 3 inches, a 3-inch pipe to 4
inches and a 4-inch pipe to 5 inches.
(Remember, most *drainage* pipes aren't manufactured in ½ pipe sizes but are used only in sizing pipes for the combination system.)

▪ Some codes will accept fixture traps and drains
in the normal code size if they're installed above
the floor.

▪ Traps and vertical waste pipes must be two pipe
sizes larger than the fixture tailpiece they serve.

▪ The fixture tailpiece should be as short as possible, but should never exceed 2 feet.

▪ Dishwashers and grease-producing sinks must
drain through a grease interceptor.

▪ To ensure adequate venting, appurtenances that
deliver large quantities of water (such as pumps)
should not discharge through a combination
system.

▪ Branch lines more than 15 feet long must be separately vented. See the footnote in Figure 6-10.

▪ Install one vent upstream and one vent downstream from all fixtures in the system.

▪ The minimum area of any vent must be at least
one-half the size of the waste pipe served. If a
waste pipe is 4 inches, the vent pipe must be at
least 2 inches.

▪ Each vent stack must have an accessible
cleanout.

▪ Long mains must have additional relief vents
installed at a minimum of 100 feet apart.

▪ Cleanouts aren't usually required on any wet-
vented branch serving a single trap when the fixture tailpiece is 2 inches and provides ready
access for cleaning through the trap.

▪ Fixtures producing 7½ gallons of waste per
minute must be considered as one fixture unit of
load value.

Laundry Interceptors

Lint interceptors aren't required in single-family
houses or apartment buildings if there's a washer in
each unit. Apartment complexes with a central
laundry room are considered commercial and do
require lint interceptors.

Commercial and self-service laundries discharge
solids (such as lint, string and buttons) along with the
liquid waste. They must have lint interceptors to

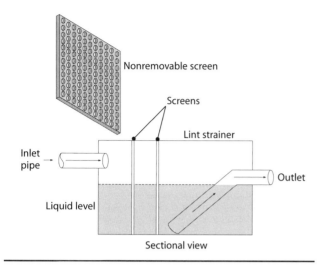

Figure 6-11
Lint interceptor detail

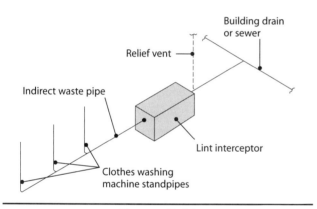

Figure 6-12
*Lint interceptor installed in drainage pipe
from laundry*

prevent these solids from entering the drainage system. Install a nonremovable screen or a $1/2$-inch mesh screen metal basket to collect the solids. The screen construction must allow for easy cleaning. See Figure 6-11.

Some codes label the horizontal drainage pipes which serve self-service clothes washing machines as indirect waste pipes (Figure 6-12). This unique method of piping is economical and practical for this particular application.

The indirect waste system doesn't have to be trapped or vented, like most other plumbing fixtures. The washing machine standpipes are open-ended, 3- or 4-inch diameter pipes extending to about 26 inches above the finished floor. These drain pipes receive the discharge from washers through flexible hoses. A 3-inch standpipe can accommodate two machines; a 4-inch standpipe will serve four machines. Horizontal drain pipes collect waste from the standpipes and convey that waste to a lint interceptor. You don't have to vent this system because the standpipes permit free circulation of air.

Other codes require each washing machine to discharge into individual traps, be vented and connect separately to the horizontal drain pipe. See Figure 6-13. Check with local authorities for the system that's acceptable in the area where you work.

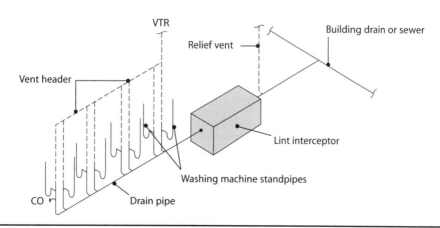

Figure 6-13
Clothes washing machines individually trapped and vented (required by some codes)

No. clothes washing machines	×	2 cycles per hour	×	Waste flow rate	×	Retention time	×	Storage factor	=	Interceptor size (liquid capacity)
10	×	2	×	50	×	2.0	×	1.5	=	3,000

Figure 6-14
Sizing criteria for lint interceptor for self-service laundries

The outlet pipe from the lint interceptor is connected to the regular sanitary drainage system serving other fixtures within the building. The code requires that a vent be installed on the horizontal discharge (outlet) pipe as close as possible to the lint interceptor. The vent pipe serves the drainage pipe between the lint interceptor and the building drainage system or building sewer. It supplies and removes air as needed. This ensures a free flow of waste water by preventing the lint interceptor from becoming air locked. See Figures 6-12 and 6-13.

Sizing Lint Interceptors

It's usually the responsibility of the architect to size lint interceptors. But you may have to calculate the lint interceptor size when an architect's seal isn't required — when a self-service laundry locates in an existing building, for example. Unfortunately most codes don't provide established sizing methods. They just recommend design criteria, or leave it to the discretion of local plumbing officials. Commercial laundries rely on the manufacturer's specifications.

Lint interceptors for self-service laundries are sized according to the number of washing machines installed and other factors like the number of cycles, waste flow rate, retention time and storage factor (Figure 6-14). Some codes require a retention period of at least 2.0 hours. Other codes don't require a retention period but depend solely on the screen to collect the solids. Let's see how two codes compare in the sizing of a lint interceptor for a self-service laundry.

A local code provides a fairly small interceptor unit, located either inside or outside a building. Under that code, the minimum size lint interceptor approved for a self-service laundry with up to ten washing machines has a liquid capacity of 450 gallons. This is the formula the local code uses:

10 washers × 2 cycles per hour × 3 fixture units per washer × 7.5 gallons flow per fixture unit = 450 gallons storage capacity

This flow-through interceptor has very little retention time. It counts on the built-in screens to collect the solids as the water rushes by. Because of its small capacity, it requires regular maintenance to clear the intercepting compartment and screens. See Figure 6-11.

For self-service laundries that have either more or fewer than ten machines, you'd multiply each machine by 45 gallons:

1 machine × 2 cycles × 3 fixture units per washer × 7.5 gallons per fixture unit = 45 gallons

The *Uniform Plumbing Code* provides guidelines for a large unit that must be located outside the building (Figure 6-14). Let's use the *UPC* to size the lint interceptor for the same ten washer self-service laundry that's open 16 hours a day. The *UPC* table that estimates waste/sewage flow rates for self-service laundries allows 50 gallons per wash cycle per washing machine. It also allows a 2.0 hour retention time and a 1.5 storage factor, which the local code doesn't require. Here's the calculation:

10 washers × 2 cycles per hour × 50 gallons per cycle flow rate × 2.0 hours retention time × 1.5 storage factor = 3,000 gallon liquid capacity

As you can see, there may be a wide variation between codes. Always verify the sizing method used in your particular area.

Most codes leave the sizing of lint interceptors for commercial laundries to the manufacturer's specifications. It's very different from the sizing of self-service laundries.

Lint Interceptor Construction

Lint interceptors may be made of precast concrete or poured-in-place concrete, steel or other approved materials. No matter what the material, a lint interceptor must have nonremovable baskets or screens to trap materials that would harm the building drain pipe. Larger units require a special basket to trap the solids from waste water entering the lint interceptor. But some codes allow removable screens, usually on small units. Where they're allowed, the screens or basket must be easy to clean.

In most cases, you'll install the inlet pipe one pipe size above the liquid level. The outlet or discharge pipe generally extends downward on a 45-degree angle to within approximately 2 inches of the bottom of the lint interceptor. This achieves two purposes:

1) It prevents lighter objects that pass through the screen and float on the water from sinking into the outlet pipe and then getting into the sanitary drainage system.

2) It creates a liquid seal that serves as a trap, preventing sewer gases from entering the building through the washer standpipes.

Gasoline, Oil and Sand Interceptors

Interceptors are absolutely essential in some settings to prevent gasoline, grease, oil or sand from getting into the drainage system. In the following places they're *always* required:

▮ Repair garages where motor vehicles are serviced and repaired, and where floor drainage is provided

▮ Commercial motor vehicle washing facilities

▮ Gasoline stations with grease racks, grease pits or wash racks

▮ Factories which have oily and/or flammable wastes as a result of manufacturing, storage, maintenance, repair or testing

▮ Public storage garages with floor drainage

▮ Anyplace where sand, oil, gasoline or other volatile liquids can enter the drainage system

Although the codes agree that certain kinds of businesses need interceptors or separators, they don't provide much direction about the size or type of interceptor to install. The code wording usually goes something like this: *The size, type and location of each interceptor or separator shall be approved by the local administrative authority, in accordance with its standards.*

The amount of volatile liquids in a system determines the sizing and design methods for interceptors handling them. Some systems have only small concentrations, while others have larger amounts. We'll look at the guidelines for both types.

Small Concentration of Volatile Liquids

Commercial garages that service or store fewer than ten vehicles generate small amounts of volatile liquids and sand. Service stations and repair shops that service but don't store vehicles are also included here.

Figure 6-15 shows a typical poured-in-place or precast interceptor. This oil interceptor with a bucket-type floor drain may be approved for small installations without a separate sand interceptor.

Here are some typical criteria that meet the requirements of most codes:

▮ The interceptor should have a minimum liquid capacity of 18 cubic feet per 20 gallons of design flow.

▮ Bucket floor drains are generally required. The floor drain outlet should be a minimum of 4 inches (3 inches, in some codes). The bucket, made of the same material as the floor drain, is removable for cleaning. The bottom portion of the bucket is solid to retain sand or other debris. Drainage holes near the top of the bucket let liquid waste pass out of the bucket and into the pipe or pipes leading to the interceptor.

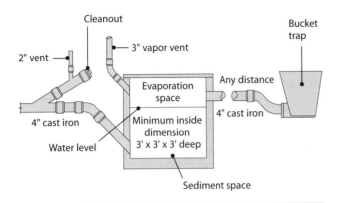

Figure 6-15
Sand and oil "combined" interceptor
for small businesses generating small amounts
of volatile liquids

▌ The inlet drain pipe should enter the interceptor above the liquid level line. Some codes may require the pipe to be vented. Others consider that the evaporation space provides adequate venting.

▌ There must be a minimum 3-inch (2-inch, in some codes) vapor vent in the evaporation space of the interceptor, venting into the open air. It must terminate in an approved location (most often the outside of a building) at least 12 feet above grade.

▌ The discharge pipe should enter the interceptor on a 45-degree angle and terminate near the bottom of interceptor. The liquid depth should be at least 2 feet below the invert of the discharge pipe.

▌ The discharge pipe should never be smaller than the inlet pipe. It should have a full-size cleanout brought to grade. The vent must be at least 2 inches in diameter.

▌ Some codes don't require the discharge pipe to be vented *if* it discharges into a catch basin or into a vented building sewer or building drain and the discharge pipe doesn't exceed a 15-foot developed length.

▌ The interceptor must be located outside the building.

▌ Each interceptor must be installed so it's readily accessible for removal of the cover, servicing and maintenance.

Large Concentration of Volatile Liquid

Businesses that generate large amounts of volatile liquids must have a waste oil storage tank. This receives and stores excess oil from the interceptor through a gravity overflow line. See Figure 6-16.

As a rule, codes require locating the waste oil storage tank outside the building. The designer will determine its size, capacity and location, with local official approval. Some codes require a minimum 500-gallon capacity. The tank must be UL (Underwriters Laboratories, Inc.) approved. You'll have to provide at least a $1^1/_2$-inch vent ending in the open air 12 feet above grade or through the building roof. You must also install a 2-inch minimum pumpout pipe to grade for removing the oil.

Install a gravity overflow line from the oil interceptor (separator) and connect it to the oil storage tank. Then use a 2-inch vent to remove any accumulation of explosive vapors.

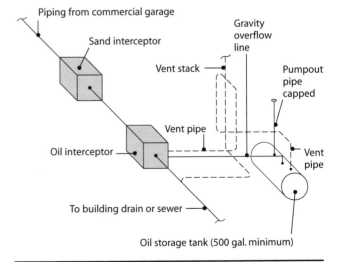

Figure 6-16
Sand interceptor, oil interceptor and oil storage tank
for businesses generating large amounts
of volatile liquids

Sand Interceptors

Most codes require floor drainage systems for all commercial garage buildings. Large numbers of automobiles produce a lot of sand, grit, grease and oil drippings that are regularly washed down the drain. It's essential that you install a sand interceptor that the waste will pass through before it discharges through the oil interceptor. See Figure 6-16.

There are a few other places where sand interceptors are required. These include stairwells, planter drains, beach cabana showers, public pool showers, public pool deck drains and play areas.

The sand interceptor holds sand and grit in the sediment section, as shown in Figure 6-17. The waste water, once free of sand and other debris, may then be discharged into the building sanitary or storm drainage system.

Bottling Plants

Bottling plants must discharge their processed wastes into an interceptor that separates out broken glass or other solids before the liquid waste discharges into the building drainage system. A properly-sized sand interceptor would do the job. See Figure 6-17.

Other Types of Interceptors or Separators

There are many businesses that require special interceptors or separators before their waste waters can be discharged into the building sanitary or storm drainage systems.

Hair Interceptors

The code may require a hair interceptor for barber shop sinks, beauty salon sinks and commercial fixtures used for bathing animals. Any such interceptor must have a removable, easy-to-clean screen basket to intercept hair. See Figure 6-18.

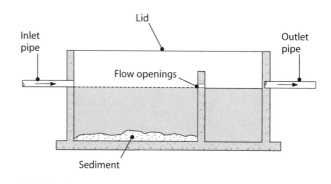

Figure 6-17
Sectional view of sand interceptor (trap)

Slaughterhouses

The code requires interceptors (separators) for all drains in slaughtering rooms and meat dressing rooms. They must exclude feathers, scales, entrails and other unacceptable materials from the building drainage system. Interceptors must be approved by local authorities. The architect will provide the type, size and locations on the approved blueprints.

Animal Boarding Businesses

Where animals are confined (animal kennels and zoos), most codes require the special handling of waste before it's discharged to a legal point of disposal (private or public).

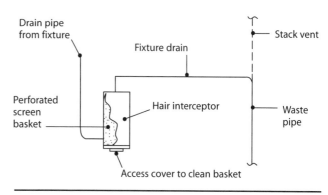

Figure 6-18
Hair interceptor installed in fixture drain pipe

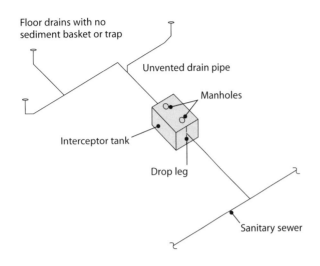

Figure 6-19
Interceptor tank and drainage piping for animal boarding establishments

The wash-down for such a system has a perforated or slotted grate but no sediment basket. The floor drain isn't usually trapped. The drainage pipe empties into an interceptor tank, which is usually similar in design to a septic tank.

This tank separates the solids, retains them and allows the liquid waste to pass on to a legal point of disposal. The bacterial process that's so effective in digesting human waste isn't very effective with animal waste. Since these animal solids have to be removed frequently, you'll have to install separator tanks that have adequate manholes. Be sure you bring the manholes to grade. See Figure 6-19. As always, your local authority can give you the criteria for an acceptable system.

Transformer Oil Spill Holding Tank

The code requires floor drainage for transformer vault rooms located inside a building. The usual installation is a 3-inch floor drain, at floor level, without a sediment basket or trap. You must install a drain pipe that connects to an oil spill holding tank. This serves as an indirect waste pipe and doesn't require a vent. You'll also need a pumpout pipe (installed to grade) for removing oil spills.

Your local power company is the authority on sizing the tank, which must be large enough to hold the amount of oil contained in the transformer in case of a rupture. Figure 6-20 shows a detailed drawing for a typical transformer oil spill holding tank.

Plaster Work Sinks

The code requires an interceptor trap in drain lines leading from dental and orthopedic sinks (Figure 6-21). Objectionable substances like wax and plaster can't discharge into the drainage system.

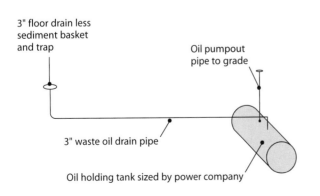

Figure 6-20
Transformer oil spill holding tank and piping detail

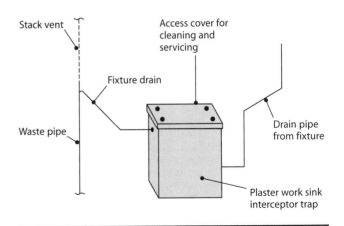

Figure 6-21
Plaster work sink interceptor trap and piping detail

Neutralizing Tanks

Corrosive liquids, spent acids and other chemicals present a distinct danger. They could damage or destroy a drainage, waste and vent system or create noxious or toxic fumes. To prevent such a possibility, these substances must be neutralized by dilution or other approved treatment before discharging into a building drainage system. Figure 6-22 shows the installation of a typical separate and independent system. Your local authority will want to know the type and amount of corrosive waste generated. It's up to them to approve the size and type of neutralizing tank you use.

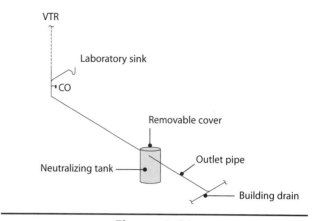

Figure 6-22
Typical acid waste system with neutralizing tank

Review Questions (answers are on page 283)

1. Name at least three types of waste that the code considers objectionable and harmful to the building drainage system.

2. What are the three options for dealing with objectionable and harmful waste before it can enter the drainage system?

3. What is the primary purpose of an interceptor or separator trap?

4. When an interceptor is used, what must your working blueprints show?

5. What type of waste must not go through an interceptor?

6. Name three types of commercial buildings that require the installation of a grease interceptor.

7. Name two types of buildings where grease interceptors are *not* generally required.

8. What are the two types of grease interceptor installations that are usually approved by code?

9. What information about inside grease interceptor installations is omitted by most codes?

10. What's your source for obtaining established sizing methods for grease interceptor installations?

11. What kind of restaurant will generally use an inside grease interceptor?

12. What is the maximum grease capacity permitted for an inside grease interceptor?

13. What are the two typical installation methods for inside grease interceptors?

14. Why must an approved flow control fitting be installed on a small inside grease interceptor?

15. If you omit the fixture trap for a pot sink, what distance requirement must be met?

16. What is the maximum vertical fixture tailpiece drop for a fixture connected to an inside grease interceptor?

17. Why must grease interceptors be easily accessible?

18. In what part of a building is the installation of an inside grease interceptor prohibited?

19. Why do most codes prohibit the installation of a water-cooled grease interceptor?

20. Most codes prohibit a food waste disposal from discharging through what device?

21. What is the mandated grease retention capacity of a grease interceptor?

22. What is the minimum retention time required for a grease interceptor in a small single-service kitchen?

23. What are the two major considerations when sizing a grease interceptor for a fully-equipped commercial restaurant?

24. What is the grease interceptor retention time required for a fully-equipped commercial restaurant?

25. Most codes don't spell out the sizing methods for commercial grease interceptors. Who do they defer to for the sizing methods?

26. Why might identical restaurants in different geographical areas require grease interceptors of different sizes?

27. For sizing commercial grease interceptors, the *Uniform Plumbing Code* considers 16 hours of operation to include how many peak hours?

28. What is the most commonly accepted construction material for outside grease interceptors?

29. Most codes require the inlet invert in an outside grease interceptor to discharge a minimum of how many inches above the liquid level line?

30. What is the minimum size cleanout manhole for an outside grease interceptor?

31. Outside grease interceptors must be designed and installed to avoid what potential problem?

32. Most codes require that outside grease interceptors have how many compartments?

33. How many manholes are required for an outside grease interceptor 21 feet long?

34. How close to a building foundation can you install an outside grease interceptor?

35. How close to a private property line can you install an outside grease interceptor?

36. The greasy waste line system is designed and installed for what purpose?

37. When is a greasy waste line allowed to connect to the building sewer?

38. What are the two code-approved greasy waste systems?

39. On which greasy waste system will you most often be working?

40. What are the guidelines for sizing pipes for a conventional greasy waste system?

41. What is the difference between a conventional greasy waste system and a combination waste and vent system?

42. Name three areas where some codes might permit the use of a combination waste and vent system.

43. Why do some codes *not* recommend connecting grease-producing restaurant kitchen equipment to a combination waste and vent system?

44. Why are pipes in a greasy combination waste and vent system sized two pipe sizes larger than a conventional greasy waste system?

45. A fixture tailpiece should be as short as possible, never exceeding what length?

46. Why can't large quantities of water (from pumps, for example) discharge into a combination waste and vent system?

47. What must be provided if a branch line in a greasy combination waste and vent system exceeds 15 feet?

48. In a greasy combination waste and vent system, how does the minimum area of a vent compare to the size of the waste pipe it serves?

49. What does the code require that each vent stack in a greasy combination waste and vent system have?

50. What businesses require lint interceptors in their drainage system?

51. What type of strainer is usually required on a commercial or self-service lint interceptor?

52. What do some codes call the horizontal drainage pipes serving commercial or self-service clothes washing machines?

53. What is the advantage of an indirect waste system for a commercial or self-service laundry?

54. How many clothes washers can a 3-inch standpipe accommodate in a self-service laundry?

55. When a commercial or self-service lint interceptor connects to a building drainage system, where on the horizontal discharge pipe would you correctly locate the vent?

56. Where are lint interceptors *not* required by code?

57. Most codes don't provide established sizing methods for self-service laundries. Who does?

58. Who sets the design criteria for commercial laundries?

59. What determines the size of the lint interceptor in a self-service laundry?

60. What is the usual code-required retention period for a lint interceptor?

61. Using the local code formula in our example, what size lint interceptor would be required for a self-service laundry with eight clothes washing machines?

62. Using the *UPC* formula, what size lint interceptor would be required for a self-service laundry with eight clothes washing machines?

63. What is the general definition of areas where gasoline, oil and sand interceptors are required by the code?

64. Name three types of establishments where the code would require a gasoline, oil and sand interceptor.

65. What governs the sizing and design of gasoline and oil interceptors that handle volatile liquids?

66. Give an example of an establishment that, according to the code, generates small amounts of volatile liquids and sand.

67. What type of floor drain is usually required for an automobile repair shop?

68. What is the minimum liquid capacity for an oil interceptor in the floor drainage system of a service station?

69. In a commercial garage that services or stores fewer than ten vehicles, where should the inlet drain pipe enter the oil interceptor?

70. If the inlet pipe to an oil interceptor is 4 inches, what is the minimum size required for the discharge pipe?

71. Under what circumstances can you omit the vent for the discharge pipe in an oil interceptor?

72. Where must you locate an oil interceptor for a service station?

73. In addition to an oil interceptor, what is required for businesses that generate large amounts of volatile liquids?

74. According to most codes, what is the minimum size vent required for a waste oil storage tank for a business that generates large amounts of volatile liquids?

75. Codes require that waste oil storage tanks be UL approved. What does the abbreviation UL stand for?

76. What's the minimum height above grade permitted for a vent serving a waste oil storage tank?

77. Before it enters an oil interceptor, floor drainage for a commercial garage building must first discharge through what device?

78. Before the liquid wastes can discharge into the building drainage system, bottling plants must discharge their processed wastes into what device?

79. What special device must be installed in the drainage line of commercial fixtures used for bathing animals?

80. The code requires interceptors or separators for all drain lines in slaughtering rooms and meat dressing rooms. Who must approve the blueprints that indicate the type, size and location of these interceptors or separators?

81. Before discharging to a legal point of disposal, most codes require that drainage pipe wastes from animal boarding businesses pass through what device?

82. Where must the waste from transformer vault room floor drainage discharge?

83. Who has the authority to size a transformer oil spill holding tank?

84. Why must dental and orthopedic sinks be equipped with an interceptor trap?

85. What is the purpose of a neutralizing tank?

Indirect Waste Piping, Receptors and Special Wastes

Any fixture, appliance or device that's not classified as a plumbing fixture may be drained by indirect means if it has a drip or drainage outlet. The indirect drainage prevents sewage from backing up into a special fixture and contaminating its contents in case there's a stoppage in the sanitary drainage system.

Special fixtures or appliances include refrigerators, ice boxes, bar sinks, hand sinks, cooling or refrigerating coils, extractors, steam tables, egg boilers, coffee urns, stills, sterilizers, commercial dishwashers, water stations, water lifts, expansion tanks, cooling jackets, drip or overflow pans, air conditioning condensate drains, drains from overflows, relief vents from the water supply system, and similar applications.

Overflow and relief pipes on a water supply system and relief pipes on expansion tanks, sprinkler systems and cooling jackets must always connect *indirectly* to the sanitary drainage system. See Figure 7-1. This prevents any possibility of contaminating the potable water supply through cross connection. To avoid contamination, there must also be a positive separation by indirect means between the waste outlet and the drainage inlet of hospital equipment, and food storage and preparation establishments. This unique method of piping is very practical for fixtures with low discharge rates.

Certain restrictions apply as to *maximum length of indirect waste piping* in hand sinks, bar sinks and similar fixtures. Figure 7-2 shows a hand sink trap that

illustrates the restrictions. If the length of the indirect waste pipe (X) exceeds 5 feet (2 feet in some codes) but is less than 15 feet (25 feet in some codes), it must be directly trapped. You don't have to vent these traps.

All indirect waste receptors must have adequate capacity and a design that prevents splashing or flooding. And any required vent in indirect waste piping must extend separately to the outside air. *Never connect it to a vent which serves a sanitary system.*

Piping Material and Sizes

Indirect waste piping materials must meet local code requirements. Acid and chemical waste pipes must be made of materials that aren't affected by these wastes. We'll cover the materials in detail in Chapter 8.

The sizing of indirect waste piping is fairly simple. It should never be smaller than the drain outlet or tailpiece of the fixture, appliance or equipment served, and in any case, never smaller than $1/2$ inch.

Types of Indirect Waste Piping

There are two types of indirect waste piping — *air break* and *air gap*. Most codes accept one or both of these types.

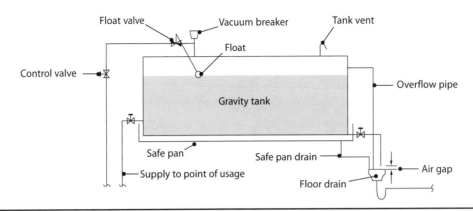

Figure 7-1
Typical water gravity tank showing required air gap

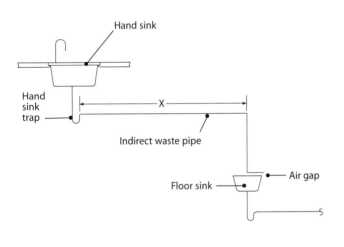

Figure 7-2
Hand sink trap

Air break is a piping arrangement in which a drain pipe from a fixture, appliance or device discharges indirectly into a fixture or receptor at a point below the flood level rim of the receptor. There are two indirect *air break* waste methods illustrated in Figure 7-3. Some local authorities will allow an *air break* installation to terminate below the floor or rim of the floor sink or receptor.

In an *air gap* arrangement, there must be an unobstructed vertical distance through the free atmosphere between the drain pipe outlet from a fixture, appliance or device and the flood level rim of the receptor into which it discharges. There must be a minimum separation of 2 inches (or twice the drain pipe size) between the plumbing fixture, appliance, or appurtenance outlet and the rim of the floor sink or receptor. See Figure 7-4.

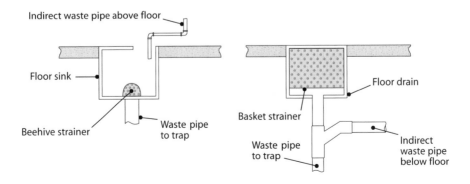

Figure 7-3
Two indirect (air break) waste methods

Receptors

All receptors must always be accessible for inspection and cleaning. We'll look at the three types of receptors you may be installing.

Floor Sink

The first receptor type, the *floor sink*, is used with an air gap indirect waste pipe, mostly in restaurants. The top of the receptor, which may be rectangular or round, is open to receive the indirect waste pipe. The type of floor sink you'll install depends on whether or not the area carries pedestrian traffic. One type has a grate that makes it safe for pedestrian traffic. The second type of floor sink, used in areas with no pedestrian traffic, doesn't require a covering grate. It's used in hidden areas where appearance isn't important. Figure 7-5 shows both types.

The code requires that the indirect waste receptor for a floor sink must have a beehive strainer at least 4 inches high. Indirect waste pipes that convey only clear water waste don't need a beehive strainer. And all indirect waste receptors must have adequate capacity and a design that prevents splashing or flooding.

Floor Drain and Clothes Washer Standpipe

A floor drain is commonly used with *air break* indirect waste pipe. It may also serve as a receptor to receive floor drainage. Look back at Figure 7-3.

Most codes categorize an *automatic clothes washer standpipe* as an indirect waste receptor. It must be no smaller than 2 inches in diameter and be installed above floor level. The *UPC* says the standpipe can never extend more than 30 inches nor less than 18 inches above the trap. The *SPC* says the standpipe can't extend more than 34 inches nor less than 24 inches above the trap. The range in the *International Plumbing Code* is from 42 to 18 inches. Rough it in at least 6 inches (but not over 18 inches) above the floor. Make sure the standpipe receptor is properly trapped and vented (Figure 7-6).

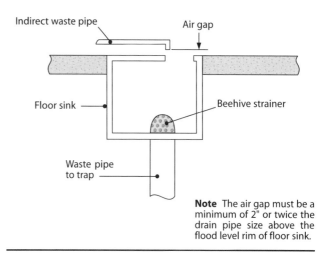

Note The air gap must be a minimum of 2" or twice the drain pipe size above the flood level rim of floor sink.

Figure 7-4
Typical air gap indirect drainage method

A Floor sink receives indirect wastes under equipment while grate provides safety for pedestrian traffic

B Floor sink receives indirect wastes in nontraffic areas

Figure 7-5
Floor sinks

The code prohibits installation of indirect waste receptors in any toilet room, closet, cupboard or storeroom — with this exception. You may install the standpipe for an automatic clothes washer in toilet and bathroom areas *if* the clothes washing machine is installed in the same room.

Sumps

When the location of a fixture, appliance or equipment prevents drainage by gravity, you'll have to use a sump. The sump pump can empty clear water waste from a basement floor, subsoil drain, air conditioning condensate drain, and similar applications. The pump will lift the liquid and convey it to an approved place of disposal. See Figure 7-7.

Special Wastes

Certain clear water wastes (water lifts, expansion tanks, stills, sterilizers, cooling jackets, sprinkler systems, drip or overflow pans, or similar devices) contain insignificant amounts of impurities. But according to the *UPC*, they still have to discharge through an indirect waste pipe with an *air gap* before they can empty into the building drainage system.

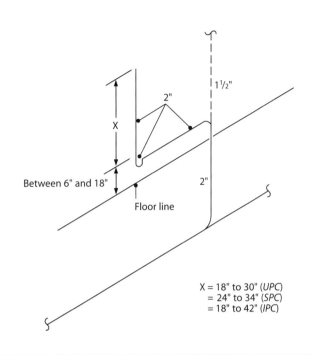

Figure 7-6
Automatic clothes washer standpipe detail

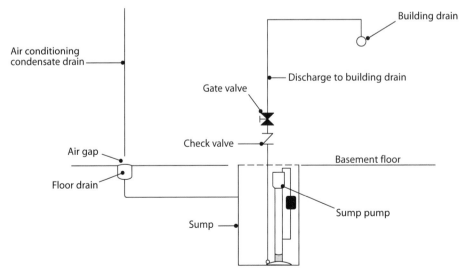

Note The sump need not be vented.

Figure 7-7
Clear water waste sump and piping detail

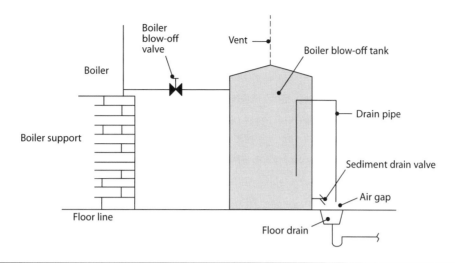

Figure 7-8
Typical boiler blow-off and blow-off tank installation

Air Conditioning Condensate Drains

Pipe carrying the special waste from air conditioning units usually connects to the building sanitary or storm system. In such cases, you must install it as indirect waste. The pipe must discharge through an *air gap* or *air break* type connection. See Figure 7-7.

Local authorities will generally approve one of the following:

▌ An approved receptor or other suitable fixture

▌ A sump

▌ A building storm or sanitary drain

▌ A building's inside rain leader

▌ A waste and overflow or lavatory tailpiece. (Note: Not all codes accept this particular method for disposing of air conditioning waste.)

In Chapter 8, we'll cover materials, installation specifics, and other disposal methods for air conditioning waste.

High Temperature Waste Water

No steam pipe or water with a temperature above 140 degrees F should discharge directly into any part of a drainage system. The pipe must first discharge into an approved intercepting sump or boiler blow-off tank. See Figure 7-8. It must connect to drainage systems by the *air gap* method. Closed tanks must be vented from the top. The vent must extend separately, full size, above the roof.

Swimming Pool Wastes

Commercial Pools

Swimming pools, wading pools and spas have special objectionable wastes. Ordinary public sewer systems and treatment plants simply can't handle huge amounts of waste suddenly discharged into the system. The health department is concerned about the potential for health hazards when the combined wastes are treated at the treatment plant. That's why piping carrying waste from swimming pools, wading pools and spas that's connected to building sanitary or storm systems has to be installed as an indirect waste. You'll always have to get advance approval from your authority.

This also includes other types of pool drainage such as backwash from filters, water from scum gutter drains and deck drains. If the pool's indirect waste line is below the sewer grade, you may be required to install a circulation pump to remove the contents.

Roof drain   Deck drain

A Two rainwater drains

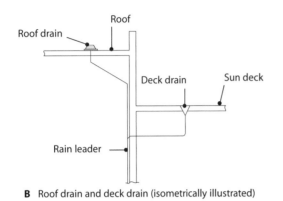

B Roof drain and deck drain (isometrically illustrated)

Figure 7-9
Roof and deck drains

Private Pools

If you obtain advance approval from the local authority, there are several approved methods to dispose of backwash water and to empty private pools. Because the quantity is less, you seldom have to connect this waste to a sanitary or storm drainage system.

And remember this: *Pool wastes should never discharge into a sanitary drainage system connected to a septic tank.* It can harm the septic tank and/or the subsoil leach field (drainfield). Sudden large amounts of pool water can create a reverse flow in a drainage system, causing the plumbing fixtures inside to overflow. Also, the chlorine in pool water can destroy the bacterial process that makes the septic tank and leach field function. See Chapter 17 for more detailed information on waste water disposal.

Storm Water Drainage Systems

Rainwater is one of the major *special wastes* that's generated on commercial properties. Take care that the piping you install receives and conveys rainwater to a point where discharge can legally occur — not always a simple task.

Storm drainage includes roof drains, area drains, catch basins, gutters, leaders, building storm drains, building storm sewers, footing drains and ground surface storm sewers.

Rainwater that's not properly collected and disposed of becomes a nuisance and a health hazard. Parking lots and large residential and commercial buildings can collect stagnant pools of storm water. Stagnant pools, besides producing offensive odors, are ideal breeding grounds for mosquitoes.

The code recognizes the importance of collecting and disposing of storm water from commercial properties. Though it's rare nowadays, you might find that some older cities near lakes, rivers or the ocean use a combined sewer system to carry both storm water and partially-treated (or untreated) sewage. These archaic systems end in outfall lines that dump waste into the water at a considerable distance from the shoreline or city. These practices are increasingly unacceptable to informed citizens and politicians. Cities are establishing stricter controls in an effort to correct such environmental assaults.

You'll need advance approval to connect any new construction into an old dual system. Most local codes now require two separate systems: one for sewage and one for storm water collection.

Roof Drains and Special Drains

You must equip *roof drains* with strainers, with the strainer cover extending at least 4 inches above the roof surface. It must have a minimum inlet area of $1^1/_2$ times the area of the pipe to which it connects ($2^1/_2$ in some codes).

Equip *deck drains* on sun decks, parking decks and similar public areas with an approved flat-surface strainer. This reduces the chance that pedestrians will trip, become injured and possibly file a lawsuit. The strainer must have an inlet area at least 2 times the area of the pipe to which it connects ($2^1/_2$ in some codes). Figure 7-9 shows both types of drains.

Install *area drains* at the foot of a stairwell or similar area where rainwater may accumulate. The pipe's diameter must be at least 2 inches, and it can't drain an area that exceeds 100 square feet. If the area is greater, use regularly-spaced floor drains to drain the entire area.

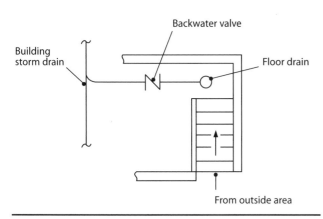

Figure 7-10
Area drain connection to storm drain

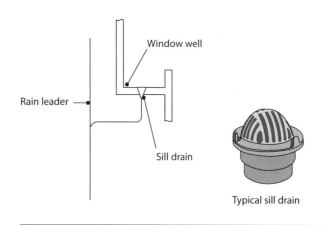

Figure 7-11
Sill drain connection to rain leader

When you connect an area drain to a building storm drain, you'll have to install a backwater valve that's accessible for inspection and maintenance. Floor drains are generally used for this purpose, although flat-surface type strainers like those used on deck drains are also permitted. See Figure 7-10.

Sill drains are drains located where there's a flat surface between a window and an outside wall. You can usually connect the wastes from these drains to an inside rain leader or to the storm water drain (Figure 7-11).

Planter drains may be required for commercial buildings with an indoor planter area. Your local authority may require that the excess waste water pass through a sand interceptor before it's discharged to a legal point of disposal. See Figure 7-12.

Subsoil drains (sometimes called *footing drains*) intercept surface water before it reaches the building's foundation wall or footings. Most codes require subsoil drains for building construction in low or depressed areas where surface water accumulation from rainstorms can present a serious problem. Some codes require subsoil drains for buildings with basements, rooms, cellars or crawl spaces below grade. You can install them inside or outside of the footings. See Figure 7-13. Most subsoil drains have to pass through a sand interceptor before being pumped to an approved legal point of disposal. See Chapter 8 for more detail on special drains.

Figure 7-12
Typical planter drain approved by code

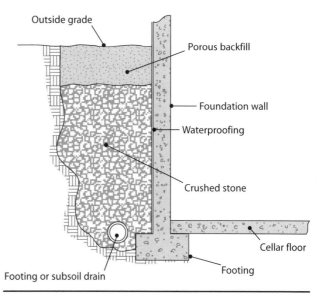

Figure 7-13
Detail of footing drain installation

Sizing Storm Water Drainage Systems

Mechanical engineers are usually responsible for figuring out sizing and disposal methods for storm water drainage pipes. Calculations are based on two factors:

1) The square footage of impervious areas (roofs, parking lots and similar surfaces)

2) The maximum anticipated inches of rainfall in any one hour

If engineers do most calculations, why should you concern yourself with the principles of sizing commercial storm drainage and disposal systems? There are a couple of good reasons. First, it's plumbers, not engineers, who actually install this kind of work. If you install a system that isn't right, engineer's fault or not, you're still going to look bad. Second, you'll have to interpret storm drainage tables and understand regulations if you expect to pass the journeyman and master plumber's examination.

You'll find that there are wide differences between codes when it comes to sizing storm drains. They use different formulas, and there are great variations in maximum anticipated rainfall in any one hour. Rainfall in this country can vary from less than 2 inches to more than 8 inches. Naturally, the less rainfall, the smaller the pipe required. The greater the rainfall, the larger the pipe required.

You need to provide good spacing of the roof drains to eliminate puddling on roof surfaces. If you don't have proper spacing, you won't pass inspection. Of course, the more roof drains you use, the fewer puddles you'll have. Remember, too, that smaller roof drains and leader pipes are less expensive.

When you know the maximum anticipated rainfall and the square feet of roof surface, you can begin your calculations. Use your local code book tables to determine the size of your leaders and horizontal drain pipes. The examples in this book are based on the *Uniform Plumbing Code.*

Every code book has rainfall tables that indicate the diameter of vertical leaders and horizontal piping based on the square footage of the roof area. Figure 7-14 shows the tables for vertical leaders from the *UPC* and *IPC.* Figure 7-15 is the horizontal table; it's identical in both codes.

Uniform Plumbing Code — Horizontal projected roof area in square feet						
Size of leader pipe (in)	Rainfall rate (inches per hour)					
	1	2	3	4	5	6
2	2,176	1,088	725	544	435	363
3	6,440	3,220	2,147	1,610	1,288	1,073
4	13,840	6,920	4,613	3,460	2,768	2,307
5	25,120	12,560	8,373	6,280	5,024	4,187
6	40,800	20,400	13,600	10,200	8,160	6,800
8	88,000	44,000	29,333	22,000	17,600	14,667

Sizes indicated in the above table are the diameter of circular piping. Square or rectangular piping may be used, provided the cross-sectional area fully encloses a circle of the pipe diameter indicated in this table.

From the UPC® with permission of the IAMPO© 1997

International Plumbing Code — Horizontal projected roof area in square feet						
Size of leader pipe (in)	Rainfall rate (inches per hour)					
	1	2	3	4	5	6
2	2,880	1,440	960	720	575	480
3	8,800	4,400	2,930	2,200	1,760	1,470
4	18,400	9,200	6,130	4,600	3,680	3,070
5	34,600	17,300	11,530	8,650	6,920	5,765
6	54,000	27,000	27,995	13,500	10,800	9,000
8	116,000	58,000	38,660	29,000	23,320	19,315

Sizes indicated are the diamter of circular piping. The above table is applicable to square or rectangular piping, provided the cross-sectional area (shape) fully encloses a circle of the pipe diameter indicated in this table.

From the International Plumbing Code®

Figure 7-14
*Sizing roof drains and **vertical** rainwater piping*

Size of pipe in inches	Maximum rainfall in inches per year					
¹/₈"/ft. slope	**1**	**2**	**3**	**4**	**5**	**6**
3	3,288	1,644	1,096	822	657	548
4	7,520	3,760	2,506	1,880	1,504	1,253
5	13,360	6,680	4,453	3,340	2,672	2,227
6	21,400	10,700	7,133	5,350	4,280	3,566
8	46,000	23,000	15,330	11,500	9,200	7,670
10	82,800	41,400	27,600	20,700	16,580	13,800
12	133,200	66,600	44,400	33,300	26,650	22,200
15	238,000	119,000	79,333	59,500	47,600	39,650
¹/₄"/ft. slope	**1**	**2**	**3**	**4**	**5**	**6**
3	4,640	2,320	1,546	1,160	928	773
4	10,600	5,300	3,533	2,650	2,120	1,766
5	18,880	9,440	6,293	4,720	3,776	3,146
6	30,200	15,100	10,066	7,550	6,040	5,033
8	65,200	32,600	21,733	16,300	13,040	10,866
10	116,800	58,400	38,950	29,200	23,350	19,450
12	188,000	94,000	62,600	47,000	37,600	31,350
15	336,000	168,000	112,000	84,000	67,250	56,000
¹/₂"/ft. slope	**1**	**2**	**3**	**4**	**5**	**6**
3	6,576	3,288	2,192	1,644	1,310	1,096
4	15,040	7,520	5,010	3,760	3,010	2,500
5	26,720	13,360	8,900	6,680	5,320	4,450
6	42,800	21,400	14,267	10,700	8,580	7,140
8	92,000	46,000	30,650	23,000	18,400	15,320
10	165,600	82,800	55,200	41,400	33,150	27,600
12	266,400	133,200	88,800	66,600	53,200	44,400
15	476,000	238,000	158,700	119,000	95,200	79,300

From the UPC® with permission of the IAMPO© 1997

Figure 7-15
*Sizing **horizontal** rainwater piping*

First, you need to find the total square feet to be drained. Then check to see if there are any vertical walls that permit storm water to drain onto the roof area. Be sure to review the vertical wall section of your code carefully. Depending on the code you use, you'll add from about 30 to 50 percent of the vertical wall area to the roof area. When you add vertical wall areas to the flat surfaces, it can dramatically increase the size of leaders and horizontal pipes.

Figure 7-16 suggests one way to arrange roof drains and size the storm water pipes for a typical roof with a vertical wall. You can follow this example for draining any type or size roof. In this case, the main roof is 5,000 square feet (100' × 50' = 5,000 SF) with an attached 900 square foot vertical wall (100' × 9' = 900 SF).

Assume the code in this area calls for adding half of the vertical wall surface. So we'll add 450 square feet to the 5,000 square foot roof area, for a total of 5,450 square feet to be drained. You can see that two roof drains (each one with a drain capacity of 2,725 SF) will be adequate to avoid puddling on this roof.

Use the *UPC* table in Figure 7-14 to size the rain leaders. The first column shows the leader pipe size in inches. Columns 2 through 7 show the maximum number of square feet that each pipe size can drain. The number at the top of each column is the rainfall rate.

Let's do a sample roof in Montgomery, Alabama. According to the maximum rainfall rates in Figure 7-17, it has a maximum anticipated rainfall of 3.8 inches. Look at the 4-inch column in Figure 7-14. As you move

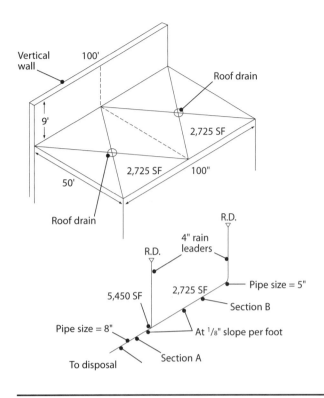

Figure 7-16
Storm water drainage system

down the column, you see 544 and 1,610. Neither of these is large enough to drain the 2,725 square foot area. You'll have to move to the next line and use a leader large enough to drain 3,460 square feet. That's a 4-inch vertical leader pipe.

You'll need Figure 7-15 for the horizontal rainwater pipes. It's divided into three sections for slopes of $^1/_8$ inch, $^1/_4$ inch and $^1/_2$ inch. The greater the slope, the larger the area that each pipe size can drain. The left column shows the slope and the pipe size in inches. The other columns show the maximum area that can be drained by that size pipe during a rainfall of 1 to 6 inches per hour.

The slope shown in Figure 7-16 is $^1/_8$ inch per foot, so you'll use the top section of Figure 7-15. Note that the horizontal pipe in the drawing is divided into two sections, A and B. Section A carries the flow for both leaders, or 5,450 square feet. At 4 inches of maximum rainfall, you'll install an 8-inch pipe for Section A and a 5-inch pipe for Section B.

Again, the greater the slope, the larger the area that each pipe size can drain. This also means that each pipe size can be smaller.

Now, from the drawing in Figure 7-16, let's size Sections A and B, using the other two slopes listed in Figure 7-15 and the same rainfall rate (4 inches per hour.) At $^1/_4$ inch slope per foot, you can still use a 5-inch pipe for Section B (an area of 2,725 square feet). Section A (an area of 5,450 square feet) needs a 6-inch pipe, which can handle up to 7,550 square feet.

You'll see that at $^1/_2$ inch slope per foot, a 4-inch pipe will work for Section B and a 5-inch pipe for Section A. A smaller pipe can accommodate both Sections A and B.

Generally speaking, you'll find that the tables in code books for sizing storm water drains and leaders are already calculated for the particular geographic area involved. But what if there's more rainfall or less rainfall than Figure 7-15 shows?

The *Uniform Plumbing Code* lets you adjust the figures in Figure 7-15 for varying amounts of rainfall. Here's how it works. In the 1-inch rainfall column, adjust the figures by dividing the square feet in the 1 inch/hour column by the desired rainfall rate in inches/hour. Let's work out an example. Using each of the three acceptable horizontal pipe slopes, we'll figure the number of square feet a 4-inch pipe will carry in a geographic area with a maximum 8 inches/hour rainfall rate.

$^1/_8$ inch slope: In Figure 7-15, the 1-inch rainfall column opposite the 4-inch pipe shows 7,520 square feet.

$$\frac{7,520 \ SF}{8"/hour} \ = \ 940 \ SF$$

$^1/_4$ inch slope: In Figure 7-15, the 1-inch rainfall column opposite the 4-inch pipe shows 10,600 square feet.

$$\frac{10,600 \ SF}{8"/hour} \ = \ 1,325 \ SF$$

$^1/_2$ inch slope: In Figure 7-15, the 1-inch rainfall column opposite the 4-inch pipe shows 15,040 square feet.

$$\frac{15,040 \ SF}{8"/hour} \ = \ 1,880 \ SF$$

You can repeat this formula with differing pipe sizes and maximum rates of rainfall that aren't listed in Figure 7-15.

Alabama:
Birmingham 3.7
Mobile 4.5
Montgomery 3.8

Alaska:
Fairbanks 1.0
Juneau 0.6

Arizona:
Phoenix 2.2

Arkansas:
Ft. Smith 3.9
Little Rock 3.7

California:
Eureka 1.5
Lake Tahoe 1.3
Los Angeles 2.0
Lucerne Valley 2.5
San Diego 1.5
San Francisco 1.5
San Luis Obispo 1.5

Colorado:
Denver 2.2
Durango 1.8
Stratton 3.0

Connecticut:
Hartford 2.8
New Haven 3.0

District of Columbia:
Washington 4.0

Florida:
Daytona Beach 4.0
Jacksonville 4.3
Melbourne 4.0
Miami 4.5
Tampa 4.2

Georgia:
Atlanta 3.5
Brunswick 4.0
Macon 3.7
Savannah 4.0
Thomasville 4.0

Hawaii:
Honolulu . . . . . . .Consult local
data

Idaho:
Boise 1.0
Lewiston 1.0
Twin Falls 1.1

Illinois:
Chicago 2.7
Peoria 2.9
Springfield 3.0

Indiana:
Evansville 3.0
Indianapolis 2.8
South Bend 2.7

Iowa:
Davenport 3.0
Des Moines 3.4
Sioux City 3.6

Kansas:
Topeka 3.8
Wichita 3.9

Kentucky:
Lexington 2.9
Lousville 2.8

Louisiana:
New Orleans 4.5
Shreveport 4.0

Maine:
Eastport 2.0
Portland 2.4

Maryland:
Baltimore 3.6

Massachusetts:
Boston 2.7
Nantucket 2.7

Michigan:
Detroit 2.5
Grand Rapids 2.6
Port Huron 2.7

Minnesota:
Duluth 2.6
Minneapolis 3.0

Mississippi:
Meridian 4.2
Vicksburg 4.5

Missouri:
Kansas City 3.9
St. Louis 3.2
Springfield 3.7

Montana:
Havre 2.0
Kalispell 1.5
Missoula 1.3

Nebraska:
Lincoln 2.7
North Platte 3.5
Omaha 3.7

Nevada:
Reno 1.2
Tonopah 1.3
Winnemucca 1.0

New Hampshire:
Concord 2.5

New Jersey:
Atlantic City 3.4
Trenton 3.2

New Mexico:
Albuquerque 2.0
Roswell 2.6
Santa Fe 2.1

New York:
Binghamton 2.4
Buffalo 2.3
New York 3.1
Syracuse 2.4

North Carolina:
Asheville 3.2
Charlotte 3.4
Raleigh 4.0
Wilmington 4.4

North Dakota:
Bismarck 2.7
Devils Lake 2.8
Williston 2.5

Ohio:
Cincinnati 2.8
Cleveland 2.4
Columbus 2.7
Toledo 2.4

Oklahoma:
Oklahoma City 4.1

Oregon:
Baker 1.5
Portland 1.3

Pennsylvania:
Erie 2.4
Harrisburg 2.9
Philadelphia 3.2
Pittsburgh 2.5
Scranton 2.8

Puerto Rico:
San Juan 5.7

Rhode Island:
Block Island 3.0
Providence 2.9

South Carolina:
Charleston 4.1
Columbia 3.5
Greenville 3.3

South Dakota:
Huron 2.5
Rapid City 2.7
Yankton 2.0

Tennessee:
Chattanooga 3.2
Knoxville 3.1
Memphis 3.5
Nashville 3.0

Texas:
Austin 4.5
Corpus Christi 4.6
Dallas 4.2
El Paso 2.0
Galveston 4.7
Houston 4.6
San Antonio 4.4

Utah:
Modena 2.1
Salt Lake City 1.3

Vermont:
Burlington 2.3
Northfield 2.5

Virginia:
Lynchburg 3.3
Norfolk 4.0
Richmond 4.0

Washington:
Seattle 1.0
Spokane 1.0
Walla Walla 1.0

West Virginia:
Elkins 2.9
Parkersburg 3.0

Wisconsin:
Green Bay 2.5
LaCrosse 2.9
Milwaukee 2.7

Wyoming:
Casper 1.9
Cheyenne 2.5
Yellowstone Park 2.5

Rates given in this table are based on U.S. Weather Bureau Technical Paper No. 40, Chart 14/ 100-year one-hour rainfall (inches)

Figure 7-17

Maximum rainfall rates in inches per hour

Review Questions (answers are on page 287)

1. By what means may wastes from fixtures, appliances and devices not regularly classed as plumbing fixtures be drained?

2. What is the purpose of the indirect drainage method for special fixtures?

3. Name three plumbing fixtures or appliances that can be drained by indirect means.

4. Name two devices that may be drained by indirect means.

5. Why must overflow pipes on the water supply system always be indirectly connected to the sanitary drainage system?

6. What is it called when there's the possibility of a potable water supply system becoming contaminated through an unsafe source?

7. What restriction applies to any indirect waste piping that exceeds 5 feet (2 feet in some codes) but is less than 15 feet (25 feet in some codes)?

8. When a vent is required in indirect waste piping, how must it be installed?

9. What is the minimum size for indirect waste pipes?

10. What are the two types of indirect waste piping?

11. What type of indirect waste piping arrangement is required to form an *air break* installation?

12. What type of indirect waste piping arrangement is used for an air gap installation?

13. What is the minimum separation required between the fixture outlet and the rim of the receptor for an *air gap* installation?

14. What is always required as part of an indirect waste receptor installation to allow for cleaning and inspection?

15. What factor determines the type receptor to use for *air gap* indirect waste pipe?

16. What kind of strainer does the code require for a floor sink?

17. What type receptor is commonly used where *air break* indirect waste pipe is installed?

18. How do most codes categorize automatic clothes washer standpipes?

19. What must be included in the installation of a standpipe receptor for an automatic clothes washer?

20. Name one place the code prohibits the installation of an indirect waste receptor.

21. In places where drainage by gravity isn't possible, what may you use to receive indirect waste pipe?

22. What equipment is needed to lift liquids from a sump to a place of disposal?

23. How must clear water wastes empty into the building drainage system?

24. In cases where air conditioning waste is connected to the building storm system, what type connection should you use?

25. Name two acceptable methods of disposing of air conditioning wastes.

26. What's the maximum water temperature that can be discharged directly into a drainage system?

27. How must wastes from swimming pools, wading pools and spas be connected to the building sanitary system?

28. What type of waste disposal system can never be used to receive waste from a swimming pool?

29. What might be the result of improperly collecting and disposing of rainwater?

30. In previous years, what type of drainage system was commonly used in older cities located near lakes, rivers or the ocean?

31. What would you need in order to connect new construction into an existing combined sewer system?

32. What must roof drains be equipped with?

33. In what locations do most codes require installation of deck drains?

34. What's the maximum area most codes permit an area drain to handle?

35. What type drain must you use where there's a flat surface between a window and an outside wall?

36. When a planter drain is used, what may be required by your local authority?

37. What's the purpose of subsoil drains?

38. Who usually determines the sizing of storm water drainage pipes?

39. What are the two determining factors in sizing storm water drainage pipes?

40. As a plumber, why should you learn all you can about sizing commercial storm drainage and disposal systems?

41. Why do storm drainage tables differ in various local codes?

42. Once you know the maximum anticipated rainfall where you work and the square feet to be drained, what can you use to determine the sizing of storm drainage pipes?

43. How does increasing the slope affect the sizing of horizontal storm drainage pipes?

Installation of Drainage and Vent Pipes

Plumbing codes exist to protect public health, welfare and safety. They outline the acceptable procedures for the safe design, installation and maintenance of plumbing systems. Although installation details may vary from code to code, the basic principles of sanitation and safety remain the same.

This chapter explains the installation requirements for most drain, vent and waste systems. Of course, it won't cover every situation, but it will make the intent of the code clear and understandable. These basics are absolutely essential to every plumber. If you violate even one of these principles, it's a safe bet the plumbing inspector won't pass your work.

Plumbing fixtures mark the end of the potable (drinkable) water supply system and the beginning of the sewage system. It's crucial that your drainage system design prevents any kind of fouling or solids being deposited along the walls of drainage pipes. Provide adequate venting to allow for free air circulation. (See Chapter 3.) This prevents siphonage or forcing of trap seals.

But even if your drainage design is fault-free, stoppages can occur. That's why you have to provide enough cleanouts so that all portions of the drainage system are accessible to cleaning equipment. (See Chapter 5.)

Make changes of direction in drainage pipes with 45-degree wyes, long or short sweeps, quarter bends, sixth, eighth or sixteenth bends, or a combination of these (or other approved fittings). You can use single and double sanitary tees, quarter bends and one-fifth bends in vertical sections of drainage pipes only where the direction of flow is from the *horizontal to the vertical.*

Never use a tee branch or a fitting that has a hub in the opposite direction of flow as a drainage fitting. There are also two other prohibitions you should be aware of. You can't have running threads, bands or saddles in the drainage system. And drainage and vent piping can't be drilled or tapped.

Install your plumbing drainage pipes in open trenches and keep them open until the plumbing inspector has tested, inspected and accepted them. You may have to remove cleanout plugs or caps so the inspector can make sure the pressure has reached all parts of the system.

Standards for Plumbing Materials

In the plumbing code, the accent is always on protection. But even if your system has an excellent design and admirable workmanship, it may not do its job — unless your materials provide satisfactory service. That's why the code sets minimum standards for plumbing system materials.

In general, the materials must be free from defects and meet the standards of the building department. All pipe, fittings and fixtures have to be listed or labeled by one or more organizations listed in Figure 8-1.

AHAM	Association of Home Appliance Manufacturers 20 North Wacker Drive Chicago, IL 60606
ANSI	American National Standards Institute 11 West 42nd Street, 13th Floor New York, NY 10036
ASME	American Society of Mechanical Engineers United Engineering Center/345 E. 47th Street New York, NY 10017
ASSE	American Society of Sanitary Engineering 100 Bar Harbor Drive West Conshohocken, PA 19428-2959
ASTM	American Society for Testing and Materials 1916 Race Street Philadelphia, PA 19103
AWS	American Welding Society 550 Northwest LeJeune Road Miami, FL 33126
AWWA	American Water Works Association 6666 W. Quincy Avenue Denver, CO 80235
CISPI	Cast Iron Soil Pipe Institute 5959 Shallowford Road, Suite 419 Chattanooga, TN 37421
CMI	Cultured Marble Institute 435 N. Michigan Avenue Chicago, IL 06011
CS and PS	Commercial Standards and Product Standards Supt. of Documents, U. S. Government Printing Office Washington, DC 20402
FM	Factory Mutual Standards — Laboratories Department 1151 Boston-Providence Turnpike, P. O. Box 9102 Norwood, MA 02062
FS	Federal Specifications, Federal Supply Service Standards Division General Services Administration 7th and D Streets Washington, DC 20407
IAPMO	International Association of Plumbing and Mechanical Officials 20001 S. Walnut Drive Walnut, CA 91789-2825
MSS	Manufacturers Standardization Society of the Valve and Fittings Industry 127 Park Street NE Vienna, VA, 22180
NSF	National Sanitation Foundation 3475 Plymouth Road Ann Arbor, MI 48106
PDI	Plumbing and Drainage Institute 1106 W. 77th Street, South Drive Indianapolis, IN 46260-3318
UL	Underwriters' Laboratories, Inc. 333 Pfingsten Road Northbrook, IL 60062
WQA	Water Quality Association 4151 Naperville Road Lisle, IL 60532

These organizations issue standards and specifications for materials as listed in Figure 8-2 (Table 14-1, *Uniform Plumbing Code*). All standards and specifications are subject to change, so designations indicating year of issue may become obsolete.

Figure 8-1
Abbreviations for plumbing-related organizations

A material is considered approved if it meets one or more of the standards cited in Figure 8-2. (Look in Table 14-1, *Uniform Plumbing Code*; Table 703, *Standard Plumbing Code;* and Chapter 14, *International Plumbing Code.*) Of course, you must be able to decipher the standards table in your code. If you can't get the listed materials, you can substitute materials only if the building department gives you the green light. Your local plumbing supplier should stock only materials that meet code standards.

You need to be familiar with the abbreviations in Figure 8-1 and the standards in Figure 8-2. Most journeyman and master examinations include at least one question from these tables.

Permitted Uses for Materials

The most common materials for drainage, waste and vent systems are listed in Figure 8-3. This chart covers sanitary and storm drainage, vent piping, and chemical and acid system materials. You'll see in the footnotes that you can't use certain materials in some locations, and that the use of other materials is limited or restricted. So how do you make your decision? Actual conditions determine your final choice: the building type and location, the type of fill material, the traffic expected over piping, and the waste conveyed.

Plumbing material standards are always subject to change. There's a steady stream of new material coming on the market. Check with your local authority about code changes that occur from time to time. Always buy the updated code supplement sheets and insert them in your code book.

Building Sewers

Cast iron You can use tar-coated, centrifugal-spun service weight or extra-heavy cast iron soil pipe. Use pipe that's the same weight as the underground pipe inside the building. If your job calls for extra-heavy underground soil pipe in the building, it's likely your code will require extra-heavy building sewer pipe, also.

Vitrified clay pipe You can also use standard or extra-strength vitrified clay pipe and fittings.

Plastic pipe The building sewer may be ABS or PVC Schedule 40 plastic pipe and fittings. Under heavy traffic areas, some codes require that the plastic pipe strength be increased to Schedule 80.

Concrete drain pipe The building sewer may be concrete drain pipe and fittings. However, since concrete pipe is highly susceptible to corrosion from acids and sewer gases, it's not usually recommended for ordinary building sewers.

Asbestos-cement sewer pipe Contrary to some environmentalists' belief, asbestos-cement pipe is still listed as an approved piping material by some codes. Its use is generally restricted to outside drainage systems.

Bituminous fiber pipe You won't find this kind of pipe listed in the standards of most codes. It's just too fragile. The government housing authority (HUD) sometimes specifies it, however, since it's less expensive than most other sewer materials. Chances are you'll never have to install it.

Copper pipe The building sewer may be copper pipe and fittings, Type DWV, K, L and M. Because of its cost and fragility, you'll seldom have to install it.

Underground Drainage Piping Within a Building

You can use any of these materials for underground drainage piping inside a building: ABS or PVC Schedule 40 plastic, brass, cast iron soil pipe, copper Type DWV, K, L or M, lead or vitrified clay pipe extra strength. Some codes also list ductible iron and heavy schedule borosilicate glass pipe.

As a rule, codes allow the use of plastic pipe in only the first three floors of a building. Underground piping for a building higher than three stories must be of one of the other materials listed above.

You'll find other special requirements for buildings in areas originally below high tide or where the fill produces hydrogen sulfide gas. You shouldn't use materials subject to corrosion. But if you do install them in corrosive soils, take care to protect them by coating or wrapping them with approved material. This type of material and material protection must extend out to the point of disposal (a public sewer or septic tank).

Materials	ANSI	ASTM	FS	IAPMO	Other standards	Footnote remarks
Ferrous pipe and fittings						
Cast iron soil pipe and fittings		A74-87				Note 2
Cast iron soil pipe and fittings for hubless cast iron sanitary systems					CISPI301-85	Note 2
Ductile-iron gravity sewer pipe		A746				
Hubless cast iron sanitary and rainwater systems (installation)				CISPI301-90		
Shielded couplings for use with hubless cast iron soil pipe and fittings				PS35-92		
Special cast iron fittings				PS 5-84		
Threaded cast iron pipe for drainage, waste and vent services	A40.5-43					
Nonferrous pipe and fittings						
Cast bronze solder-joint drainage fittings	B16.23-84					
Copper drainage tube (DWV)		B306-88				
High silicon iron pipe and fittings		A861-86 (R91)				
Lead pipe, bends and traps			FSWW -P-325			
Wrought copper and wrought copper alloy solder-joint drainage fitting- DWV	ANSI/ ASME B16-29-86					
Wrought copper and copper alloy solder joint fittings for DWV systems	B16-43					
Glass pipe						
Borosilicate glass pipe and fittings for DWV systems		C1053-90				
Nonmetallic pipe						
Asbestos-cement sewer pipe		C428-81(R85)				
Asbestos-cement storm sewer pipe		C663				
Concrete drain tile		C412-90				Note 1
Concrete sewer and storm drain pipe		C 14-90				
Plastic pipe and fittings						
ABS Schedule 40 DWV pipe and fittings		D2661-91				Note 2
ABS Schedule 40 DWV pipe with cellular core		F 628-91				Notes 2 & 3
ABS sewer pipe and fittings		F 628-91				Note 2
Coextruded PVC pipe with cellular core		F 891-91				Notes 2 & 3
Clay drain pipe		C4				
Clay sewer pipe		C700				
Concrete perforated pipe		C444				
Concrete reinforced sewer pipe		C361				
Extra strength vitrified clay pipe in building drains (installation)				IS 18-85		
DWV plastic fitting patterns		D3311-91				Note 2
PVC corrugated sewer pipe		F949-91A				Note 2
PVC plastic DWV pipe and fittings		D2665-91B				
PVC sewer pipe and fittings (thin wall)		D2729				
PVC sewer pipe and fittings (Type PSM)		D3034-89				
PVC gravity flow sewer pipe and fittings (Type PS-46)		F 789-89				
PVC sewer pipe and fittings (Type PSM)		D3034-89				

Notes 1) Type II only 2) Although this standard may be referenced in some code tables for some of the pipe, tube or fittings, it is not always acceptable under every code 3) Additional requirements for inner and outer layers

Figure 8-2
Plumbing materials standards for drainage systems

Pipe material	Building sewer	Underground within buildings	Aboveground within buildings
Sanitary drainage and waste systems			
Asbestos-cement sewer	X		
ABS Schedule 40	X	X	X
Borosilicate glass			X
Brass			X
Cast iron soil	X	X	X
Concrete drain	X		
Copper, DWV	X	X	X
Copper, Type K, L, M	X	X	X
Galvanized steel			X
Wrought iron			X
PVC Schedule 40	X	X	X
Vitrified clay, standard	X		
Vitrified clay, extra strength	X	X	
Lead		X	X
Vent pipes for drainage and waste systems			
ABS Schedule 40	N/A	X	X
Borosilicate glass	N/A		X
Brass	N/A		X
Cast iron soil	N/A	X	X
Copper, DWV	N/A	X	X
Copper, Type K, L, M	N/A	X	X
Galvanized steel	N/A		X
Lead	N/A		X
PVC Schedule 40	N/A	X	X
Vitrified clay, extra strength	N/A	X	
Wrought iron	N/A		X
Storm drainage systems			
Asbestos-cement pipe	X		
ABS Schedule 40	X	X	X
Brass			X
Cast iron soil	X	X	X
Concrete drain	X		
Copper, DWV	X	X	X
Copper, Type K, L, M	X	X	X
Galvanized steel			X
Lead			X
PVC Schedule 40	X	X	X
Vitrified clay, extra strength	X	X	
Wrought iron			X
Chemical waste and acid systems			
ABS Schedule 40	N/A	X	X
Borosilicate glass	N/A		X
High silicon content iron	N/A	X	X
Plastic lined	N/A	X	X
PVC Schedule 40	N/A	X	X
Lead	N/A		X
Vitrified clay, extra strength	N/A	X	

Notes Other codes may list materials that don't appear in above charts. Some codes limit plastic pipe to first three floors of the building. Fragile sewer piping materials, like asbestos, plastics, concrete and clay, generally require special protection. See Figure 8-22.

Figure 8-3
Common materials and locations where they can be used

The fittings you use must match the material and type of piping in the drainage system. Joints must be gastight, watertight and root-proof.

Footing and Subsoil Drains

When the design calls for subsoil drains under a cellar or basement floor, or around the outer walls of a building, they must be at least 4 inches in diameter. The most common materials are:

- clay drain tile
- concrete drain tile
- perforated or horizontally split concrete pipe
- corrugated polyethylene tube
- perforated or horizontally split SR plastic drain pipe
- PVC sewer pipe
- vitrified clay pipe (standard or extra strength)

Some codes also accept perforated asbestos cement pipe. See Figure 8-3.

Aboveground Drainage Within a Building

Aboveground drainage pipes may be ABS or PVC Schedule 40 plastic, borosilicate glass, brass, cast iron soil pipe, Type DWV, K, L or M copper, galvanized steel, wrought iron or lead (Figure 8-3).

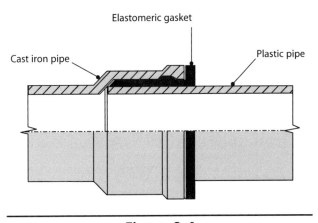

Figure 8-4
Typical transition joint — plastic to cast iron pipe

Your fittings in drainage systems must be compatible with the pipe you've used. Make sure they have no ledges, shoulders or reductions that could restrict or obstruct flow. Check your threaded fittings to see that they're the recessed drainage type.

Underground and Aboveground Vents Within a Building

Vent pipes can be made of ABS or PVC Schedule 40 plastic, borosilicate glass, brass, cast iron soil pipe, copper Type DWV, K, L or M, galvanized steel, wrought iron or lead pipe. See Figure 8-3.

All your fittings must match the materials and type of piping used. When you use threaded pipe, your fittings may be either the drainage or pressure type, galvanized or black.

In general, codes place these restrictions on drainage, waste and vent piping within a building, aboveground or underground:

- Most codes limit plastic pipe to no more than three floors above grade.
- Plastic pipe or fittings can't support the weight of any plumbing fixture.
- You can't mix or combine different types of plastic materials (ABS or PVC) in any plumbing system.
- To extend, relocate or add to any existing soil, waste or vent pipe, use material of like grade and quality.
- When you've got to join together different piping materials in new work, you must do it with an approved transition fitting. Figure 8-4 shows plastic pipe joined to cast iron pipe with a compression gasket.
- You can't use galvanized steel or galvanized wrought iron pipe underground in a drainage or vent system. Install it at least 6 inches aboveground.
- You can't use vitrified clay pipe above ground or where pressurized by a pump or sewage ejector. Keep it a minimum of 12 inches below ground.

Chemical Waste and Acid Systems

All piping for chemical waste and acid systems must be made of corrosion-resistant materials approved by your local authorities. These materials are usually acceptable: borosilicate glass, high silicon content cast iron, vitrified clay, ABS and PVC Schedule 40 plastic, plastic-lined and lead pipe. See Figure 8-3. And again, match the fittings to the material and pipe you're using.

Indirect Waste Piping

For indirect waste piping, you can use any materials approved for potable water, or sanitary or storm drainage. The most common materials are cast iron soil pipe, galvanized steel, wrought iron, brass, copper Type DWV, K, L or M, ABS or PVC plastic and lead pipe. Figure 8-3 shows where you can use each material, above or below ground.

Storm Drainage Systems

Most codes accept the materials listed in Figure 8-3 for storm drainage systems.

For the building sewer, you can use asbestos-cement, ABS and PVC Schedule 40 plastic, cast iron, concrete, vitrified clay, or copper Type DWV, K, L and M. But unless there are extraordinary circumstances, you'll never install copper drains. They're just too expensive.

For underground storm drain piping within a building, choose from ABS and PVC Schedule 40 plastic, cast iron soil pipe, copper Type DWV, K, L and M, and vitrified clay extra strength. Some of these materials require advance approval by your plumbing official. And *some codes will accept only cast iron or ferrous-alloy piping*. Always check the local code for acceptable materials in your area.

Acceptable materials for aboveground storm drains and leaders within a building are ABS and PVC Schedule 40, brass, cast iron, copper Type DWV, K, L and M, galvanized steel, lead and wrought iron pipe. Some codes also accept black steel.

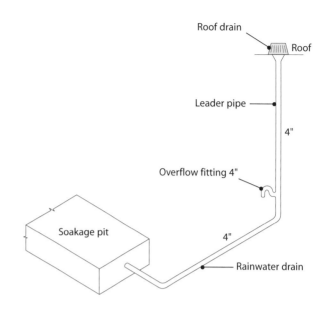

Figure 8-5
Rainwater overflow

Rainwater drains conveying runoff from leaders that discharge directly into a soakage pit must have an overflow fitting. In fact, each rainwater leader requires an overflow fitting at its base. It must be the same size as the leader pipe, up to 4 inches. See Figure 8-5.

Plan to protect all exposed rainwater leaders you install in parking garages, alleys, driveways or any place subject to damage. The conventional protection is a 3-inch galvanized steel pipe supported in a concrete base, placed in front of the leader. You can protect fragile materials (sheet metal conductors or plastics) by installing a cast iron pipe boot on the lower 5-foot portion of the leaders. And use fittings that conform to the materials and pipe you're using.

Standards for Plumbing Materials

All plumbing materials used in the construction, installation, alteration or repair of any plumbing or drainage system must comply with the standards in your code. For easier reference, I've organized the standards for plumbing materials to match the

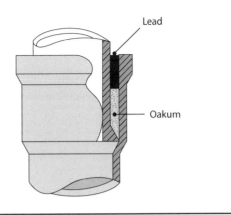

Figure 8-6
Lead and oakum joint

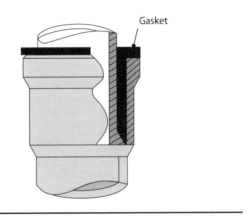

Figure 8-7
Compression joint

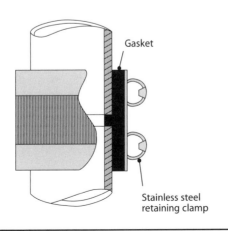

Figure 8-8
No-hub joint

categories of materials you'll use in a plumbing system. That means that in this chapter I'll cover only those materials you use in drainage, waste and vent systems.

Joints and Connections

Joints and connections in the DWV system must be gastight and watertight at the pressure required for testing or for the intended use. The code makes an exception for those parts of a drainage system which use perforated or open-joint piping designed to collect and convey underground water to a legal point of disposal.

Types of Joints

Cast Iron Caulked Joints

Every lead caulked joint for cast iron bell-and-spigot soil pipe must be firmly packed with oakum or hemp and filled with molten lead at least 1 inch deep (Figure 8-6). The lead can't extend more than $1/8$ inch below the rim of the hub. You should run the lead in one pouring and caulk it tight. Don't use paint, varnish or other coatings on the jointing material until after the joint's tested and approved.

Cast Iron Compression Joints

Neoprene rubber compression gaskets for bell-and-spigot cast iron soil pipe and fittings must be compressed when the spigot is inserted into the hub of the pipe. You can use this kind of joint as an alternate for lead and oakum joints. See Figure 8-7.

Cast Iron Hubless Joints

Make joints for hubless cast iron soil pipe and fittings with an approved elastomeric sealing sleeve and stainless steel clamp (Figure 8-8). Any clamp assembly you use to join a hubless cast iron soil pipe for DWV and building sewer lines must meet the standards of the Cast Iron Soil Pipe Institute (CISPI), or the recommendation of the manufacturer.

Asbestos-Cement Sewer Pipe Joints

Make all joints in asbestos-cement pipe with sleeve couplings of the same composition as the pipe, sealed with preformed rubber rings (Figure 8-9). Make joints between asbestos-cement pipe and metal pipe with an adapter coupling caulked with lead and oakum. Joints between asbestos-cement pipe and clay pipe need an adapter coupling with approved rubber rings or a preformed bituminous ring approved by your local authority. Joints between asbestos-cement pipe and plastic pipe should always have an approved adapter coupling with approved rubber rings.

Join asbestos-cement pipes used for subsoil or storm sewer pipe with plastic couplings designed to fit snugly over the ends of the pipe.

When you need to cut asbestos-cement pipe for new installations, use a tapering tool designed to taper the pipe so it'll make a watertight joint. Tapered couplings of the same material as the pipe will provide a tight joint. See Figure 8-10.

Plastic Pipe Joints

For joints connecting plastic pipe to plastic fittings, you'll usually use solvent-cemented or heat-joined connections. If you choose ABS or PVC threaded joints, use only approved thread tape or lubricant seal recommended by the manufacturer. Never use regular pipe dope to seal plastic threaded joints.

You can also use a stainless steel clamp and elastomeric gasket joint or a push-on elastomeric gasket joint (Figure 8-11). Make this joint by pushing the pipe end into the hub which contains the gasket. It's a common joint for building drains and sewers.

When you need a connection between plastic pipe and any other material, use only approved transition fittings. The manufacturer will probably recommend the type of fitting and installation method. And finally, remember that you can't mix ABS or PVC plastic pipe and fittings in the same system without the correct transition fittings.

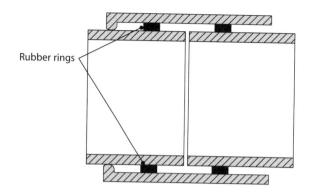

Figure 8-9
Asbestos-cement pipe coupling

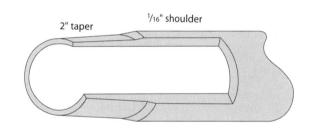

Figure 8-10
Tapered pipe end

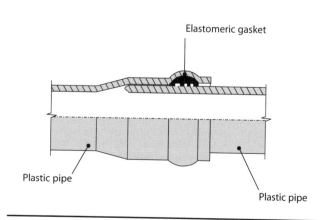

Figure 8-11
Typical push-on elastomeric gasket joint

Bituminous Fiber Pipe Joints

When you're allowed to use bituminized fiber pipe joints, make them with tapered couplings and fittings of the same material as the pipe. Be sure to use the specially-designed tapering tool when you need to cut the pipe. See Figure 8-10. Make your joints between bituminized fiber pipe and metal pipe with an adapter coupling caulked with lead and oakum. You'll need to use approved transition fittings to make the joints between bituminous fiber pipe and other materials.

Vitrified Clay Sewer Pipe Joints

Use neoprene preformed elastomeric rings to join vitrified clay pipe with bell-and-spigot connections. Join plain ends with a flexible coupling. It has an approved oil-resistant gasket attached to the pipe with adjustable stainless steel clamps and bolts (Figure 8-8).

Some codes let you use this method to seal vitrified clay pipe with bell-and-spigot joints: Use hot poured bitumastic compound if it has a bond strength of at least 100 psi in shear. Fill approximately 25 percent of the joint space at the base of the socket with jute or hemp. Pour each joint in one operation until it's filled. Don't try to test the joint for at least one hour after pouring. This joint is about like a lead and oakum joint (Figure 8-6), but with less hemp and more bitumastic compound. Vitrified clay sewer pipe joints have to be flexible to avoid breakage.

Concrete Sewer Pipe Joints

For new construction, most codes prohibit installing concrete pipe and fittings with cement mortar joints. But you can use mortar joints and connections for repairs or for connections to existing lines that have mortar joints.

The code is very specific on how to make this joint. First, pack a layer of jute or hemp into the base of the cement mortar joint space. Then dip the jute or hemp into a slurry of portland cement before inserting it into the bell. This prevents mortar from getting into the interior of the pipe. Fill only 25 percent of the joint space with jute or hemp. Fill the remaining space in one continuous operation with a thoroughly-mixed

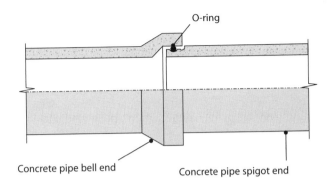

Figure 8-12
Concrete pipe joint made with O-ring

mortar made of one part cement and two parts sand. Use only enough water to make the mixture workable by hand. Apply additional mortar to form a 1 to 1 slope with the barrel of the pipe. Swab the interior of the pipe to remove any material that dripped into the pipe. Trowel more of the same mortar on your joint to form a 45-degree taper with the barrel of the pipe.

You can make a joint in concrete pipe using approved compression elastomeric materials. That's simply an O-ring that's located in the bell that makes a tight joint when the spigot end is forced into the hub. See Figure 8-12.

Trap Slip Joints

In a drainage system slip joint, you can use connectors (nuts and washers) on both sides of the trap and in the trap seal. However, some codes prohibit the use of slip joint connectors on the outlet side of the trap above the water seal. Why? Over a period of time the washer may decay or the seal of the joint may break. This would permit sewer gases to enter the building. You can see some approved traps back in Chapter 4.

Wiped Joints

The use of wiped joints is rare today. But be aware that wiped joints are required for these connections: lead pipe or fittings with lead, and lead pipe with brass or copper (includes ferrules, solder nipples or traps). Be sure your wiped joint has an exposed surface on each side of the joint no less than $3/4$ inch and at least as thick as the material being joined. Joints between lead pipe and cast iron, steel or wrought iron pipe require a caulking ferrule, soldering nipple or bushing. The minimum length of lead from wiped joint to fixture connection is 4 inches.

Borosilicate Glass Joints

Make glass-to-glass connections with a bolted compression-type stainless steel coupling that has a contoured acid-resistant elastomer compression ring and fluorocarbon polymer inner seal ring.

When you connect glass pipe joints to other types of piping material, use only approved adapters with a TFE seal, installed to manufacturer's recommendations.

Make your glass caulked joints like cast iron caulked joints, with this exception: you must use acid-resistant oakum or hemp rope and acid-proof cement. Check your local code for verification.

Burned Joints

Lap burned (welded) lead joints and fuse the sections together uniformly to make one continuous piece. Make the weld as thick as the lead pipe you're joining, using the same welding material as the material being joined.

Ductile Iron Gravity Sewer Pipe, Bell-and-Spigot Joints

Always join ductile iron bell-and-spigot joint gravity sewer pipe with a push-on type single oil-resistant gasket. The specially-designed bell end of the pipe is shaped so the gasket can lock into place. This prevents displacement and avoids leakage.

Plain-End Ductile Iron Gravity Sewer Pipe

You can join plain-end ductile iron gravity sewer pipe the same way as no-hub cast iron pipe — with an elastomeric gasket joint and stainless steel retaining sleeve. See Figure 8-8.

Threaded Joints

Any threaded joints you use in DWV systems must conform to approved standard pipe threads shown in Chapter 14. Always use recessed drainage fittings. The manufacturer taps the threads to allow $1/4$ inch fall per foot. Fittings must have a smooth interior waterway.

Copper Soldered Joints

Use approved cast brass or wrought copper fittings when joining copper pipe in a DWV system. Make your joints with approved solder containing no more than 0.2 percent lead. See Figure 8-13.

Special Joints in DWV Systems

You need to have a working knowledge of several special joints used in a DWV system, including:

- *Slip joints* — They're located in fixture drains and traps. Use only approved materials.

- *Expansion joints* — They're permitted in vent piping or drainage stacks when necessary for expansion and contraction of pipes. They don't need to be accessible.

- *Joints of different types of materials* — Various materials for drainage piping can be joined together by using approved adapters or prefabricated sealing rings or sleeves.

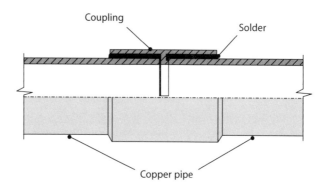

Figure 8-13
Typical copper soldered joint

∎ *Joints with different outside diameters* — You can join pipe materials with different outside diameters with an approved elastomeric sealing sleeve and clamping device. Provide continuous support between the two pipes.

∎ *Plastic threaded joints* — You can't thread Schedule 40 plastic pipe on the job site. For transitions, use only approved male or female threaded adapter fittings.

Increasers and Reducers

When you connect different size pipes or different size pipes and fittings, take care to select the proper size increaser, reducer or reducing fitting.

Changes in Direction

Make changes in direction in horizontal or horizontal-to-vertical drainage systems with 45-degree wyes, long or short sweeps, quarter bends, sixth, eighth or sixteenth bends, or with a combination of these or other approved fittings. When the direction of flow is from horizontal to vertical, use sanitary tees, quarter bends and one-fifth bends. All the fittings in Figure 8-14 are the no-hub type.

Prohibited Joints and Connections in Drainage Systems

Never use fittings or connections with an enlarged chamber, or a ledge, shoulder or reduction of pipe area which might obstruct the drain flow.

Installation Methods

New materials are constantly being developed, approved and accepted by local authorities for use in DWV systems. New concepts in pipe and fitting materials often have advantages in performance, versatility, low-cost installation and product availability. There are drawbacks, however. These pipes and fittings are manufactured of fragile substances. You've got to strictly follow the manufacturer's installation instructions.

Always get advance approval from your local authority before you use materials that aren't yet included in the standards cited in your code.

Securely support your piping in trenches and above ground to avoid sagging, misalignment and breaking. We'll take a closer look at support later in this chapter.

Building Sewers

Cast iron soil pipe is the preferred material for sewers because of its strength, durability and resistance to trench loads. There are three types of cast iron pipe and fittings:

1) Lead caulked joints for hub and spigot pipe

2) Compression gasket joints for hub and spigot pipe

3) Stainless steel shield with elastomeric gasket joints for no-hub pipe.

The two grades of piping material still in use today are centrifugally-spun service weight and extra heavy cast iron.

Sewer Installation

When installing cast iron soil pipe, be sure to keep the pipe barrel in firm contact with solid ground. To do that, you have to excavate for the hub (bell) or coupling. That distributes the weight evenly along the full length of the pipe. See Figure 8-15. The depth of the excavation isn't too important because cast iron has a high resistance to trench loads. But don't backfill with large boulders, rocks, cinder or other materials which could damage or corrode the pipe.

Vitrified clay, plastic, asbestos-cement, copper and bituminous fiber pipe are considered fragile sewer materials. Where they're permitted, these pipes require special installation methods. Here are some precautionary measures that most codes require:

∎ Make sure the trench base continuously and uniformly supports the bottom quarter of these pipes.

∎ Support the pipe with fine material free of stones.

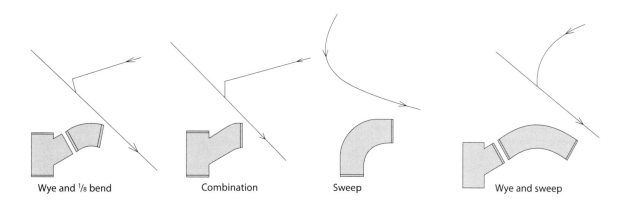

A Horizontal to horizontal change of direction

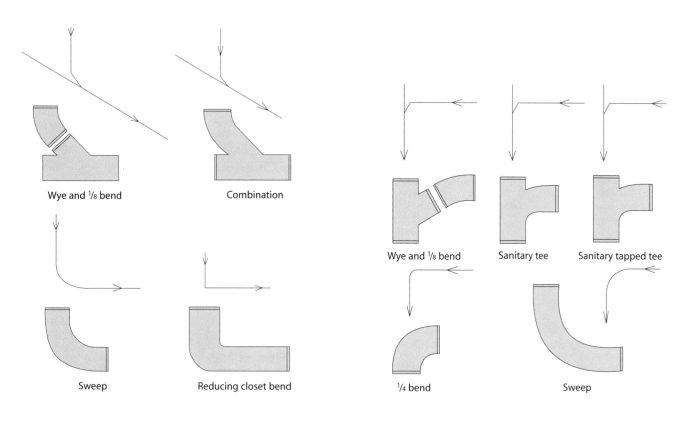

B Vertical to horizontal change of direction

C Horizontal to vertical change of direction

Figure 8-14
Changes of direction

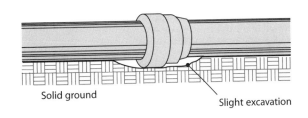

Figure 8-15
Excavation for piping hub projection

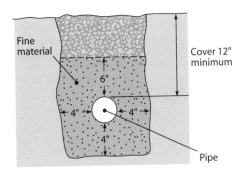

Figure 8-16
Protecting fragile piping materials

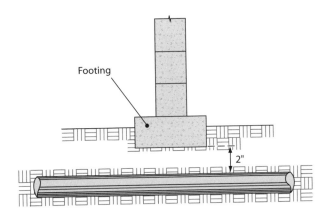

Figure 8-17
Clearance for building drains beneath foundations

❚ Excavate for hub and coupling projections so no part of the pipe load is supported by the hub or coupling. The supporting material must extend 4 inches on each side of the pipe.

❚ Firmly compact the backfill with selected fine material. Extend it from the trench bottom to a point 6 inches over the top of the pipe. Make sure your pipe is at least 12 inches below ground. See Figure 8-16.

❚ Make sewer pipe trenches wide enough to provide adequate work space for placing and joining the pipe.

❚ Prepare the trench bottom so that in final position, the pipe is true to line and grade.

Drainage, Waste and Vent Piping Within a Building

For underground or aboveground installation within a building, there's a greater variety of piping materials to choose from. But there are restrictions on the use of certain pipe and fitting materials, especially in regard to the building height. Here are some of the restrictions:

❚ Underground or horizontal drainage, waste and vent piping must be adequately supported. It's possible you'll need approved hangers or masonry supports to keep the pipe in alignment and to prevent sagging.

❚ Support the bases of stacks with masonry or concrete.

❚ When drainage pipe passes through cast-in-place concrete, provide sleeves that create $1/2$ inch of annular space around the entire circumference of the pipe. Tightly caulk the space between the sleeve and the pipe with coal tar or asphaltum compound, lead or other approved material.

❚ Building drains that pass beneath foundations need a clearance of at least 2 inches from the top of the pipe to the bottom of the footing. See Figure 8-17.

❚ Building drains and vent pipe passing through a fire wall or penetrating poured-in-place concrete floors must be sleeved (Figure 8-18). Seal the

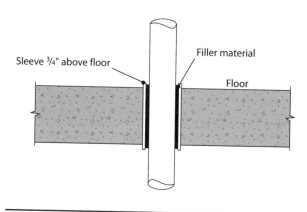

Figure 8-18
Approved floor penetration

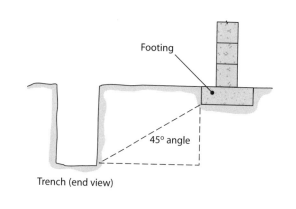

Figure 8-19
45-degree angle of pressure

annular space between sleeve and pipe with fire-resistant material so it's watertight. The filler material must meet fire and building code requirements.

▌ Unless the building official approves a special design, you can't place any excavation for drainage piping within a 45-degree angle of pressure from the base of an existing structure to the sides of the trench. In other words, you can't dig a drainage pipe trench that's deeper than the base of the structure and parallel to the structure's foundation within the 45-degree angle illustrated in Figure 8-19.

▌ When installing drainage piping in contact with cinders, concrete or other corrosive materials, use sleeves, coating, wrapping or other approved methods to protect it.

▌ Close any openings for pipes through walls, floors or ceilings by fastening approved metal collars securely to the structure.

▌ All waste pipes installed in exterior walls and other areas must be protected from freezing.

▌ Never install used drainage piping which doesn't conform to your code's specifications.

▌ Make all joints and connections gastight and watertight to withstand any required pressure testing.

▌ Don't install fittings or connections that present any abnormal obstruction to the flow of waste.

▌ Never drill or tap any waste or vent pipe for the purpose of rodding or making a connection to it.

▌ Never make a waste connection to a closet bend, stub of a water closet or similar fixture.

▌ Never use a vent pipe to convey any kind of waste and never use a soil or waste stack as a vent.

▌ When joining pipe or pipe and fittings of different sizes, always use increasers or reducers.

Pay special attention to these two limitations: First, most codes limit the use of plastic DWV pipe to buildings not over three stories high. Check local code requirements. Second, buildings that exceed three stories may use centrifugally-spun service weight or extra-heavy cast iron, copper or galvanized steel pipe.

Drainage, Waste and Vent Piping Supports

Make certain that all hangers and anchors that support horizontal and vertical piping are secure enough to maintain pipe alignment and prevent sagging. Figure 8-20 shows five common pipe supports for horizontal and vertical DWV pipe. You can check your hanger rod sizes by looking at Figure 8-21. They won't be acceptable if they're smaller than the sizes shown in that table.

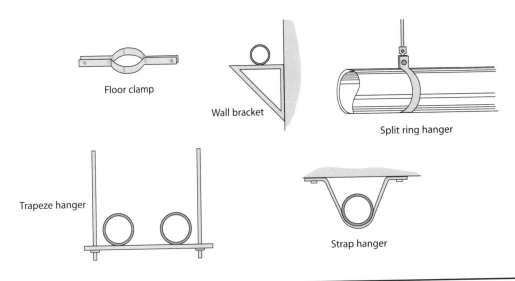

Figure 8-20
Five common horizontal and vertical pipe supports

We'll look at the specific requirements for supporting *horizontal* and *vertical* piping constructed of different materials.

Horizontal DWV Piping Supports

Always support cast iron soil pipe that has lead and oakum joints with hangers or other approved means at every hub. If you're using 5-foot lengths of pipe, that's at least every 5 feet. When you use pipe lengths over 5 feet, your support hanger intervals can be 10 feet. Your supports must be within 18 inches of the hub or joint. Support cast iron soil pipe with hubless or compression gasket joints every other joint, unless the developed length exceeds 4 feet, then support at each joint.

Pipe and tube size (in)	Rod size (in)
½ to 4	³⁄8
5 to 8	½
10 to 12	⁵⁄8

Figure 8-21
Hanger rod sizes

- *Screwed pipe:* For ³⁄4 inch and smaller, you need 10-foot interval supports. Pipe this small is used mostly as indirect waste piping. Pipe 1 inch and larger requires 12-foot interval supports.

- *Copper pipe or tubing:* Provide supports at about 6-foot intervals for copper pipe or tubing that's 1½ inch or smaller. When using 2-inch and larger copper pipe or tubing, you'll need supports at 10-foot intervals.

- *Schedule 40 PVC and ABS solvent-cemented pipe:* Regardless of pipe size, support it every 4 feet.

- *Lead pipe:* This must be supported for its entire length.

- *Borosilicate glass piping (or similar material):* Provide supports every 8 to 10 feet. Hangers for this material must be padded.

Figure 8-22 shows a summary of the horizontal piping support intervals.

Vertical DWV Piping Supports

For all metal DWV piping, the size of the pipe isn't a factor in the support requirements. Here are the supports you'll need to provide for these vertical pipes:

- *Cast iron DWV pipe:* Support at its base and at each floor, at intervals that don't exceed 15 feet.

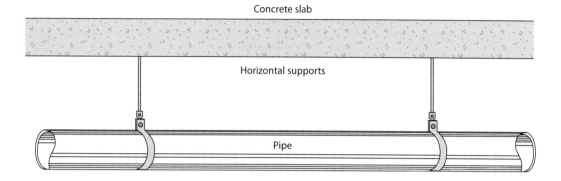

Distance between hangers for:

A Cast iron lead and oakum joints, 5' pipe . . . . .5'

B Cast iron lead and oakum joints, 10' pipe . . . . .10'

C Cast iron compression joints, every other joint, unless over 4', then each joint . . . . —

D Cast iron hubless joints (same as C above) . . . . —

E Copper pipe or tubing 1½" and smaller . . . . .6'

F Copper pipe or tubing 2" and larger . . . . .10'

G Screwed or welded pipe 1" and larger . . . . .12'

H Schedule 40 plastic pipe ABS and PVC all sizes . . . . .4'

I Lead pipe continuous support . . . . .Entire length

J Borosilicate glass pipe . . . . .8' to 10'

Notes Support adjacent to joint, not to exceed 18". To prevent horizontal movement, brace every 40'. Support at each horizontal branch connection. Do not place hangers on the coupling.

Figure 8-22
Horizontal piping support intervals

■ *Screwed vertical DWV pipe:* Support at least every other story, never exceeding 25 feet.

■ *Copper pipe or tubing:* Support at each floor, but intervals must never be more than 10 feet apart.

■ *Schedule 40 PVC and ABS pipe:* Install supports at the base and at each floor. Provide mid-story guides and expansion fittings each 30 feet.

■ *Lead DWV pipe:* Your support intervals can't exceed 4 feet. The size of the pipe doesn't affect the support requirements.

■ *Borosilicate glass pipe:* Sizes 2 inches and smaller require supports at every other story height. Sizes 3 inches and larger need support at each story height. You must use padded hangers.

Figure 8-23 shows a summary of the support requirements for vertical DWV pipe.

Vent Piping

Install horizontal vent piping with enough slope that gravity will drain it to the soil or waste pipe. Inadequate slope or sags in the pipe allow condensation to collect in low places and restrict air circulation. That reduces venting capacity.

When your horizontal vent pipe connects to a stack vent or vent stack, make sure you have an upward slope. This prevents entrapment of warm, moist air which restricts free air circulation. Trapped moist air

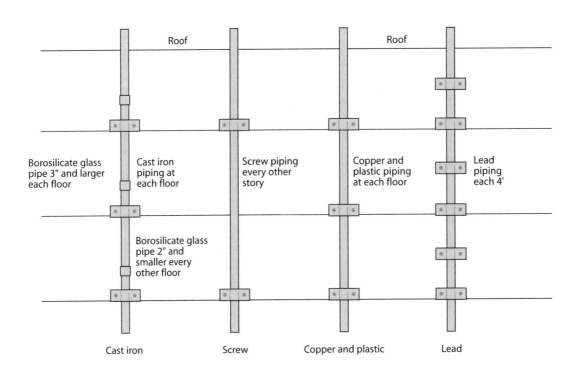

Figure 8-23
Vertical piping supports

Indirect Waste Piping and Special Waste

Size and install all indirect waste piping to accommodate the outlet drainage of any fixture or appliance. You don't need to vent indirect waste piping that doesn't have its own trap, but never connect it directly to the sanitary drainage system.

Special indirect waste piping can carry some offensive odors to the receiving fixture floor sink or floor drain. The liquid content, flowing slowly, may create slime deposits. Install all indirect drain piping without sags to eliminate the potential for slime buildup and stoppage. Gravity should cause it to drain dry. You can

accelerates metal pipe corrosion and greatly reduces its life. Review Chapter 3 for additional vent installation requirements.

further reduce slime formation by locating the receiving fixture in a well-ventilated area. Always provide accessible cleanouts for cleaning and flushing purposes.

When practical, install indirect waste pipe below the floor. Install any indirect waste pipe through the receiving fixture above the water seal of the trap. This is known as an *air break* installation. See Figure 8-24.

When an above-floor installation is required, install the pipe at least 3 inches above the floor to allow room for floor cleaning. Terminate the outlet with an *air gap* of 1 or 2 inches above the receiving fixture. See Figure 8-25.

Use pipe that's at least $3/4$ inch diameter, but never smaller than the outlet drains of fixtures or appliances served. If you're installing indirect waste pipe below the floor, most codes require $1^1/4$-inch pipe.

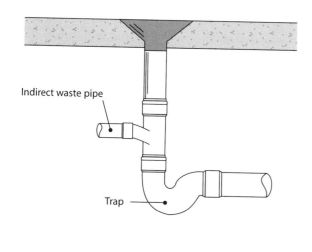

Figure 8-24
Air break type indirect waste pipe connection beneath floor

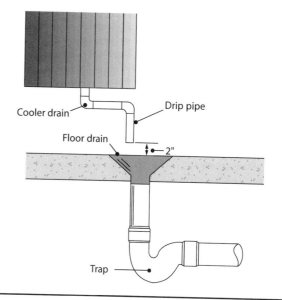

Figure 8-25
Air gap

When your installation serves *special fixtures rated more than 1 fixture unit*, use these minimums for the pipe size:

- 2 fixture units: $1^{1}/_{2}$ inch minimum

- 3 fixture units: 2 inch minimum

Place receiving fixtures in accessible locations for inspection and cleaning. Make sure workers can get to them without moving or disconnecting equipment or heavy objects. Don't put them in a closet, cupboard or storeroom.

You'll usually have to directly trap indirect waste pipe that's between 5 and 15 feet long, depending on your code, but you don't need to vent the traps. Some local codes don't require trapping or venting on indirect waste pipes of any length, as long as they're properly graded to drain dry.

Always connect a commercial dishwashing machine directly to the building waste drain line. And many codes require an *air gap* assembly device on a domestic dishwashing machine when it's connected to the sink tail piece. See Figures 8-26 and 8-27.

Never install a floor drain or floor sink which serves as indirect waste pipe in a toilet room.

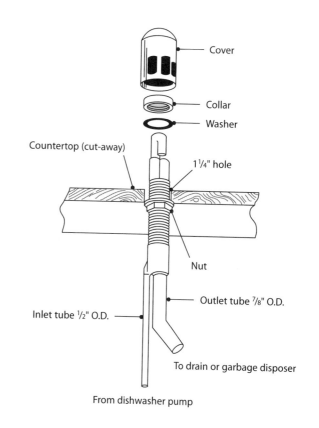

Figure 8-26
Domestic dishwasher air gap assembly device

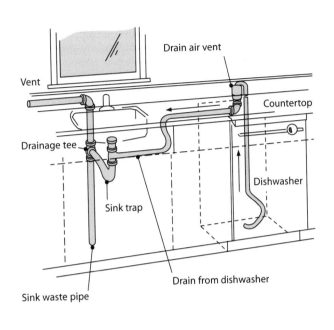

Figure 8-27
Domestic dishwasher with air gap device

Install the floor drainage from a walk-in cooler, freezer or food storage room indirectly to the floor drain installed outside of the room. The indirect waste pipe must have a flap check valve, and the room floor must be 2 inches above the receiving fixture (Figure 8-28).

When a drinking fountain is located outside a toilet room, you can indirectly connect it to a floor drain, for the purpose of resealing the trap (providing a source of water so the trap doesn't evaporate).

Air Conditioning Condensate Drains

Always locate indirect waste piping for air conditioning units at least 2 inches below the bottom of the floor slab. Don't install it until *after* fill and compaction are complete. Lay it on a firm base for its entire length, backfilled with 2 inches of sand. Be sure to protect all risers passing through the slab with sleeves.

Most codes require an air conditioning unit's vertical condensate drain line to be vented and trapped. Some codes will accept a perforated cap on the riser pipe about 2 feet above the topmost unit. Others

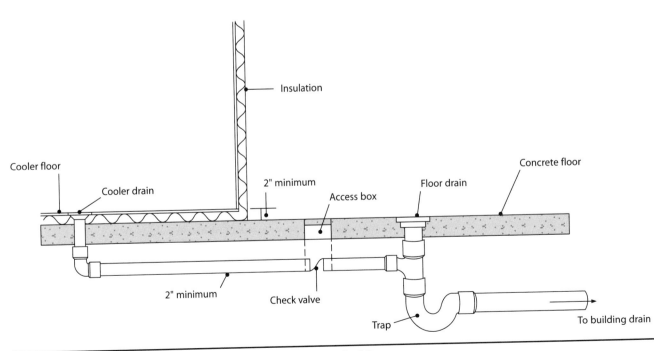

Figure 8-28
Special cooler floor drainage by air break method

require the riser to extend through the building roof. See Figure 8-29. Always check your local code requirements.

Note that the waste or condensate from an air conditioning unit is classified as a plumbing fixture only if it's connected to the plumbing drainage system. For a continuous flow like this, allow 2 fixture units per gallon per minute. Look back to Figures 2-10 and 2-11 in Chapter 2 to size the fixture drain or trap that will accommodate those fixture units.

Never use the air conditioning condensate waste to reseal the trap of the floor drain.

Any air handling equipment located in a room must be indirectly connected to the floor drain located outside the room. See Figure 8-30.

Sometimes sanitary or storm water drainage pipes aren't readily available. When that happens, your code may permit one of the following for air conditioning condensate waste disposal:

▌ Air conditioning units not exceeding 5 ton capacity may discharge their waste onto a pervious area such as bare soil.

▌ Air conditioning units over 5 tons but less than 10 tons may discharge their waste into a buried pipe filled with ³/₄-inch rock. The pipe must have a minimum 10-inch diameter and be 24 inches long. No cover is required.

▌ Air conditioning units 10 tons or larger may discharge condensate into a drainage well, storm water system, adequate-size soakage pit, drainfield or the building sanitary drainage system.

▌ When an air conditioning unit is centrally located below the roof of a building, it may indirectly connect to a rain leader pipe. See Figure 8-31. You can never use this method on a sanitary drainage, waste or vent pipe.

Remember these restrictions on air conditioning equipment or condensate drains: They can never discharge onto the roof of a building, across a sidewalk to the curb gutter, or onto any impervious area such as a parking lot.

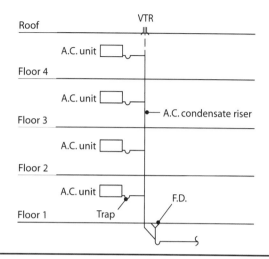

Figure 8-29
Typical air conditioning riser condensate drain connected to building drain by air break method

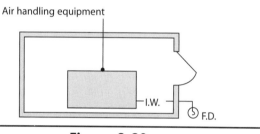

Figure 8-30
Indirect waste pipe connected to floor drain located outside of room

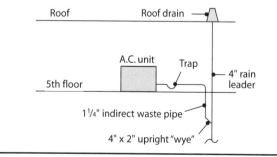

Figure 8-31
Indirect waste connection to inside rain leader

Review Questions (answers are on page 289)

1. What is the purpose of a written plumbing code?

2. Give a general description of what the plumbing code covers.

3. Where does the potable water supply system end and the sewage system begin?

4. Why is it important to design a proper drainage system?

5. Why must a drainage system be properly vented?

6. What does a properly vented drainage system prevent?

7. Why must you provide adequate cleanouts on a drainage system?

8. Name three fittings that are acceptable for making changes in direction in drainage pipes.

9. Name two of the five fittings that aren't acceptable for use in a drainage system.

10. How long after the installation of drainage pipes in open trenches must the trenches remain open?

11. Who sets the minimum standards for plumbing system materials?

12. If listed or labeled materials aren't available, when can you use a substitute material?

13. There are certain organizations that approve various plumbing materials. If you see the abbreviation ASTM on a cast iron soil pipe and fitting, you would know that these materials were approved by what organization?

14. Why is it important that you familiarize yourself with the abbreviations of organizations that approve plumbing materials?

15. What organization approves most standards for drainage materials?

16. Why is it important to keep your copy of the code updated?

17. If you're using extra heavy cast iron pipe underground within a building, what weight cast iron sewer pipe will most codes require?

18. Under what conditions do some codes require Schedule 80-strength plastic sewer pipe?

19. Why is concrete pipe not generally recommended for ordinary building sewers?

20. The use of asbestos-cement pipe in a building drainage system is limited to what area?

21. In what situations can bituminous fiber pipe be used as a building sewer?

22. Where, other than in building sewers, can you use extra strength vitrified clay pipe?

23. On which floors in a multistory building can you install plastic drainage piping?

24. What kind of piping material should you not use where the fill is known to be deleterious?

25. According to the code, fittings used in a drainage system must conform to what?

26. The joints in a drainage system must meet what three conditions?

27. What is the minimum size required for subsoil drains?

28. Name three of the approved materials most commonly used for a building subsoil drain.

29. Besides being compatible with the pipe used, what other standards must fittings in a drainage system meet?

30. What type of threaded fittings must be used in a drainage system?

31. What types of fittings may be used with threaded pipe in a vent system?

32. The code prohibits the mixing or combining of what type of piping materials in the same plumbing system?

33. What type of material must you use when adding to an existing DWV system?

34. When joining different piping materials together in new work, what kind of fitting must you use?

35. Galvanized steel pipe can't be used underground in a DWV system. How far aboveground must it be kept?

36. What is the minimum depth below grade that vitrified clay pipe must be kept?

37. Name two types of materials that are acceptable for use in a chemical or acid system.

38. The same materials are approved for use in indirect waste piping as are approved for what other systems?

39. What must you install at the base of a rain leader when it discharges directly into a soakage pit?

40. What is the conventional protection for exposed rainwater leaders located in areas where they may be subject to damage?

41. For what purpose are joints and connections in DWV systems pressure tested?

42. What type of packing is used in every lead-caulked joint with cast iron bell-and-spigot soil pipe?

43. How, specifically, should you pour a lead joint?

44. What material can you use besides lead and oakum to seal bell-and-spigot cast iron soil pipe joints?

45. The clamp assembly used in joining hubless cast iron soil pipe and fittings for DWV systems must comply with the standards set by what organization?

46. How should joints between asbestos-cement pipe and plastic pipe be made?

47. When it's necessary to cut asbestos-cement pipe for a new installation, what tool must be used to dress the pipe so that the fittings will be watertight?

48. What two types of joints are used most often to connect plastic pipe and plastic fittings?

49. What type of sealant should never be used to seal plastic threaded joints?

50. What type fitting must be used when joining bituminous fiber pipe with other types of materials?

51. What kind of fitting is used to join the plain ends of vitrified clay sewer pipe?

52. What's the only type of construction in which you can use cement mortar joints and connections for concrete sewer pipe and fittings?

53. On what part of a regular fixture trap do some codes prohibit the use of slip joint connectors (nuts and washers)?

54. Joints between lead pipe and cast iron pipe may be made with which kinds of fittings?

55. Caulked glass joints are made in the same manner as caulked cast iron joints with what exception?

56. What type materials should you use to make a burned (welded) lead joint?

57. Plain end ductile-iron gravity sewer pipe may be joined in the same manner as what other type pipe?

58. What type fittings must be used in a threaded DWV system?

59. What types of fittings are approved for copper DWV systems?

60. Expansion joints may be used in vent piping or drainage stacks for what purpose?

61. When piping materials have different outside diameters, how may they be joined together?

62. How are pipes and fittings of different sizes connected in a plumbing system?

63. When you need to make changes of direction in horizontal systems or in horizontal to vertical drainage systems, what are two of the acceptable fittings that you can use to accomplish these changes?

64. Name the three fittings that are acceptable for use where the direction of flow is from the horizontal to the vertical.

65. Under what condition may you use materials not covered by the standards cited in your code?

66. There are three types of cast iron soil pipe and fittings approved for building sewers. Name two of these.

67. What are the two grades of cast iron soil pipe used today?

68. What characteristics make cast iron soil pipe superior for building sewers?

69. When you lay pipe with hubs or couplings, what must you do to protect it from damage in the trench?

70. What is the minimum code-required depth for installing plastic pipe or any one of the other fragile pipes that are approved for building sewers?

71. Why must underground or horizontal drainage, waste and vent pipe be adequately supported?

72. How must the bases of stacks be supported?

73. How should drainage pipe passing through cast-in-place concrete be protected?

74. What is the required clearance from the top of a drainage pipe to the bottom of the footing?

75. How should drainage piping be protected when it's installed in corrosive materials?

76. What must you ensure that used drainage piping conforms to before installing it in any plumbing system?

77. Under what circumstances will the code permit you to drill or tap a waste or vent pipe for the purpose of rodding?

78. Under what circumstance might the code permit a lavatory waste to connect to a water closet stub?

79. What is the maximum number of stories in which a plastic DWV system may be installed?

80. What are the two types of supporting methods you must consider in drainage, waste and vent systems?

81. When pipe lengths exceed 5 feet, what's the maximum distance you should allow between hangers for horizontal cast iron soil pipe with lead and oakum joints?

82. What's the maximum distance you should allow between hangers for horizontal copper pipe $1^{1}/_{2}$ inches and smaller?

83. What special requirements are there for the hangers used with horizontal borosilicate glass piping?

84. What maximum distance is allowed between supports for vertical copper piping?

85. What two requirements must be met when placing hangers for support of horizontal and vertical piping?

86. How should horizontal vent piping be installed and sloped?

87. Why must you not allow moist air to be trapped in a horizontal vent pipe?

88. What must you always accommodate when you size and install indirect waste piping?

89. When practical, where should indirect waste pipe be installed?

90. What do you call the type of indirect waste pipe installation that is made below the floor and connects through the receiving fixture above the water seal of the trap?

91. The outlet for above the floor indirect waste pipe should terminate in what kind of installation?

92. What minimum size indirect waste pipe should you use for an above-floor installation?

93. What is the minimum indirect waste pipe size most codes require for a below-floor installation?

94. Where should you locate receiving fixtures for indirect waste piping?

95. By what means must commercial dishwashing machines be connected to the building greasy waste drain line?

96. For what purpose may a drinking fountain be indirectly connected to a floor drain?

97. How far below the bottom of the floor slab must indirect waste piping for A/C units be installed?

98. How is an A/C unit classified when its waste or condensate connects to the plumbing drainage system?

99. Where may an air conditioning unit that's over 5 tons but less than 10 tons discharge its waste?

100. Name three of the five acceptable areas into which air conditioning units 10 tons and larger can discharge waste.

Septic Tanks and Drainfields

9

There was a time when cesspools and outhouses were the most common means of human waste disposal in this country. But they just weren't adequate as population densities increased. In rural areas, where drinking water comes from wells, contamination from sewage is a real possibility. That can spread diseases like cholera and typhoid. Local authorities no longer accept outhouses or cesspools as an acceptable method of waste disposal.

If public sewers aren't available, the only acceptable method for sewage disposal is the septic tank system. And there's still some controversy about whether the septic tank is truly safe. The Septic Tank Association contends that there's never been a proven case of a septic tank contaminating drinking water. But the Department of Environmental Resource Management (DERM) has questioned septic tank safety. The controversy and field testing continue. DERM established strict guidelines that have been adopted by most model code organizations.

Few plumbing contractors handle septic tank and drainfield installation any more. Licensed septic tank contractors usually do the installation and maintenance work for these systems. The most you're likely to do is connect the building drainage system to the inlet tee of a septic tank. But you still need to know the basic principles about septic tanks and drainfields, since you're sure to find questions about them on the journeyman's and master's examinations.

Definition of the Septic Tank

A septic tank is a watertight receptacle that receives the sewage discharge of a drainage system. It's designed to separate solids from liquid wastes. (There's usually about 3/4 pound of solids in each 100 gallons of water.) The heavier portions settle to the bottom of the tank while lighter particles and grease rise to the top.

The tank's capacity should hold approximately 24 hours of anticipated flow. This retention period allows the bacterial action to digest most of the solids. That transforms the sewage waste into gases and harmless liquids. As new sewage enters the septic tank, it forces the gases up through the drainage vent pipes and into the atmosphere above the building roof. An equal amount of treated liquid is forced out through the tee of the septic tank as *effluent*. This effluent flows into the drainfield, a subsurface system of open-joint or perforated piping installed on a bed of washed rock. As the effluent seeps out between the joints or holes in the perforated piping, it oxidizes and finally evaporates.

When the bacterial process is complete, a small amount of solids remains in the tank and settles to the bottom as sludge. Lighter undigested particles rise and form scum on top of the liquid. Periodically the tank has to be cleaned of these undigested materials by a certified professional with the proper equipment.

Septic Tank Construction

The code prohibits sectional septic tanks or tanks made of blocks, brick or wood. Precast concrete or cast-in-place septic tanks are most common. In some areas, local authorities will approve steel or fiberglass septic tanks. The more fragile fiberglass tanks must meet strict installation guidelines to protect them from damage.

Concrete septic tanks must be protected from corrosion with an approved bituminous coating or other acceptable means. You can't have voids, pits or protuberances on the inner wall of a septic tank. All septic tanks must be watertight and produce a clarified effluent that meets code standards. They must be large enough to accommodate sludge and scum accumulations.

Figure 9-1 shows the criteria that most codes require for designing septic tanks. Here are the main points:

- Septic tanks must have at least two compartments (although some codes accept one).

- The inlet compartment should be at least two-thirds of the total capacity of the tank and must retain a minimum of 500 gallons of liquid. (The smallest approved tank size has a 750 gallon liquid capacity.)

- The secondary compartment (outlet compartment) should be at least one-third of the total capacity of the tank, with a minimum 250 gallon liquid capacity.

- Minimum septic tank size is 3 feet by 5 feet. Liquid depth must be between $2^1/2$ feet and 6 feet. (Some codes require a liquid depth of 4 feet.)

- There must be a 9-inch air space above the liquid level in the septic tank (8 inches in some codes).

- The secondary compartment (outlet compartment) of a septic tank holding over 1,500 gallons must be at least 5 feet in length.

- A cover slab in removable sections is acceptable for cleaning a residential septic tank.

- Commercial septic tank manholes must be brought to grade for cleaning access. One manhole must be located over the inlet (inlet tee) and one over the outlet (outlet tee).

- When the first compartment is longer than 12 feet, add an additional manhole over the baffle wall.

- Manhole sizes vary from 20 to 24 inches in diameter, depending on the code used.

- A septic tank inlet and outlet pipe should never be smaller in diameter than the connecting sewer pipe.

- The vertical legs of the inlet and outlet tees must be as large as the connecting building sewer pipe, at least 4 inches in diameter.

- When you use a baffle-type fitting, its cross-sectional area should be the same as the connecting building sewer pipe.

- The inlet and outlet pipe or baffle must extend at least 4 inches above and at least 12 inches below the liquid level of the septic tank.

- The inlet pipe should be at least 2 inches higher than the outlet pipe.

- The inlet and outlet pipe or baffle must allow free ventilation above the liquid level. This provides circulation of air from the disposal field and septic tank through the building drainage and vent system.

- Partitions or baffles between compartments must be of solid, durable materials and extend 4 inches above the liquid level. To allow liquids to pass from the inlet to the secondary compartment, there's an inverted fitting midway in the liquid depth. (Some codes accept a slot.) The opening must be equivalent in size to the connecting building sewer.

- Septic tanks must be strong enough to withstand all anticipated earth or other loads. Septic tank covers must be able to support an earth load of at least 300 pounds per square foot.

- Septic tanks in parking lots or other areas with vehicular traffic must have a traffic cover acceptable to the building department.

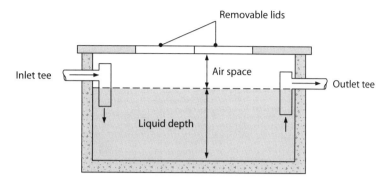

A Cross section of small septic tank

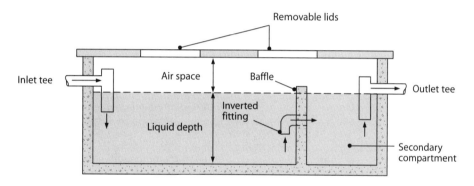

B Cross section of septic tank with secondary compartment

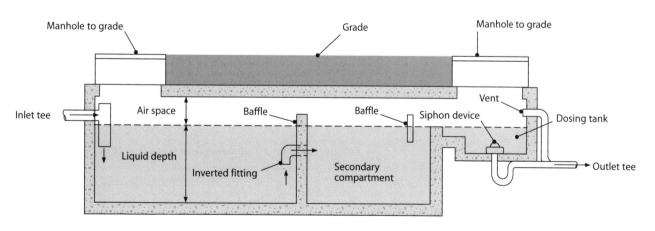

C Cross section of large septic tank with secondary compartment and dosing tank

Figure 9-1
Typical septic tank construction

Single-family residences — number of bedrooms*	Minimum capacity (gal)
2 or less	750
3	1,000
4	1,200
5 or 6	1,500

* For extra bedrooms, add 150 gallons each. Septic tank sizes in this figure include sludge storage capacity and the connection of domestic food waste disposal units without further volume increase. The minimum size of any septic tank is 750 gallons.

Figure 9-2
Required capacity of septic tanks for single-family residences

Multiple residential units — one bedroom each	Minimum capacity (gal)
2 units	1,200
3 units	1,500
4 units	2,000
5 units	2,250
6 units	2,500
7 units	2,750
8 units	3,000
9 units	3,250
10 units	3,500

For each extra bedroom within any unit, add 150 gallons. For extra dwelling units over 10, add 250 gallons each. Septic tank sizes in this figure include sludge storage capacity and the connection of domestic food waste disposal units without further volume increase.

Figure 9-3
Required capacity of septic tanks for multiple residential units

Sizing Septic Tanks

For single-family or multiple *residential* units, the minimum capacity is based on the number of bedrooms (Figures 9-2 and 9-3). For *commercial* buildings the septic tank capacity is based on the maximum fixture units for public use (Figure 9-4). Both the *Uniform Plumbing Code* and the *Standard Plumbing Code* agree on the minimum capacities.

But there are fixture unit code differences when applied to *public* use. This can affect your septic tank sizing. In Figure 9-5, notice that the *UPC* assigns urinals a fixture unit value of 5. The *SPC* uses a fixture unit value of 4 for the same type of urinal. Variations like this can affect septic tank sizing, particularly on large jobs.

The septic tank sizes listed in Figures 9-2 and 9-3 include sludge storage capacity. They also allow for the use of garbage disposals.

Let's work through some examples, both residential and commercial, based on the *UPC*. We'll consider several types of construction and compute the septic tank liquid capacity for each.

Example 1, Single-Family Residence

Let's figure the capacity for a seven-bedroom house. Check Figure 9-2 for the required tank capacity for up to six bedrooms. For each additional bedroom, just add 150 gallons to the six-bedroom capacity. So for a seven-bedroom house, add 150 gallons to the 1,500 gallon tank. You'll need a 1,650 gallon tank.

Example 2, Multiple Residential Unit

Figure 9-3 shows the required liquid tank capacity for up to 10 one-bedroom units. For each bedroom above one per unit, add 150 gallons to the minimum liquid tank capacity. Let's look at a four-unit apartment building with a total of six bedrooms. For the two extra bedrooms at 150 gallons each, add 300 gallons to the septic tank liquid capacity for four one-bedroom units (2,000 gallons). You'll need a 2,300 gallon tank.

Maximum fixture units, commercial use	Minimum septic tank (gal)
15	750
20	1,000
25	1,200
33	1,500
45	2,000
55	2,250
60*	2,500

*For fixture units exceeding 60, see your local code for septic tank capacities in gallons.

Figure 9-4
Maximum fixture units and minimum septic tank capacities for commercial use

Fixture type*	Number of fixture units for public use
Lavatories (wash basins)	1
Urinals (wall-mounted)	5
Water closet	6

* These fixtures are for the purpose of illustrating Figure 9-6, only. More complete tables are listed in your code. Fixture units may vary slightly from one code to another. Check local code.

Note Water closets must be computed as 6 fixture units when determining septic tank size.

Figure 9-5
Fixture units for commercial use

Example 3, Multiple Residential Unit with More Than 10 Units

According to Figure 9-3, you must add 250 gallons for each unit above 10. Let's look at a unit with 14 one-bedroom apartments. Four extra units at 250 gallons each equals 1,000 gallons. Add 1,000 gallons to the septic tank liquid capacity opposite the 10 units (3,500 gallons). You'll need a 4,500 gallon septic tank. If any unit has more than one bedroom, add an additional 150 gallons for each extra bedroom.

Example 4, Commercial Building

Minimum septic tank capacities for normal use are shown in Figure 9-4. These sizes will generally meet the requirements of the DERM and the local health department. Always check with local authorities if there's any doubt.

The minimum septic tank size is 750 gallons for any commercial use up to 15 fixture units. Of course, the septic tank capacity increases as the type and number of fixtures increase. Figure 9-5 shows the fixture units assigned by the *UPC* for some fixtures.

In our example, we'll use a restaurant with men's and women's bathrooms. Figure 9-6 shows the total fixture units (54). For commercial use, a septic tank that handles 54 fixture units needs a 2,250 gallon tank. That's the number opposite 55 F.U. in Figure 9-4. For fixture units exceeding 60, you'll need to check your local code.

Installing Septic Tanks

You can install cast-in-place septic tanks where they're subject to overburden loads, such as parking lots. The tank lid must be designed and installed to support an anticipated load equal to the weight of a 10-ton truck. You can't use precast septic tanks in parking lots or any area where vehicular traffic is anticipated unless it's designed with a traffic lid that's supported by the soil, not the tank itself. If you're installing a septic tank under any kind of paving, all cast-in-place lids must have 20- or 24-inch manholes located directly above the inlet and outlet tees. The manhole cover must be brought up to grade level.

No matter where you're installing a septic tank, it's essential that it be installed level.

Minimum septic tank size for eat and drink establishments with men's and women's toilet rooms		
Fixture type	Fixture units	Min. septic tank capacity (gal)
Men's		
4 lavatories (public) 1 F.U. each	4	
2 water closets (public) 6 F.U. each	12	
2 urinals (wall mounted) 5 F.U. each	10	
Women's		
4 lavatories (public) 1 F.U. each	4	
4 water closets (public) 6 F.U. each	24	
Total fixture units	54	2,250

Figure 9-6
*Required septic tank capacity in gallons
for commercial use*

Restrictions for Septic Tanks

Here are some locations where you *can't* install septic tanks:

▪ Under or within 5 feet of any building.

▪ Within 5 feet of any water supply lines.

▪ Within 5 feet of property lines other than public streets, alleys or sidewalks.

▪ Within 50 feet of the shoreline of open bodies of water.

▪ Within 50 feet of a private water supply well which provides water for human consumption, bathing or swimming.

You can't excavate for a septic tank within the angle of pressure as transferred from the base of an existing structure to the sides of an excavation on a 45-degree angle. (Look back to Figure 8-19 in Chapter 8.)

Air must circulate within the septic tank and drainfield through the plumbing system and then through the inlet and outlet tees of the septic tank. No other circulation is permitted.

Types of Drainfields

There are several ways to distribute the effluent from septic tanks evenly throughout the drainfield bed. The most common are open-jointed or perforated drain tile, block or cradle-type drain units, and corrugated plastic perforated tubing. We'll look at the installation requirements for each type. The end of each distribution line must be sealed by capping or cementing a block to the ends.

Tile Drainfields

Drainfield tile must have a minimum inside diameter of 4 inches, installed on a slope not more than 3 inches per 100 feet. Lay drainfield tile on a bed of washed rock ($^3/_4$ to $2^1/_2$ inch) that's 12 inches deep under the tile. Then fill the trench with another 6 inches of rock. You need 2 inches of rock over the top of the tile. You'll have a total depth of 18 inches for the full width of the trench. See Figure 9-7.

Lay the tile with a space of $^1/_4$ inch between the tile ends. Use a strip of 4- by 16-inch 30-pound bituminous saturated paper to cover that gap to prevent sand or other small particles from filtering into the openings.

Each trench for drainfield tile must be at least 18 inches wide. The maximum width is 36 inches. But regardless of the width of the trench, each linear foot of drainfield tile is considered to be 1 square foot. We'll cover that later in the section on sizing drainfields.

The maximum length of a single tile drainfield trench is 100 feet. Where the job requires more than one trench, space the trenches at least 6 feet apart from center to center.

Reservoir-Type Drainfields

Block or cradle drain units are used in single excavations based on the square footage (length x width) rather than in individual trenches like drainfield tile.

Install block or cradle drain units to a maximum slope of 3 inches per 100 feet. Lay them on a bed of $^3/_4$- to $2^1/_2$-inch washed rock which extends from 12

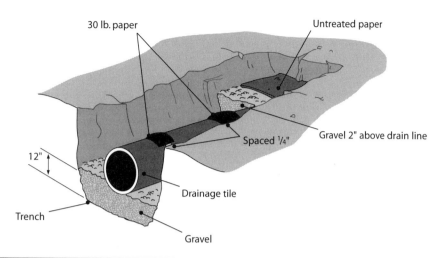

Figure 9-7
Tile drainage trench installation

inches under the drain units to 10 inches above the bottom of the units. This gives a total of 22 inches of rock for the full width of the drainfield.

Where drain units have a fixed opening to provide seepage, you can butt the units tight against each other. Where there's no fixed opening, lay the units with a ¼-inch space between the ends. Cover the gap with a strip of 30-pound bituminous saturated paper, 4 inches wide, to protect the top seam. The paper must extend down 4 inches on each side of the units. (See Figure 9-8.) Cover the entire area of washed rock

with untreated paper to prevent sand and other small particles from filtering down and through the rock.

The maximum distance between the centers of the distribution lines can't exceed 4 feet. The outside distribution line must be a minimum of 2 feet from the excavated wall of the filter bed. Each drain unit is considered to occupy 4 square feet.

Distribution lines can't exceed 100 feet in length. Where you're using two or more lines, make them as near the same length as possible, and connect them with a distribution box. A tight-jointed pipe must

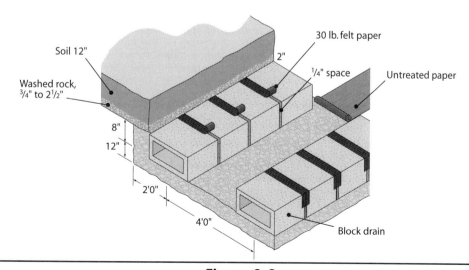

Figure 9-8
Cross section of reservoir-type drainfield

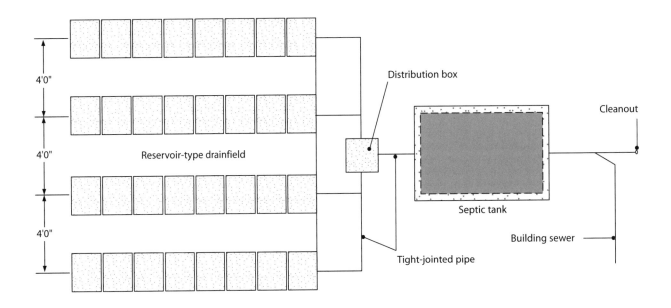

Figure 9-9
Detail of septic tank distribution box and drainfield

connect the septic tank's outlet tee to the distribution box and the distribution box to the fixed reservoir distribution lines. (See Figure 9-9.)

Corrugated Plastic Perforated Tubing Drainfield

Plastic tubing drainfields are generally used in single excavations based on the square footage (length x width) rather than in individual trenches. Like the others, it must slope a maximum of 3 inches per 100 feet, on a bed of $3/4$- to $2^{1}/_{2}$-inch washed rock 12 inches deep under the tubing. Cover it with 6 inches of rock over the tubing bottom, for a total depth of 18 inches for the full width of the drainfield.

Since the tubing has adequate fixed openings, all joints are tight. Cover the entire area of washed rock with untreated paper.

The distance between centers of the distribution lines must be between 48 and 72 inches. The outside distribution line should be a minimum of 36 inches from the excavated wall of the filter bed. See Figure 9-10. Don't make distribution lines more than 100 feet long. If there are two or more lines, make them as near the same length as possible. You don't need a distribution box to connect more than one plastic tubing distribution line, but you may use tees.

Sizing Drainfields

The drainfield won't work unless the soil can absorb the effluent properly. The best soils are those made up of coarse sand or gravel. They absorb water the fastest. The next best soil is fine sand. In areas where the soil has limited porosity (such as clay with some sand), a percolation test is usually required. The test will tell you how long it takes for water to be absorbed into the soil. That determines the size of the leaching area. Figure 9-11 shows the design criteria for five common soil types.

Codes vary in how to size a drainfield. *Always check your local code.*

Let's use the *Uniform Plumbing Code* to size some drainfields with different soils and different types of buildings.

Residential Drainfields

First, we'll size a drainfield for a three-bedroom, single-family house on a site with fine sand. The first column in Figure 9-2 shows the number of bedrooms for a single-family residence. The other columns are the minimum capacity for the septic tank in gallons in both the *Uniform Plumbing Code* and the *Standard Plumbing Code*. For our three-bedroom house, that's 1,000 gallons.

Now look at Figure 9-11. The second column gives the minimum square feet of leaching area per 100 gallons for each soil type. Since we have fine sand, it's 25 square feet.

Here's how to size the drainfield: Multiply the required capacity, 1,000 gallons, by the required leaching area, 25 square feet, then divide by 100 gallons:

$$\frac{1,000 \ gal. \ \times \ 25 \ SF}{100 \ gal.} = 250 \ SF \ total \ leaching \ area$$

Let's look at another example. We'll use the same three-bedroom house, but this time the site is clay with small amount of sand or gravel. According to Figure 9-11, that requires 120 square feet of leaching area for each 100 gallons. We know that a three-bedroom house requires a 1,000 gallon tank (Figure 9-2). Multiply required capacity times required leaching area, then divide by 100:

$$\frac{1,000 \ gal. \ \times \ 120 \ SF}{100 \ gal.} = 1,200 \ SF \ total \ leaching \ area$$

The percolation rate determines the size of the leaching area — and the cost. Naturally, the larger the leaching area, the more expensive it is.

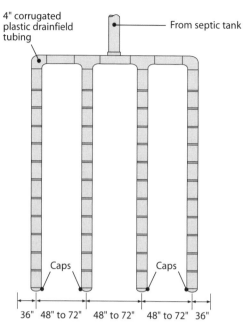

Plan view

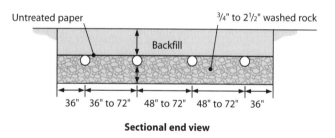

Sectional end view

Figure 9-10
Spacing the drain lines

Type of soil	Required SF of leaching area/100 gallons	Maximum absorption capacity gal/SF of leaching area for a 24-hour period
Coarse sand or gravel	20	5.0
Fine sand	25	4.0
Sandy loam or sandy clay	40	2.5
Clay with considerable sand or gravel	90	1.1
Clay with small amount of sand or gravel	120	0.8

Figure 9-11
Design criteria of five typical soils

Commercial Drainfields

Drainfields for commercial establishments are usually sized by DERM or by the local health department. Their size is based on the daily sewage flow (which you can find in a table of flow rates in your code book). Figure 9-12 is similar to the table in your code book, although your code may have more data.

We'll size the drainfield for a church building with a kitchen. The sanctuary seats 500. Sewage flow per seat according to Figure 9-12 is 7 gallons. The soil is coarse sand or gravel, so we need 20 square feet of leaching area per 100 gallons (Figure 9-11).

Using our formula, multiply 500 (seats) times 7 gallons per seat (required capacity) times 20 square feet (required SF of leaching area), and divide by 100 gallons:

$$\frac{500 \ seats \times 7 \ gal. \times 20 \ SF}{100 \ gal.} = 700 \ SF \ total \ leaching \ area$$

Now let's do the same church building on soil made up of clay with considerable sand or gravel. The only difference is that you need 90 square feet of leaching area (Figure 9-11) per 100 gallons instead of 20:

$$\frac{500 \ seats \times 7 \ gal. \times 90 \ SF}{100 \ gal.} = 3,150 \ SF \ total \ leaching \ area$$

Type of occupancy	Gallons per day
Airports	15 per employee
	5 per person
Bowling alleys (snack bar only)	75 per lane
Day camps (no meals served)	15 per person
Churches:	
Sanctuary	5 per seat
With kitchen waste	7 per seat
Factories:	
No showers	25 per employee
With showers	35 per employee
Hotels (no kitchen waste)	60 per bed (2 persons)
Nursing homes	125 per person
Motels:	
No kitchen	50 per bed space
With kitchen	60 per bed space
Offices	20 per employee
Mobile homes	250 per space
Schools:	
Staff and office	20 per person
Elementary students	15 per student
Intermediate and high	20 per student
Theaters:	
Auditorium	5 per seat
Drive-in	10 per space

Figure 9-12
Estimated sewage flow rates

Number of bedrooms	Drainfield required (SF)	Conventional block drain (LF)	Corrugated 4″ plastic tubing (LF)
2	100	25	40
3	125	32	50
4	150	38	60
5	175	44	70

Figure 9-13
Typical drainfield conversion table

We need to look at Figure 9-4 to size the drainfield for a commercial building, since the septic tank size is based on the fixture units. We'll try a restaurant, using the fixture units from Figure 9-6. The total fixture units are 54, requiring a 2,250 gallon septic tank. Here's how to size the drainfield using the septic tank size in gallons and a soil of fine sand (25 SF):

$$\frac{2{,}250 \text{ gal.} \times 25 \text{ SF}}{100 \text{ gal.}} = 562.5 \text{ SF total leaching area}$$

When drainfields are located under pavement, you have to increase the absorption area by these percentages:

Percolation rate	Area increase (%)
0 to 5	10
6 to 10	17
11 to 15	25

For example, if the required drainfield for a three-bedroom house is 300 square feet and the percolation rate is 3, you'd have to increase the drainfield area by 10 percent, an additional 30 square feet.

Drainfield Restrictions

Here are the locations where you can't locate a drainfield:

▮ Under any building or within 8 feet of any building (*UPC*).

▮ Within 5 feet of a seepage pit.

▮ Within 8 feet of a basement wall or terraced area.

▮ Within 5 feet of water supply pipe lines.

▮ Within 5 feet of property lines other than public streets, alleys or sidewalks.

▮ Within 50 feet of shorelines of open bodies of water.

▮ Within 100 feet of any private water supply well which provides water for human consumption, bathing or swimming.

A drainfield must have a minimum earth cover of 12 inches and a maximum of 24 inches.

Drainfield Conversion Table

In some cases drainfields become *dead* and the soil can't absorb the effluent properly. This requires replacement of the drainfield. You can use Figure 9-13 to calculate the square feet of drain block or linear feet of plastic tubing you'll need to replace a drainfield for an existing building.

Review Questions (answers are on page 294)

1. Under present code standards, what two former disposal methods are no longer acceptable as a permanent means of dealing with human waste?

2. Even today, drinking water can be contaminated by untreated sewage. Which two diseases can be spread as a result of this type of contamination?

3. When public sewers aren't available, what's the most acceptable method for sewage disposal?

4. What governmental department has established guidelines for septic tank safety?

5. Although septic tank and drainfield installations are important areas of plumbing work, who usually does the installation and maintenance work for these systems?

6. Since you probably won't do the work, why is it important that you be informed about the basic principles of septic tank and drainfield installation?

7. What are septic tanks designed to accomplish?

8. What is the approximate amount of solid waste for each 100 gallons of water in a septic tank?

9. A septic tank is sized to have a capacity equal to approximately how many hours of anticipated flow?

10. How are the solids in a septic tank digested?

11. What happens to the effluent when it enters the drainfield?

12. What remains when the bacterial process is completed within the septic tank?

13. What do you call the lighter, undigested particles that rise to the top of the liquid after the bacterial process in the septic tank is complete?

14. Who cleans out the undigested materials remaining in the septic tank?

15. Name two materials that are *not* code-approved for septic tank construction.

16. What are the two most common types of septic tanks?

17. How are concrete septic tanks protected from corrosion?

18. In order to prevent contamination by leaking sewage, what does the code require of all septic tanks?

19. What must be taken into consideration when sizing a septic tank?

20. How many compartments do most codes require septic tanks to have?

21. What is the liquid capacity of the smallest approved septic tank?

22. The secondary compartment (outlet compartment) of the tank should have a minimum liquid capacity of how many gallons?

23. On a *residential* septic tank, what type of cover slab is acceptable for cleaning purposes?

24. How many manholes are needed for a 1,500-gallon capacity commercial septic tank?

25. When the first compartment of a *commercial* septic tank exceeds 12 feet in length, how many manholes are required?

26. What is the minimum size requirement for the vertical legs of the inlet and outlet tees of a septic tank?

27. At least how many inches higher than the outlet pipe should a septic tank inlet pipe be?

28. How much air space must be provided above the liquid level in a septic tank?

29. When a septic tank is located in a parking lot, what does the building department require it to have?

30. What determines the size capacity for septic tanks installed for single-family or multiple *residential* units?

31. What determines the capacity requirements for septic tanks for *commercial* buildings?

32. How do you determine the capacity requirement of a septic tank for a single-family residence that has more bedrooms than there are listed in your code table?

33. How do you determine the capacity requirement of a septic tank for multiple *residential* units when the number of units exceeds those listed in your code table?

34. A septic tank can't be located within how many feet of any water supply line?

35. A septic tank can't be located within how many feet of the shoreline of an open body of water?

36. Name two drainfield materials commonly used to distribute the effluent from septic tanks.

37. What is the minimum inside diameter of drainfield tile?

38. What is the minimum slope for drainfield tile per 100 feet?

39. What is the minimum width for each drainfield trench?

40. What is the maximum length of a single tile drainfield trench?

41. What is the required depth of washed rock under the drain units of a reservoir type drainfield?

42. What is the maximum distance between centers of the distribution lines of a reservoir type drainfield?

43. What are the best soils for absorbing drainfield effluent?

44. What is determined by a percolation test?

45. Other than the type of soil, what else is used to determine the sizing of a *residential* drainfield?

10

Mobile Home and Travel Trailer Parks

Local agencies regulate plumbing systems for mobile home parks that serve the growing number of mobile homes and recreational vehicles. They set specific minimum sanitary plumbing facilities and installation methods for the parks. These standards vary considerably from the requirements for conventional permanent structures. You'll notice this immediately when you study the standards for designing and installing sanitary collection and water distribution systems for these parks. And study them you should. The odds are that you'll do this kind of work sooner or later.

Some of the terms used in this chapter are unique to this kind of work. I'll define any that aren't self-explanatory as we come to them, but look in the glossary at the end of the book for the complete list.

A *mobile home* or *RV trailer park* is land designated and improved to accommodate one or more trailers, used either for temporary or permanent living quarters. There are two kinds of parks under the code, depending on whether the trailers served are *dependent* or *independent*.

Dependent Trailers

Recreational vehicles (RVs) and motorized homes are generally defined in the code as *dependent trailers*. That's any motorized vehicle used as a temporary dwelling unit for travel, vacation and recreation.

They're portable structures built on a chassis for traveling public roads. As a rule, they have limited built-in sanitary facilities (usually, but not always, including a toilet). They don't have a plumbing system that can permanently connect to a park sewage and water supply system. Most of today's motor homes are equipped with a water storage tank to operate the plumbing fixtures and a holding tank to retain the waste water.

Independent Trailers

A mobile home is defined by code as an *independent trailer coach*. That's any trailer coach designed for permanent occupancy, equipped with kitchen and bathroom facilities and a plumbing system that can connect to the park sewage, water and gas supply system. It's built on a permanent chassis and designed as a dwelling. Most mobile home owners lease or buy space in a mobile home park. The park provides each space with a water, electrical, gas and sewer connection, and is responsible for maintaining these services.

Toilet Facilities for Trailer Parks

Let's look at the code requirements for toilet facilities, the DWV system, and the water distributing system for both kinds of parks.

Dependent Trailer Parks

The code doesn't require a *dependent trailer park* to provide individual water/sewer connections, but it must have a separate *service building*. It's required to have toilet facilities and a waste disposal station equipped to empty intermediate waste holding tanks. Some codes even require limited laundry facilities. The service building must be located within 500 feet (200 feet in some codes) of the most distant trailer site and must have an individual sewer connection.

Any park's service building or buildings must provide a minimum number of fixtures for each sex. Here's what the *Uniform Plumbing Code* requires:

■ For the first 25 sites, there must be one toilet for each sex. They need an additional toilet for each additional 25 sites (when they don't have individual sewer connections).

■ Each toilet room must have a floor drain and be protected with an automatic trap seal.

■ For each toilet up to six, the service building must have an equal number of lavatories. For each two toilets above six, they must provide one lavatory. For example:

Eight toilets require seven lavatories (6 + 1 = 7)
Ten toilets require eight lavatories (6 + 1 + 1 = 8)

■ In men's facilities, the park can replace one-third of the required toilets with urinals. For example, if six toilets are required, you may install four toilets and two urinals. When nine toilets are required, you may install six toilets and three urinals. When only two toilets are required, the code permits one toilet to be replaced with a urinal.

■ Toilets must be the elongated type with open front seats.

■ Toilets must have separate stalls with a door and latch for privacy.

■ Urinals must be either the individual stall or wall-hung type.

■ Showers, if provided, must have a floor area at least 36 by 36 inches. Each shower stall and dressing area must be screened from view. Shower water can't flow into the dressing area.

■ If laundry facilities are required, they must be installed in a room separate from the toilet rooms.

■ Drinking fountains aren't usually required. If they're provided, they must be an approved type.

Independent Trailer Parks

Some codes require minimum toilet facilities for independent trailer parks, while others don't. The *Standard Plumbing Code* is one that does. It requires a service building even when each trailer has its own facilities. Here are the *SPC* requirements:

1) The service building must have one laundry tray.

2) For women, there must be one toilet, one lavatory, and one shower or bathtub.

3) For men, there must be one toilet, one lavatory, and one shower or bathtub.

The Park Drainage and Vent System

All drainage materials must comply with the code standards. (Look back to Figures 8-2 and 8-3 in Chapter 8.) In Figure 8-3, you'll find a list of materials you can use in a park drainage system in the column labeled *Building sewer.*

Sizing a Park Drainage and Vent System

There's a big difference between fixture unit loads for a park drainage system and a conventional building. In a conventional building you always total each plumbing fixture separately. In a trailer park, your local code assigns the total fixture units for each site drainage inlet. The *Uniform Plumbing Code* assigns 12 units, while the *Standard Plumbing Code* assigns 15 units to each site.

You'll find tables in your code for sizing park drainage systems. Figure 10-1 shows the maximum number of fixture units permitted on each size pipe.

Size of drainage pipe (in)	Maximum fixture units
2*	8
3	35
4	256
5	428
6	720
8	2640

* Special fixtures and equipment which are rated 6 F.U. or more can't use a 2-inch pipe. (More complete tables are in your code)

Figure 10-1
Drainage pipe sizes and number of fixture units permitted

Pipe size (in)	Slope per 100 feet (in)
2	25
3	25
4	15
5	11
6	8
8	4

The larger the pipe, the less slope required per 100 feet. (More complete tables are in your code)

Figure 10-2
Minimum pipe size and slope of drainage pipe

Here's an example that illustrates the differences in two of the major codes. Let's use a 4-inch drainage pipe. According to Figure 10-1, it will carry a maximum of 256 fixture units.

Uniform Plumbing Code:

Total fixture units (256) divided by 12 F.U. per site = 21 trailer sites.

Standard Plumbing Code:

Total fixture units (256) divided by 15 F.U. per site = 17 trailer sites

You can see that the *Uniform Plumbing Code* accommodates four more trailer sites than the *Standard Plumbing Code*, even though the number of fixture units and pipe sizes are identical.

Figure 10-2 shows the minimum slope per 100 feet (in inches) for each size pipe. While it's important, it's not a factor in determining the number of sites you can connect to the park's drainage system.

Installing the Drainage and Vent System

You'll install, backfill and use the same type of materials for this system as for building sewers. Provide each trailer site with a sewer lateral no smaller than a 3-inch diameter pipe (4 inches in some codes). Cap the line when it's not in use. Locate vent pipes on building drainage systems at least 10 feet from adjoining property lines, and extend them 10 feet above ground level. Securely anchor the vents to the equivalent of a 4 × 4 post driven into the ground. Be sure your supports are rot- and deterioration-resistant.

Make your first vent 3 or 4 inches in diameter and install it not more than 5 feet downstream from the first sewer lateral. The park sewer main must be re-vented at intervals of not more than 200 feet. See Figure 10-3.

Install cleanouts at no less than 100-foot intervals. They'll have to be the same nominal size as the pipe they serve but no larger than 3½ inches (4 inches in some codes).

Terminate your sewer laterals at least 4 feet outside the left wheel and within the rear third of the trailer coach. This should allow a short trailer drain connection between the trailer outlet and sewer inlet.

For trailers that are properly trapped and vented, make the lateral terminate with a sweep that has a 3- or 4-inch sanitary tee caulked into it. The tee should end 4 to 6 inches above grade. Be sure you caulk a cleanout in the top of the sanitary tee.

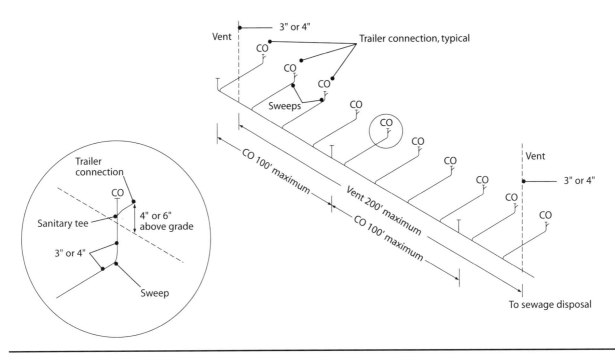

Figure 10-3
Sewage collection system for properly trapped and vented trailers

Trailers that aren't properly trapped and vented should have a lateral that terminates with a 3-or 4-inch P-trap into which is caulked a sanitary tee that ends 4 to 6 inches above grade. Caulk a cleanout in the top of the sanitary tee as shown in Figure 10-4. Since a P-trap is required at the end of the lateral (branch line), the measured horizontal distance from a vented sewer without a re-vent can never exceed 15 feet. Don't place the P-trap more than 24 inches below grade unless you get specific permission from local authorities.

The Park Water Distributing System

A $3/4$-inch branch service line connected to the park main supplies potable water to each trailer site. The line terminates on the same side of the trailer site as the trailer sewer lateral.

The park's distributing system should provide a minimum pressure of 20 psi at each trailer site. The

service connection to each trailer must be a minimum of $1/2$ inch diameter.

When a backflow preventer is required, install a pressure relief valve on its discharge side. Manufacturers must ensure that pressure relief valves release pressure at a maximum of 150 psi. Locate the backflow preventer and pressure relief valve at least 12 inches above grade. See Figure 10-5.

On each branch service line, connect the trailer to the park's water distributing system with a separate shutoff valve and a spring-loaded, soft-seat check valve. Locate the valve near the service connection for each trailer.

Use an approved flexible tubing with a quick disconnect fitting at either end for each service connection. No special tools or knowledge are required to install and remove these fittings.

You'll use the same materials and installation methods here as for the water systems in Chapter 14. Be sure that any piping materials you use in the park water distributing system meet the requirements listed in Figure 14-1.

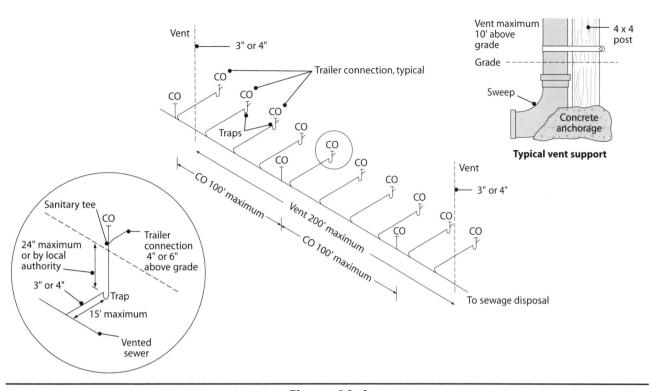

Figure 10-4

Sewage collection system for trailers not properly trapped and vented

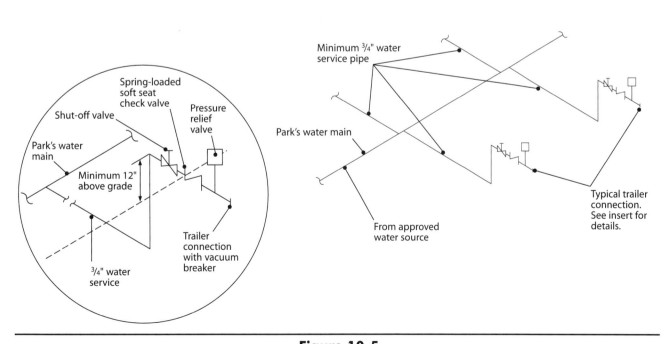

Figure 10-5

Typical trailer water service outlet with approved controls

Review Questions (answers are on page 296)

1. Who regulates the plumbing standards and installation methods for mobile home and RV parks?

2. How are recreational vehicles defined in the code?

3. Recreational vehicles have limited plumbing facilities, usually not including which fixture?

4. How is a mobile home defined in the code?

5. What does the code *not* require dependent trailer parks to provide?

6. What type of building facility must a dependent trailer park provide?

7. What type of individual connection must a dependent trailer park service building have?

8. What particular segregated facilities must a dependent trailer park service building provide a minimum number of?

9. According to the *Uniform Plumbing Code*, how many sites in a dependent trailer park may be served by one toilet for each sex?

10. What must each toilet room in the service building of a dependent trailer park have?

11. In a dependent trailer park service building, how many lavatories must be provided for each toilet, up to six toilets?

12. In the men's toilet room of a dependent trailer park service building, what portion of required toilets may be replaced with urinals?

13. What toilet design must be used in the toilet rooms of a dependent trailer park?

14. What must each toilet installed in a dependent trailer park have?

15. What is the required floor area of a shower installed in a dependent trailer park?

16. What area in a dependent trailer park's shower room must be protected from water overflow?

17. According to the *Standard Plumbing Code*, what are the minimum fixtures required in the women's area of an independent trailer park?

18. How are fixture units assigned in a trailer park?

19. What criterion is used in sizing a park drainage system?

20. How many fixture units are permitted on a 3-inch pipe in a trailer park drainage system?

21. In a trailer park, what's the minimum slope for a 4-inch pipe per 100 feet?

22. The installation and backfill for a trailer park drainage system is the same as for what other system?

23. In a trailer park drainage system, where should the first vent be located?

24. When a park lateral terminates with a 4-inch P-trap and a sanitary tee, where should a cleanout be installed?

25. What's the minimum pressure required at each trailer site for the park's water distribution system?

11

Graywater Recycling Systems for Single-Family Residences

By far, most water for public use comes from lakes, rivers and deep wells. Although many people seem to think the earth's water supply is endless, it's actually a finite resource. Without long-range planning today, America could face a critical water supply shortage tomorrow.

For years the primary water source for single-family and multifamily residential buildings has been the municipal water treatment plant. But in virtually all major cities, water treatment plants are overloaded. Some of them, because of neglect and careless maintenance, are beginning to fail. Moratoriums on new construction and other restrictions are clear indications that cities are waking up to the inadequacy of their available water sources.

Since several areas have experienced water shortages, water conservation is increasingly becoming a national issue. Plumbing codes are just now beginning to wrestle with this problem. In recent years, their major weapon has been to formulate regulations that conserve water through maximum allowable water usage for plumbing fixtures. Figure 11-1 shows the maximum allowable water usage for plumbing fixtures based on the *Standard Plumbing Code.*

Some codes are presently exploring new water recycling technologies. We'll look at older and newer solutions to the growing water shortage that our nation must resolve in the coming years.

The Older Water Conservation Methods

As I write this, many local codes still employ only *older* water conservation methods. While having been minimally helpful, they'll never be totally adequate. Using this approach, plumbing fixtures and their fittings must comply with these standards:

▌ All faucets, showerheads and their packing must be marked by the manufacturer to comply with the provisions of ANSI Standard A112.18.1M. Water closets and urinals and their packaging must be marked to meet the provisions of ANSI A112.19.2M. They're listed in Chapter 21, on plumbing fixtures.

▌ New or replacement water closets, urinals, faucets and showerheads must have a flow rate or flush volume no greater than those listed in Figure 11-1.

Besides these conservation efforts, many areas limit landscape irrigation (even on wells) to certain days of the week and/or hours of the day. But these restrictions are hard to enforce and have been largely unsuccessful.

Considerable water is lost in older buildings that have leaking pipe joints and faucets. Most people don't realize that a hidden cause like that can create such great waste. But when you consider that it takes only $2^1/2$ seconds to form one drip, one leak can waste

Maximum allowable water usage for plumbing fixtures[1]	
Water closets, flushometer tank or close-coupled 2-piece gravity-flush type	1.6 gal/flush[4]
Water closets, one-piece	1.6 gal/flush
Water closets, commercial	3.5 gal/flush
Urinals	1.0 gal/flush
Residential sink and lavatory faucets	2.2 gal/minute[5]
Commercial lavatory faucets, metering type[1]	0.25 gal/cycle
Commercial lavatory faucets, nonmetering type	0.5 gal/minute
Showerheads[2,3]	2.5 gal/minute

[1]Metering type lavatory faucets must be installed in the following public toilet rooms:
   A. in toilet rooms having 6 or more lavatories.
   B. in student-use toilet rooms in public schools.
   C. in all public-use toilet rooms in assembly type occupancies.

[2]Flow rate for showerheads as tested at 80 psi in accordance with ANSI standard A112.18.1M. See Figure 21-1.

[3]Safety showers for special purposes are exempted from maximum flow rate limitations.

[4]The average low consumption water closets must not exceed 1.6 gpf over a range of test pressures. At any one test pressure the consumption should not exceed 2.0 gpf.

[5]Measured at 60 psi.

Note Maximum water usage allowable for plumbing fixtures and fixture fittings not listed in Figure 11-1 must conform to the applicable ANSI standards listed in Figure 21-1. Blowout fixtures, panelwear, clinic sinks and service sinks are exempted from these limitations.

From the Standard Plumbing Code, ©1996

Figure 11-1
Maximum allowable water usage for plumbing fixtures

about 365 gallons of water per year. If you multiply that by all the hidden leaks throughout the country, it's a staggering loss. And unfortunately, code enforcement can't do anything to solve this problem.

A New Water Conservation Method: Graywater Recycling

One solution being explored by the *National Standard Plumbing Code,* the *Uniform Plumbing Code* and the *International Plumbing Code* is known as graywater recycling. Graywater is untreated household waste water that has had no contact with toilet waste. It includes used water from bathtubs, showers, lavatories, laundry trays and clothes washing machines, but not dishwashers or kitchen sinks. See Figure 11-2. So far its use is limited to underground landscape irrigation, permitted only in single-family residences.

Graywater piping can't connect to potable water piping. It must be installed in a way that prevents any of the graywater from surfacing. The location, the soil type and the groundwater table determine which type of system you can use. Any system must properly dispose of all graywater generated by the residential building.

A graywater recycling system consists of a holding tank which discharges the waste into underground irrigation piping or disposal fields. If installed underground, your holding tank must rest on dry, level, compacted soil. See Figure 11-3. If you install your holding tank above ground, it must rest on a 3-inch-thick concrete slab sized to give adequate support. See Figure 11-4, which shows both pumped and gravity systems installed above ground.

After you install any holding tank, fill it with water to the overflow line for inspection. Then perform a flow test through the system to the point of disposal. Make sure all lines, components, seams and joints are watertight.

Estimating Graywater Discharge

Here's the procedure in the *Uniform Plumbing Code* for estimating graywater discharge from single-family residences. The waste water is based on the number of occupants and bedrooms:

1) First bedroom: 2 occupants

2) Each additional bedroom: 1 occupant

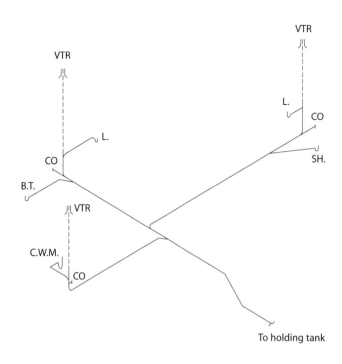

Note Kitchen sink, dishwasher and toilets on separate system

Figure 11-2
Isometric drawing of a typical two-bath residence graywater drainage system

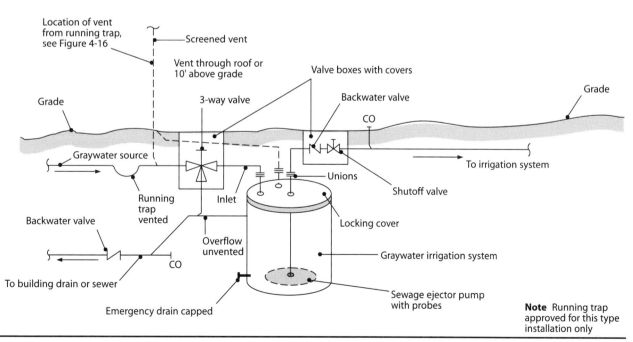

Figure 11-3
Underground graywater system tank, pumped

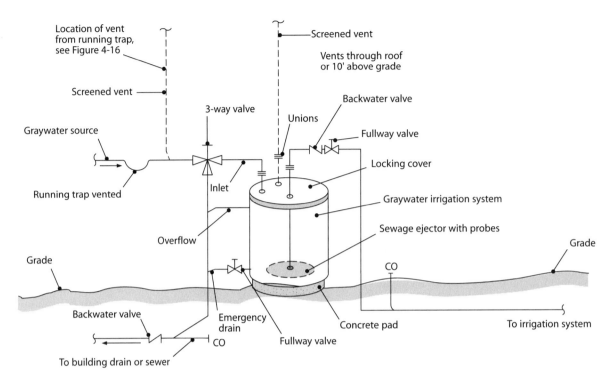

A Aboveground graywater system tank, pumped

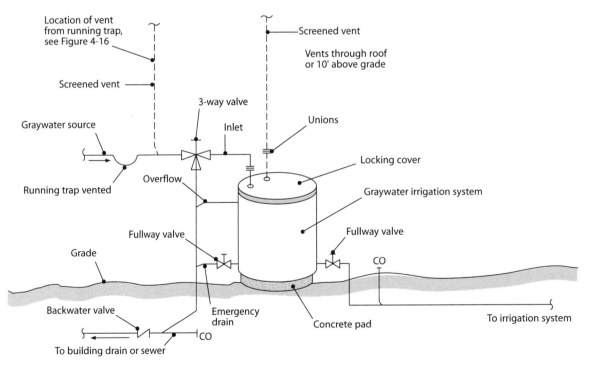

B Aboveground graywater system tank, gravity

Note Running trap approved for this type of installation only

Figure 11-4
Code-approved trap on aboveground graywater system tank

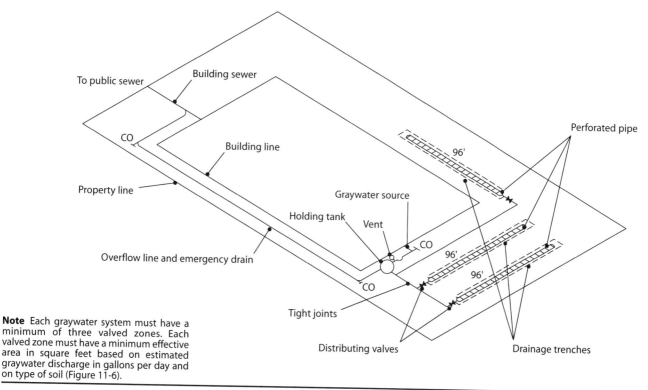

Note Each graywater system must have a minimum of three valved zones. Each valved zone must have a minimum effective area in square feet based on estimated graywater discharge in gallons per day and on type of soil (Figure 11-6).

Figure 11-5
Three valved zones, graywater system

To estimate your graywater flow, calculate the usage for each occupant using the following information:

1) Showers, bathtubs and lavatories: 25 gallons per day per occupant

2) Laundry: 15 gallons per day per occupant

Now we'll try a couple of examples, beginning with a four-bedroom house:

1) Number of occupants: 2 + 1 + 1 + 1 = 5 occupants

2) Estimated graywater flow: 5 occupants × (25 + 15) = 200 gpd

For the second example, we'll consider a three-bedroom residence with only the laundry facilities connected to the graywater system.

1) Number of occupants: 2 + 1 + 1 = 4 occupants

2) Estimated graywater flow: 4 occupants × 15 = 60 gpd

Graywater Subsurface Irrigation/Disposal Fields

Use the information above to calculate the sizing of graywater discharge for irrigation/disposal fields. Any graywater system must have at least three valved zones. See Figure 11-5. Each valved zone requires a minimum effective irrigation area (in square feet) based on the estimated graywater discharge (gpd) and soil type. Figure 11-6 shows the design criteria in the *Uniform Plumbing Code*.

Never let your excavation come within 5 vertical feet of the highest known seasonal groundwater. It can't reach a depth where graywater could contaminate the groundwater or ocean water.

Construction of Holding Tanks

▌ Each holding tank must be watertight, and constructed of materials not subject to excessive corrosion.

Type of soil	Minimum SF of leaching area per 100 gallons of estimated graywater waste per day	Maximum percolation rate in gallons per SF of leaching area for a 24-hour period
Coarse sand or gravel	20	5.0
Fine sand	25	4.0
Sandy loam	40	2.5
Sandy clay	60	1.7
Clay with considerable sand or gravel	90	1.1
Clay with small amounts of sand or gravel	120	0.8

From the UPC® with permission of the IAPMO ©1997

Figure 11-6
Design criteria of six typical soils

■ The holding tank must be vented, and have a locking, gasketed access opening which permits inspection and cleaning.

■ Each holding tank must have a permanently marked sign: *Graywater irrigation system, danger — unsafe water.*

■ The minimum capacity of each tank is 50 gallons. Its rated capacity must be permanently marked on the unit.

■ Aboveground holding tanks need an emergency drain and an overflow drain. The overflow drain can't have a shut-off valve. The tanks must be permanently connected to the building drain or sewer, and upstream from septic tanks, if any. See Figures 11-3 and 11-4.

■ Neither the emergency nor the overflow drain can be smaller than the inlet pipe.

■ A holding tank *vent* must be sized by total fixture units.

■ For replacement purposes, all connecting pipes to holding tanks must be equipped with unions or other approved effective fittings. See Figures 11-3 and 11-4.

■ An underground holding tank must be structurally designed to withstand an earth load of 300 pounds per square foot.

■ Underground holding tanks require a backwater valve to protect against sewer backups.

■ All holding tanks must be constructed of steel and protected from external and internal corrosion by an approved coating or other acceptable means.

■ To protect the building from any possible drainage gases, a vented running trap must be installed upstream from its connection to the holding tank. See Figures 11-3 and 11-4.

■ All valves, including the three-way valve, must be accessible.

Construction of Irrigation/Disposal Fields

The materials and installation methods for irrigation or disposal fields for graywater systems are similar to the requirements for septic tank drainfields. The differences are minor. The *Uniform Plumbing Code* requires the following:

■ *Piping material:* The piping can't be smaller than 3 inches in diameter. Approved materials include perforated ABS or PVC pipe, perforated high-density polyethylene pipe or other approved materials. The pipe must have enough openings for the proper distribution of the graywater into the trench area.

■ *Filter material:* The filtering material must be clean stone, gravel, slag or other approved filtering material. The size may vary from $^3/_4$ inch to no more than $2^1/_2$ inches.

■ *Installation methods:* Lay the perforated piping on a bed of properly-sized filtering material. This material must extend 3 inches beneath the pipe and at least 2 inches above the pipe, and be covered with untreated building paper, straw or similar porous material. This prevents backfill soil

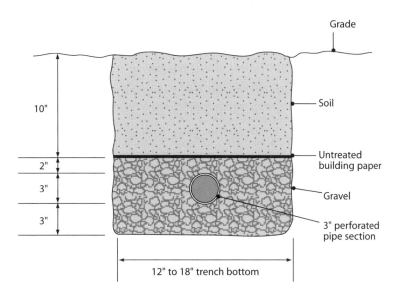

Figure 11-7
Graywater irrigation absorption trench (end view)

from filtering down and obstructing the drainage. Lay the piping with a slope of 3 inches per 100 feet. See Figure 11-7 for an end view detail of an absorption trench.

▮ *Irrigation/disposal field requirements:* When installing an irrigation or disposal field for a graywater system, make certain you follow the guidelines listed in Figure 11-8. There are times when you'll encounter steep grades in the leaching area. This might cause the graywater waste to pass too quickly through the drain lines, without enough time to seep into the soil. In those regions, the code will require that you step the irrigation or disposal lines. Make certain that the lines joining each horizontal leaching section have watertight joints. Also, they must be installed on natural or unfilled ground. See Figure 11-9.

▮ *Location of graywater systems:* If you're following the *Uniform Plumbing Code,* use the guidelines in Figure 11-10 for locating the holding tank and irrigation/disposal field.

For graywater irrigation/disposal systems in typical soils, you'll use the same design criteria as for septic tank drainfields, with one exception. Graywater systems use six typical soils, while septic tank drainfields use five typical soils.

	Minimum	Maximum
Number of drain lines per valved zone	1	—-
Length of each individual line	—	100 ft.
Width of trench bottom	12 in.	18 in.
Lines center-to-center	4 ft.	—
Earth cover over lines	10 in.	—
Filter material over lines	2 in.	—
Filter material beneath lines	3 in.	—
Slope of perforated pipe	3 in. per 100 ft.	—

From the UPC® with permission of the IAPMO©1997

Figure 11-8
Irrigation/disposal field requirements

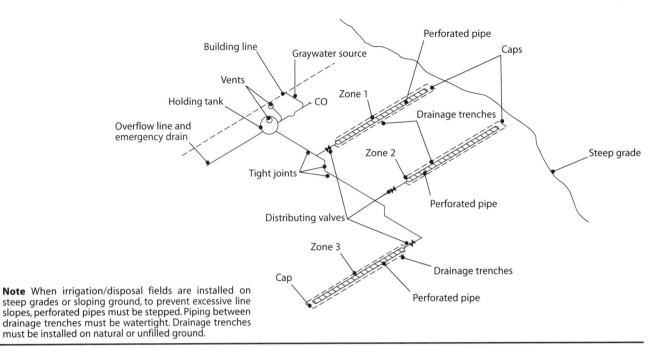

Note When irrigation/disposal fields are installed on steep grades or sloping ground, to prevent excessive line slopes, perforated pipes must be stepped. Piping between drainage trenches must be watertight. Drainage trenches must be installed on natural or unfilled ground.

Figure 11-9
Graywater drainage trenches (stepped)

Minimum horizontal distance from:	Holding tank(s) (ft)	Irrigation/ disposal field (ft)
Building structures[1]	5[2]	2[3]
Private property lines	5	5
Domestic wells[4]	50	100
Streams and lakes[4]	50	50[5]
Cesspools or sewage pits	5	5
Drainfields plus 100% expansion	5	4[6]
Septic tank	0	5
Domestic water service piping	5	5
Public water main	10	10[7]

Notes

[1]Including porches and steps. Covered or uncovered breezeways. Roofed patios or carports. Covered walks, driveways and similar structures.

[2]Distance from structures may be 0 feet for aboveground tanks if approved by local authority.

[3]Assumes a 45-degree angle from foundation. See Figure 8-19.

[4]Where special hazards are involved, the local authority may increase required distance.

[5]The minimum clear horizontal distances for streams and lakes shall also apply between the irrigation/disposal fields and the ocean high tide line.

[6]Plus 2 feet for each extra foot of depth in excess of 1 foot below the drain line bottom.

[7]The approval of the local authority is required if irrigation/disposal field is installed parallel or crosses public water main.

Figure 11-10
Location of graywater system based on the Uniform Plumbing Code

Once you've established the flow in gallons per day and you know the type of soil, you're ready to calculate the piping material and drainage trenches. Let's try an example using a five-bedroom residence with laundry facilities.

1. Total number of occupants: 2 + 1 + 1 + 1 + 1 = 6 occupants

2. Estimated graywater flow: 6 occupants × (25 + 15) = 240 gpd

The soil is sandy loam. According to Figure 11-6, sandy loam needs 40 square feet of leaching area per gallon of graywater waste.

$$\frac{240\ gpd \times 40\ SF}{100} = 96\ SF$$

Since the code requires that each graywater system have three valved zones, this example would need three 96-foot drain lines. (*Each linear foot of trench equals one SF.*) If the lot size is too small to accommodate three 96-foot drain lines, you may use several shorter lines that equal the total length of three 96-foot lines. For example, you could have three valved zones with three 32-foot lines in each zone. In this instance, you'd have to space the drains at 4 feet, center to center.

The Future of Graywater Systems

Graywater systems are designed to manage wastewater by returning it to the natural water table underground. That makes it safe to use. At the time of this writing, graywater systems are permissible in some areas *for single-family dwellings only*. But they're being considered for all types of residential, commercial, institutional and industrial buildings.

As long-term water shortages continue to increase, we can expect the use of graywater systems to expand. Instead of being restricted to landscape irrigation, it may soon be used to supply water for ornamental ponds and cooling towers, and perhaps other uses not connected to potable water. Once these systems are established nationally, the water we conserve may prove to be life-saving for generations to come.

Review Questions (answers are on page 297)

1. From what origins do we obtain water for public use?

2. What methods are being used in many large cities to extend their available water resources when they are faced with an inadequate supply of water for human consumption?

3. What has to be done to protect our present water sources and search for additional fresh water sources?

4. What is the primary source of water for single-family and multifamily residential buildings?

5. Until recently, local codes have only considered water conservation methods that involve what?

6. If the average drop of water takes only $2^{1}/_{2}$ seconds to form, how much water is wasted in a year by one leaky faucet?

7. What new water conservation method is now being tried in many areas?

8. At this time, what types of buildings are allowed to use a graywater recycling system?

9. What's the current limitation on the use of recycled graywater?

10. What is graywater?

11. What piping system cannot be connected to a graywater recycling system?

12. Name three fixtures that can connect to a gray water system.

13. Name the three things that determine the type of graywater recycling system that may be used.

14. Where must a graywater recycling system for a residential building discharge the waste?

15. When a holding tank is installed above ground, what kind of base is required to support it?

16. What test must you perform on a graywater system after you install the holding tank and piping?

17. On what do you base your estimate of the amount of graywater discharge from a single-family residence?

18. In estimating graywater discharge for a single-family residence, how many occupants are considered for the first bedroom?

19. In estimating graywater discharge for a single-family residence, how many occupants are considered for each bedroom after bedroom number one?

20. What is the allowance per occupant (excluding laundry) that you use to estimate the total number of gallons of graywater flow from a single-family residence?

21. To estimate the total number of gallons of graywater flow from a single-family residence, what is the allowance you use per occupant, including laundry facilities?

22. How many valved zones must each graywater system have?

23. When you excavate for a graywater subsurface irrigation/disposal field, you must not allow your excavation to extend within how many vertical feet of the highest known seasonal groundwater?

24. Why must each graywater holding tank have a locking gasketed access opening?

25. What must be permanently marked on each graywater holding tank?

26. What's the minimum capacity for a graywater holding tank?

27. What determines the size of a holding tank vent?

28. For replacement purposes, how must all connecting pipes to holding tanks be fitted?

29. A holding tank must be designed to withstand what amount of earth load?

30. What does an underground holding tank require to protect against sewer backups?

31. What material must all holding tanks be constructed of?

32. What's the minimum pipe size for use in graywater system irrigation/disposal fields?

33. Name two piping materials that are acceptable for graywater irrigation/disposal fields.

34. What does the code require that graywater irrigation/disposal field piping have?

35. What's the minimum size filter material required for a graywater irrigation/disposal field?

36. What's the minimum depth of filter material that's required beneath the pipe in a graywater irrigation/disposal field?

37. How far above the pipe must the filter material in a graywater irrigation/disposal field extend?

38. With what slope must graywater piping be laid?

39. In a graywater disposal system, what's the maximum length of each individual line?

40. When graywater irrigation/disposal lines are installed in areas having steep grades, what does the code require?

41. What's the minimum distance required from a private property line to a graywater system holding tank?

42. What is the minimum distance required from a graywater irrigation/disposal field to streams and lakes?

43. Who must approve the work if it's necessary to install a graywater irrigation/disposal field parallel to a public water main?

44. How does the design criteria regarding typical soils differ between graywater irrigation/disposal fields and septic tank drainfields?

45. What is a graywater system designed to do?

Public Water Supply and Distribution Systems

Two thousand years ago, the people of Rome constructed overhead aqueducts and underground tunnels to bring water into the city from mountains 50 miles away. These masonry aqueducts are ranked among the world's engineering triumphs.

To distribute the water carried in by the aqueducts, the Romans took the next logical step. They developed the first lead pipes to carry water by gravity flow to the private bathrooms of the wealthy. Some sections of these lead pipes still exist today. A recent hydraulic test was conducted on a section of ancient 4-inch diameter pipe with walls $1/4$ inch thick. It didn't fail until it reached a pressure of 250 pounds per square inch.

In 1652, the first gravity water supply system in America was built in Boston, using hollowed hemlock logs as pipes. About 1700, New York City installed a gravity water supply system with wood pipes laid under streets. They carried the water to street pumps or hydrants, where it could be sold.

Plumbing fixtures weren't generally installed in buildings until the 1800s. That's when the first pressurized water system was installed. For the first time, a safe and abundant supply of water was available at the turn of a handle. Today, water is supplied to nearly every home and every neighborhood in the U.S. Local authorities usually require users to connect to a public water supply system if one is available.

Raw water from lakes, rivers and deep wells is almost never safe for human consumption until it's filtered and treated. The treatment system removes unpleasant tastes, odors and impurities before the water is distributed through mains. The result is *potable water*, which meets the requirements of the health authority for drinking, culinary and domestic purposes. Local health departments monitor public water supply systems on a regular basis to assure pure and wholesome water is available.

Water System Components

The code uses specific definitions for most components of the water distributing system. You'll find these in the glossary at the end of this manual. In this chapter I'll describe the important parts (in simpler language than the code uses) and include some diagrams to clarify the code requirements. Figure 12-1 shows the service pipe up to the house valve. Figure 12-2 is the water distributing piping inside the structure.

Water Main

A water main is a public water distribution system located in a street, alley or dedicated easement adjacent to each owner's parcel. The main carries *public water* for *community use*. This means that the main is a common pipe installed, maintained and controlled by local authorities. For a fee, property owners can tap into this public water system after the utility company approves the connection. The service connection must have a curb stop (valve) and be connected to a water meter at the property line.

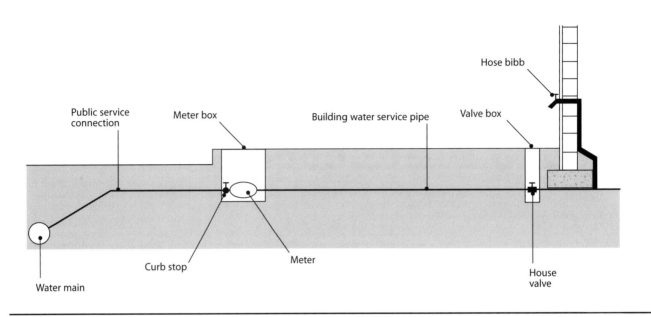

Figure 12-1
Service pipe

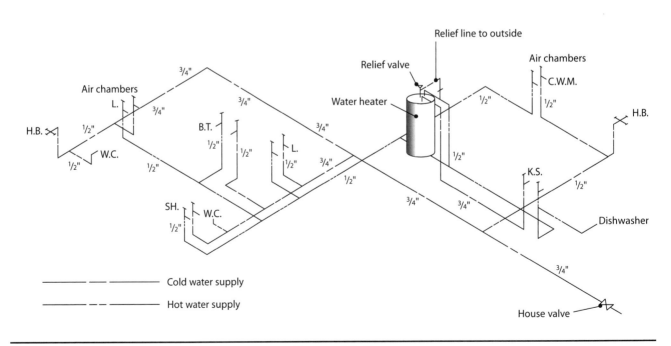

Figure 12-2
Water distributing piping diagram

Water Service Pipe

This pipe begins at the outlet side of the water meter at the property line and ends where the line reaches the first water distributing pipe. Each property owner must install and maintain his own service pipe.

Water Distributing Pipe

This is the pipe within a building that conveys water from the water service pipe to the plumbing fixtures, appliances and other water outlets.

Water Supply System

This includes the water service pipe, water distributing pipes, standpipe system, the necessary connecting pipes, fittings, control valves, and all appurtenances on private property.

Water Outlet

As defined in the code and used in connection with the water distributing system, the water outlet is the discharge opening for water:

▮ To a plumbing fixture

▮ To atmospheric pressure (except into an open tank which is part of the water supply system)

▮ To a boiler or heating system

▮ To any water-operated device or equipment not a part of the plumbing system but which requires water for operation

Sizing the Water System

You can't predict the maximum rate of flow or the demand in a building's water supply system. After all, you don't know how many fixtures the occupants will be using at the same time. But you must estimate the maximum demand as accurately as you can to provide an adequate water supply to all fixtures. Your goal is to avoid undersizing, yet to be as economical as possible in sizing the water supply piping.

In working with elaborate water supply systems, you'll need to calculate precisely to avoid any pressure loss. The physical properties of water — density, viscosity, compressibility, boiling point, minimum available pressure, friction loss and velocity flow — all have an effect on a water supply system. Most plumbers never have to figure complex systems or account for all the physical properties. But you *do* need to be familiar with the basic principles to design and install even residential systems.

There's a simplified method for sizing building water supply systems based on the demand load. It's based on water supply fixture units. Using this method, you can size the water supply system for all buildings that have adequate water pressure to supply the highest and most remote fixtures during peak demand.

When you size the supply system, you have to observe the velocity limitations in Figure 12-3. If the water pressure is at least 50 psi, you can use this method for sizing:

▮ Almost all one- and two-family dwellings

▮ Most multifamily dwellings not over four stories high

▮ Many small one- or two-story commercial and industrial buildings

In these buildings, 50 psi is usually more than enough to overcome the static head and ordinary pipe friction losses. The method is based on two factors:

1) The velocity limitations recognized as good engineering practice

2) The recommendations of piping material manufacturers

The calculations take into account the seven physical properties of water. This relatively simple sizing method is the way plumbing plan examiners and inspectors check building water supply pipe sizes. Figures 12-3, 12-4, 12-5 and 12-6 are sizing tables like those you'll find in many codes. Later, we'll work through a few examples to make sure you understand how to use them.

Fixture	Rate pressure (psi)	Flow rate (gpm)
Ordinary basin faucet	8	2.0
Self-closing basin faucet	12	2.5
Sink faucet — 3/8"	10	4.5
Sink faucet — 1/2"	8	5.0
Bathtub faucet	8	4.0
Laundry tub cock — 1/2"	8	4.0
Shower	8	2.0
Ballcock for water closet	8	3.0
Flush valve for water closet	15	25.0
Flush valve for urinal	15	15.0
Garden hose (50') and sill cock	8	5.0

Figure 12-3
Rate of flow and required pressure during flow for different fixtures

Number of bathrooms and kitchens Tank-type closets		Diameter of water service pipe	Recommended meter size	Approximate pressure loss, meter and 100' of pipe	Number of bathrooms and kitchens Flush valve closets	
Copper	Galvanized	(in)	(in)	(psi)	Copper	Galvanized
1-2	—	3/4	5/8	27	—	—
—	1-2	3/4	5/8	40	—	—
—	—	1	1	30	1	—
3-4	—	1	1	22	—	—
—	3-4	1	1	24	—	—
—	—	1 1/4	1	32	2-3	—
—	—	1 1/4	1	36	—	1-2
5-9	—	1 1/4	1	28	—	—
—	5-8	1 1/4	1	32	—	—
—	—	1 1/2	1 1/2	29	4-10	—
—	—	1 1/2	1 1/2	30	—	3-7
10-16	—	1 1/2	1 1/2	17	—	—
—	9-14	1 1/2	1 1/2	21	—	—
—	—	2	1 1/2	26	11-18	—
—	—	2	1 1/2	32	—	8-18
17-38	—	2	1 1/2	27	—	—
—	15-38	2	1 1/2	32	—	—
—	—	2	2	25	19-33	—
—	—	2	2	24	—	19-24
39-56	—	2	2	25	—	—
—	39-45	2	2	24	—	—
—	—	2 1/2	2	28	34-57	—
—	—	2 1/2	2	32	—	25-57
57-78	—	2 1/2	2	28	—	—
—	46-78	2 1/2	2	32	—	—
—	—	3	3	16	58-95	—
—	—	3	3	19	—	58-95
79-120	—	3	3	16	—	—
—	79-120	3	3	19	—	—

Figure 12-4
Minimum water service pipe size for one- and two-story buildings
(hotels, motels, and residential occupancy only)

Number of fixture units Flush tank water closet		Service size (in)	Meter size (in)	Approximate pressure loss (psi)	Number of fixture units Flush valve water closet	
Copper	Galvanized iron or steel				Copper	Galvanized iron or steel
18	—	³/₄	⁵/₈	30	—	—
—	15	³/₄	⁵/₈	30	—	—
19-55	—	1	1	30	—	—
—	16-36	1	1	30	—	—
—	—	1	1	30	9	—
56-84	—	1¹/₄	1	30	—	—
—	37-67	1¹/₄	1	30	—	—
—	—	1¹/₄	1	30	10-20	—
—	—	1¹/₄	1	30	—	14
86-255	—	1¹/₂	1¹/₂	30	—	—
—	68-175	1¹/₂	1¹/₂	30	—	—
—	—	1¹/₂	1¹/₂	30	21-77	—
—	—	1¹/₂	1¹/₂	30	—	15-52
226-350	—	2	1¹/₂	30	—	—
—	176-290	2	1¹/₂	30	—	—
—	—	2	1¹/₂	30	78-175	—
—	—	2	1¹/₂	30	—	53-122
351-550	—	2	2	30	—	—
—	291-450	2	2	30	—	—
—	—	2	2	30	176-315	—
—	—	2	2	30	—	123-227
551-640	—	2¹/₂	2	30	—	—
—	450-580	2¹/₂	2	30	—	—
—	—	2¹/₂	2	30	316-392	—
—	—	2¹/₂	2	30	—	228-343
641-1340	—	3	3	22	—	—
—	581-1125	3	3	22	—	—
—	—	3	3	22	393-940	—
—	—	3	3	22	—	344-895

Figure 12-5

Minimum water service pipe size for one- and two-story buildings (commercial use)

Type of fixture or device	Pipe size (in)	Type of fixture or device	Pipe size (in)
Bathtub	¹/₂	Laundry tray - up to three compartments	¹/₂
Combination sink and tray	¹/₂		
Drinking fountain	¹/₂	Shower (single head)	¹/₂
Dishwasher (domestic)	¹/₂	Sink (service, mop)	¹/₂
Hose bibbs	¹/₂ or ³/₄	Sink (flushing rim)	³/₄
Hot water heater	³/₄ min.	Urinal (flush tank)	¹/₂
Kitchen sink, residential	¹/₂	Urinal (direct flush valve)	³/₄
Kitchen sink, commercial, one or more compartments	³/₄	Water closet (tank type)	¹/₂
		Water closet (flush valve type)	1
Lavatory	¹/₂		

Figure 12-6

Size of fixture supply pipe

The first thing you must find is the minimum size required for normal fixture use at the highest water outlets during periods of normal flow conditions. Assume these minimum pressure requirements:

▌ For most water supply outlets at common plumbing fixtures, 8 psi

▌ For floor-mounted ball-cock-equipped water closets and urinals with flush valves, 8 psi

▌ For flush-valve-equipped water closets, 10 to 20 psi

For other types of water-supplied fixtures, refer to Figure 12-3 for flow pressure in psi and flow rate in gpm.

Even with the new water conservation measures, the code still requires the same minimum pressure (psi) for normal fixtures as those shown in Figure 12-3. *Where water conservation is in effect*, you reduce the flow rate (gpm) from Figure 12-3 to reflect the rates in Figure 11-1 in the previous chapter. But this adjustment probably won't affect the size of water service and water distributing pipes.

If the water supply system uses pipe sizes larger than those in the following figures, leave the job to a professional engineer.

It's essential that you understand how to use Figure 12-3 and the code sections that apply to sizing the piping for a water supply system.

Sizing the Water Service Pipe

The water service pipe is the first part of the water supply system you'll size. For the fixtures to operate properly, this pipe must carry sufficient volume at an acceptable velocity to the water distribution pipes throughout the building.

The water service pipe size continues within the building to become the building water main, whether it's horizontal or vertical. After you connect the water distributing branch pipes into the building main, using proper fittings, you can reduce the size of the main. As the main progresses through the building, the demand likely to be placed on the line decreases.

Figures 12-4 and 12-5 show the correct pipe sizes for residential and commercial buildings. These sizes apply only where water main pressure doesn't fall below 50 psi at any time. Figure 12-6 shows the minimum sizes for fixture supply pipe from the main or from the riser to the wall openings. The fixtures are subject to certain limitations which aren't difficult to follow when sizing water supply pipes. Keep these restrictions in mind:

▌ In buildings three and four stories high, use the next larger pipe size.

▌ Buildings located where the water main pressure falls below 50 psi should use the next larger pipe size.

▌ According to code, only two fixtures can connect to a $1/2$-inch water supply branch.

Use an automatic control pressure pump or gravity tank in buildings that are five stories or more, or where residual pressure in the system is below the minimum needed at the highest water outlet when the flow is at peak demand. The gravity tank must have the capacity to supply sections of the building which are too high to be supplied directly from the public water mains. If a tank is required, a professional engineer needs to do the design.

Note that you can install more fixtures on a copper system than on a galvanized iron or steel system of the same size, because the interior of the pipe is smoother, reducing friction. In most cases, the pressure loss is reduced or remains stable as the size of a water service pipe increases.

Some Sizing Examples

Now you can work through three examples to test your knowledge. You'll size these water supply systems using Figures 12-4, 12-5 and 12-6.

Example 1: An Apartment Building

Assume that you're sizing the cold water service piping in the apartment building shown in Figure 12-7. You don't have to consider the hot water piping to size the water service and distribution lines. The horizontal and vertical piping in Figure 12-7 serves 18 kitchens and 18 baths. The street level water pressure is 55 psi. You'll be using galvanized water piping.

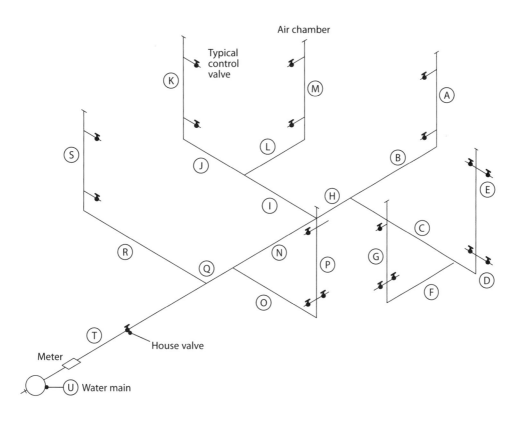

Air chamber

Typical
control
valve

House valve

Meter

Water main

Figure 12-7
Piping sizing problem (Example 1)

Size the 20 sections of pipe lettered A through T in Figure 12-7 without looking ahead to find the right answer. Start at the most remote outlet A and work back toward the water main (U), using Figure 12-4. Compute the cumulative demand for the apartment units. Remember that you're using galvanized pipe. After you've worked it out on your own, look at my answers in Figure 12-10.

Example 2: A Smaller Apartment Building

The single water supply riser in Figure 12-8 serves eight kitchens and eight baths with tank-type toilets (two on each floor). The water piping material is copper. The street level water pressure is 55 psi. Without looking ahead to find the answer, figure the sizes for sections A through D. Start at the most remote outlet (A) and work back toward the building main (E), computing the cumulative demand for the apartment units on each floor. In the *copper* section of column

one of Figure 12-4, you'll see that you have to use the next larger pipe size because the building is over two stories high.

Example 3: A Two-Bath House

Figure 12-9 shows the piping for a typical two-bath residence. The street level pressure is 50 psi. You'll use copper water piping material. There are 18 individual sections of pipe to size, A through R. Use Figure 12-4 to size the main water line and branch lines. Use Figure 12-6 to size the fixture supply pipes. Start at the most remote outlet (A) and work back toward the building water meter.

Done? Now you can look at Figure 12-10 for the answers to all three examples. If your answers don't match mine, take the time to go through the chapter again. This is the heart of the plumbing system. You've got to understand it before you go on.

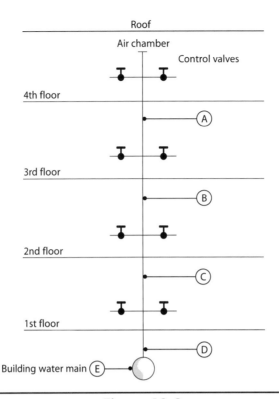

Figure 12-8
Sizing riser piping (Example 2)

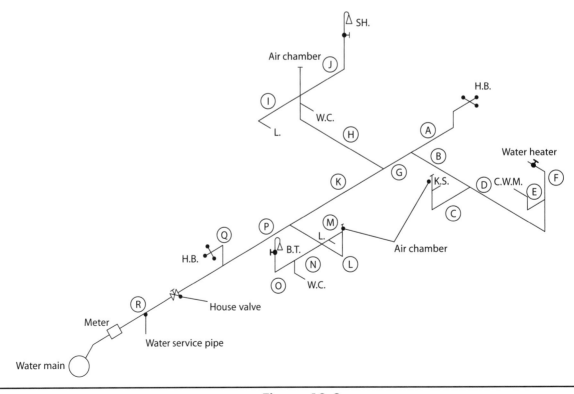

Figure 12-9
Sizing piping for typical residence (Example 3)

A	one-bath unit	3/4"	K	one-bath unit	3/4"
B	two-bath units	3/4"	L	two-bath units	3/4"
C	seven-bath units	1 1/4"	M	one-bath unit	3/4"
D	four-bath units	1"	N	thirteen units	1 1/2"
E	two-bath units	3/4"	O	three-bath units	1"
F	three-bath units	1"	P	one-bath unit	3/4"
G	one-bath unit	3/4"	Q	sixteen units	2"
H	nine-bath units	1 1/2"	R	two-bath units	3/4"
I	four-bath units	1"	S	one-bath unit	3/4"
J	two-bath units	3/4"	T	eighteen units	2"

A Pipe sizes for the building in Figure 12-7

A	2 one-bath units	1"
B	4 one-bath units	1 1/4"
C	6 one-bath units	1 1/2"
D	8 one-bath units	1 1/2"

B Pipe sizes for the building in Figure 12-8

A	hose bibb	1/2" or 3/4"	J	one fixture	1/2"
B	branch line	3/4"	K	building water main	3/4"
C	one fixture	1/2"	L	bathroom supply	3/4"
D	branch line	3/4"	M	over 2 fixtures	3/4"
E	one fixture	1/2"	N	two fixtures	1/2"
F	water heater	3/4"	O	one fixture	1/2"
G	building water main	3/4"	P	building water main	3/4"
H	bathroom supply	3/4"	Q	hose bibb	1/2" or 3/4"
I	one fixture	1/2"	R	water service pipe	3/4"

C Pipe sizes for the building in Figure 12-9

Figure 12-10
Answers for the examples

Review Questions (answers are on page 299)

1. In which American city was the first gravity water supply system built?

2. What did the installation of a pressurized water system make possible?

3. In order to make raw water safe and pleasant for drinking, name two things that are removed by treatment.

4. What is potable water?

5. What agency is generally responsible for monitoring our public water supply systems?

6. What is the purpose of a public water main?

7. What is a private service connection to a public water main required to have?

8. Where does a building water service pipe begin?

9. What are the pipes called that carry water from the water service pipe to the plumbing fixtures and other outlets?

10. Name three components of a water supply system.

11. In a water distributing system, how does the code define a discharge opening for water?

12. In sizing a water system, what won't you ever be able to predict exactly?

13. When you size water supply piping, it's important to be economical, but what cost-saving measure must you be sure to avoid?

14. Name two physical properties that can affect water in supply pipes.

15. What type of water supply system are plumbers not expected to size?

16. At what pressure psi does an ordinary faucet operate properly at ground level?

17. What's the minimum required pressure in psi for a floor-mounted ball-cock type water closet?

18. What should you do when there's a need for a water supply system using pipes larger than those in your plumbing code?

19. When a water service pipe continues within the building, what does it become?

20. As the building water main progresses through the building, what happens to the demand on the line?

21. According to code, how many fixtures can connect to a $1/2$-inch water supply branch?

22. What must you provide when a building exceeds four stories, or where a system's residual pressure is below what's required at the highest water outlet?

23. If a building requires a gravity tank, who should do the design work?

24. If each piping system is the same size, on which type system can you install the most fixtures?

25. In most cases, what happens with regard to water pressure as the size of the water service pipe increases?

Hot Water Systems

Surprisingly, even today's plumbing codes don't usually mandate hot water systems for houses. They do require hot water for buildings designed for a special occupancy like restaurants or rooming houses. But for houses, apparently the code writers consider hot water a luxury, not a necessity.

Of course, most buildings do have a hot water system, even though the code doesn't require it. And where there *is* a hot water system, the code specifically requires several safety devices. Properly installed, they'll help prevent damage to property and injury to people.

The code doesn't specify requirements for hot water distribution systems. It leaves the design to professional engineers and plumbers. The code simply states: "The sizing of the hot water distribution system shall conform to good engineering practice."

In large commercial buildings, a professional engineer usually designs the hot water supply system and the plumber just installs it. In residential and smaller commercial buildings, the installing plumber usually designs the hot water system. This chapter will help you design safe and functional hot water systems for residential and small commercial buildings.

Design Objectives

There are two principal objectives in designing a good hot water system:

1) The system must satisfy the hot water demand for the particular type of occupancy.

2) It must include safety features to guard against the hazards of excessive pressure and temperature.

I've included several sample diagrams at the end of this chapter to show you how to plan simple hot water systems.

The Water Heater

The normal design temperature of hot water in most plumbing fixtures is between 140 and 150 degrees Fahrenheit. That's hot enough for most purposes but cool enough to prevent scalding of the skin. Most heater thermostats are preset within that range at the factory and don't need additional adjustment.

Most direct heating units you'll install will use gas, oil, or electricity. Whatever the fuel, the energy transfer rate must be adequate to heat all the water stored in the tank. All commercial and residential storage tanks now have enough insulation to prevent excessive heat loss.

In today's homes, electric water heaters are the most common. They're clean and attractive enough to install nearly anywhere in a building. Gas or oil water heaters must be located in a well-ventilated area with flues to carry away combustion gases. That limits the possible locations in a building. Any confined space must have two permanent openings, one starting within 12 inches from the top and one 12 inches from the bottom (Figure 13-1). They must provide enough air for combustion, ventilation and dilution of flue gases.

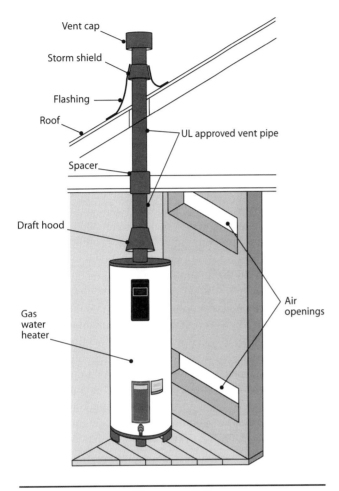

Figure 13-1
Installing a fuel-burning water heater

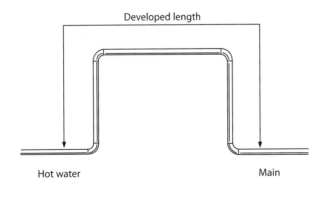

Figure 13-2
Horizontal U bend (four-elbow loop)

There's no need to constantly circulate hot water in a system designed for a one- or two-family dwelling. These small systems have small pipes and short runs. Although the water in these pipes will cool when it's not being used, it doesn't take long for hot water to reach the fixtures when needed.

In apartment buildings with a central hot water generating system, be sure to install a return line and circulating pump for greater efficiency. Without them, there's a long waiting time for hot water, and a lot of water wasted. And insulate the hot water feed and hot water return lines to prevent heat loss.

Water expands as it's heated, so your design for hot water lines in a large system must allow for the expansion and contraction. In smaller dwelling units with short pipe runs, this isn't a problem. In larger units with hot water riser supply pipes and circulating lines, however, expansion and contraction can pose a serious problem. Depending on the piping materials, 100 linear feet during a 100-degree F temperature rise may expand as little as $3/4$ inch or as much as 2 inches. Allow an expansion variance of $1^1/_2$ inches for each 100 feet of piping materials. The common four-elbow loop shown in Figure 13-2 provides adequate developed length to prevent excessive stress in a hot water piping system.

Heater Capacities

Modern electric and gas water heaters have small storage tanks. That's all they need, because they can provide enough heating capacity to keep the water at 140 to 150 degrees F during the peak draw period. The peak draw period is usually assumed to be one hour, though in many homes it's as little as 20 minutes. When sizing hot water tank capacity, here are the assumptions you should make:

▌ Assume that only 75 percent of the tank's hot water supply is available during the peak draw period of one hour. A 40-gallon water heater should provide 30 gallons at temperatures of 140 to 150 degrees F.

▌ If you think the peak demand rate may extend over a two-hour period, assume that 37.5 percent of the tank's capacity is available per hour. A 40-gallon water heater should provide 15 gallons per hour.

Fuel type	Number of bedrooms	Storage (gal)
Gas		20
Electric	1	30
Oil		30
Gas		30
Electric	2	40
Oil		30
Gas		30
Electric	3	50
Oil		30
Gas		40
Electric	4	66
Oil		30

Figure 13-3
Recommended hot water storage tank capacity

▮ If the peak demand rate could last three hours, assume that 25 percent of the tank's capacity is available per hour (75 percent over the entire three-hour period). A 40-gallon water heater should provide 10 gallons per hour.

For a storage tank heater, the heating capacity per hour determines the amount of hot water that's available during the peak demand periods. The tank storage capacities in Figure 13-3 are economical and satisfactory for the average dwelling unit *when the peak draw period doesn't exceed one hour*. If there are special demands that may last more than one hour, consider moving up to the next larger size.

The number of bedrooms (not bathrooms) is the best indication of the quantity of water needed. It determines the size of both the septic tank and the capacity of the hot water storage tank. For each additional bedroom above four, use the next larger size hot water heater. You may want to split the system and install two water heaters, especially if the layout needs long pipe runs.

For multifamily dwellings with a central hot water system, a packaged *high recovery rate* heater with these capacities is generally adequate:

▮ A 75-gallon storage tank for up to 12 living units

▮ A 100-gallon storage tank for up to 18 living units

Safety Devices

Pressure Relief Valves

The code requires a pressure relief valve to relieve excess storage tank pressure in all equipment used for heating or storing hot water. The pressure portion of the relief valve will be either 75, 100, 125 or 150 psi. Always install water heaters so the plates with the maximum working water pressure and other data are visible. Locate the heater where it's accessible for servicing or replacement without removing any permanent part of the building. See Figure 13-4.

The rate of discharge for the pressure relief valve must limit the pressure rise for any given heat input to within 10 percent of the pressure that opens the valve. Most standard hot water storage tanks are designed to withstand working pressures up to 150 psi. The pressure relief valve should open at 25 psi above the maximum working pressure, but under no circumstances should it exceed 175 psi.

Temperature Relief Valves

Temperature relief valves are required on all equipment that heats or stores domestic hot water. The temperature portion of the valve releases at a temperature of 210 degrees F. This valve must be the *reseating*

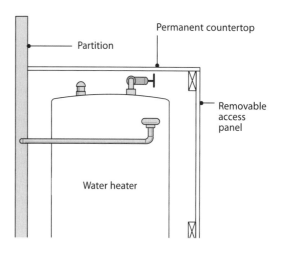

Figure 13-4
Installation of undercounter hot water heater

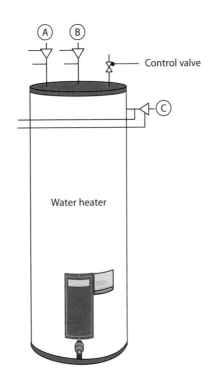

Note Positions illustrated at A, B, or C are acceptable

Figure 13-5
*Locate relief valves in the top 6 inches
of the water heater*

type, rated by its Btu capacity. The Btu rating of the temperature relief valve must always be greater than the Btu rating of the appliance it serves. This prevents premature opening of the valve.

In some cases you'll know the wattage of an electric water heater but not its Btu rating. So how do you choose the correct temperature relief valve and size the relief valve discharge line? It's simple! You can find the Btu rating with this formula:

1 watt = 3.4 Btu

Just multiply the watts by 3.4 to find the Btu rating of the appliance or combination of appliances.

Domestic packaged hot water heaters don't need separate pressure and temperature relief valves. You can use a combination pressure and temperature relief valve that meets the requirements of the American Gas Association, ASME, or other recognized authority.

Install the relief valves so the temperature sensing element is immersed in the top 6 inches of the tank (Figure 13-5). That's where the hottest water is. The

code says you can't install a check valve or shut-off valve between the relief valve and the water heater or storage tank it serves. Why? If the check valve fails to operate or if someone accidentally closes the shut-off valve, the relief valve wouldn't work. The tank could rupture, causing property damage and perhaps personal injury.

Relief Valve Drip Lines

Don't connect relief valve drip pipes (*popoff lines*) directly to any plumbing drainage or vent system. Not only could they contaminate the potable water system, they may also conceal any continuous discharge. And never terminate them above a water closet, urinal, bidet, bathtub or shower stall where anyone would be scalded if they suddenly discharge.

Drip pipes for most buildings should terminate outside the building, at an observable point within 6 inches of the ground. You can't thread the end of relief valve drip pipes. You don't want to make it easy for someone to connect something to this pipe.

The exception is when a building covers the entire lot. Then you can terminate the drip pipe in an indirect connection above a floor drain or other suitable fixture approved by local authorities. If hot water storage tanks are placed above the roof of a building, the relief valve drip pipe may discharge on the surface of the roof.

All relief pipes are sized by their inside diameter. The size of relief lines depends on the Btu rating of the appliance. If you know the total Btu rating, you can use Figure 13-6 to size relief lines for single-family

Btu rating	Inside diameter pipe size (in)
Up to 69,000 Btu	3/8
Over 69,000 to 127,000 Btu	1/2
Over 127,000 to 340,000 Btu	3/4
Over 340,000 to 600,000 Btu	1

Figure 13-6
Relief valve discharge lines

residences and multistory buildings with manifold lines (where two or more individual lines connect into one line).

Other Safety Features

Hot water heaters and storage tanks must have the drain cock in an accessible location. This allows both flushing sediment from the tank and emptying it for repairs or replacement.

The cold water supply pipe to hot water heaters must have a minimum $^3/_4$-inch shut-off valve in an accessible location. The shut-off valve must have a cross-sectional area equal to 80 percent of the nominal size of the pipe in which it's installed. And the minimum size cold water supply pipe for any hot water heater, regardless of its capacity, is $^3/_4$ inch diameter.

Water Heater Drain Pans

When water heaters or hot water storage tanks are located above the ground floor level of a building, you have to provide a drain pan. This helps prevent injury to the building occupants or damage to the building. See Figure 13-7. Here are some additional requirements for drain pans:

▪ They must be of a high-impact plastic at least 60 mil ($^1/_{16}$ inch) thick. You can also use galvanized sheet steel or other corrosive-resistant metal that's at least 24 gauge.

▪ They must be at least $1^1/_2$ inches deep (2 inches in some codes).

▪ They must have a minimum $^3/_4$-inch drain outlet (1 inch in some codes) located $^1/_2$ inch above the bottom of the pan.

▪ You must provide a minimum 2-inch clearance between the drain pan sides and the heater.

▪ In a multiple vertical installation, make sure the drain from pans runs downward a minimum of 6 inches before connecting into the main riser. See Figure 13-7.

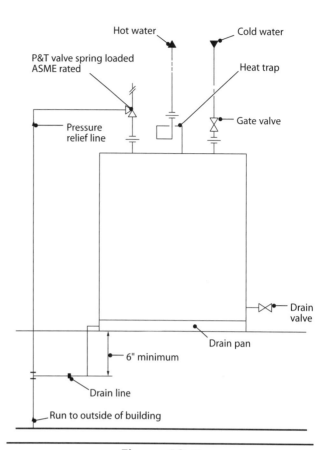

Figure 13-7
Electric water heater detail and drain pan

Sizing Drain Pan Risers

The number of drain pans you need depends on the size of the riser:

▪ For a 1-inch riser, connect a maximum of three drain pans.

▪ For a $1^1/_4$-inch riser, connect a maximum of four to ten drain pans.

▪ For a $1^1/_2$-inch riser, you can connect 11 or more pans.

Under a floor slab on grade, increase horizontal pipes that receive drain pan waste by one size. Some codes allow water heater relief lines and safe pan lines to use a common vertical riser. Use only approved materials and always check local code requirements. See Figure 13-7. Like the relief lines, drain pan lines must terminate over a suitable waste receptor or extend outside the building in a visible location 6 inches above grade.

Energy Conservation

Water heaters account for approximately 25 percent of the energy used in the average household. That's why codes have adopted stringent methods to conserve energy. These are typical of most code requirements:

■ Heat traps are required on hot water lines leading from water heaters. See Figure 13-7. The heat trap prevents hot water from rising into the hot water line, saving approximately 2 percent of the cost to generate hot water.

■ Water heaters must be equipped with an energy shut-off device. This will cut off the supply of heat energy to the water tank and prevent temperatures from exceeding 210 degrees F.

■ Water heaters are required to have efficient heating devices and controls.

There are stringent standards for water heater insulation to prevent excess heat loss from hot water storage tanks.

Three Common Hot Water Circulating Systems

Now let's look at the piping diagrams for hot water circulating supply. They illustrate the three most conventional systems: the inverted upfeed system, the looped system, and the downfeed system.

The inverted upfeed system (Figure 13-8) conveys hot water to the lowest part of the system. Then a circulating pump returns the water through the upfeed risers to the hot water generator. The hot water looped system (Figure 13-9) is simple and economical for small apartment buildings. Install the hot water feed and return directly beneath the bathrooms of each apartment. A circulating pump returns the water to a high recovery heater. In the downfeed system (Figure

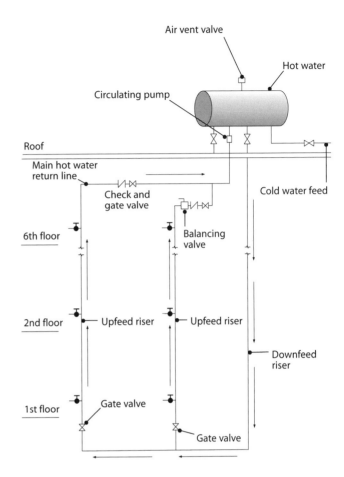

Note The inverted upfeed system conveys hot water to the lowest part of the system. The circulating pump returns the water through the upfeed risers to the hot water generator.

Figure 13-8
Inverted upfeed system

13-10), the upfeed riser from the hot water source conveys hot water to the downfeed risers. The circulating pump returns this water to the hot water generator.

Hot water piping materials and installation methods are the same as those described for water distribution in Chapter 14, *Installing Water Systems*. The only exception is that hot water piping supports must allow for the anticipated expansion.

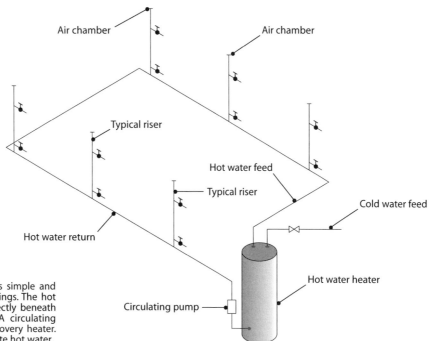

Note The hot water looped system is simple and economical for small apartment buildings. The hot water feed and return is installed directly beneath the bathrooms of each apartment. A circulating pump returns the water to a high recovery heater. This assures each apartment of adequate hot water.

Figure 13-9
Hot water looped system

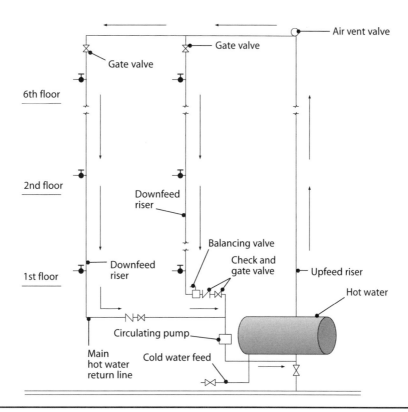

Note The upfeed riser from the hot water source conveys hot water to the downfeed risers. The circulating pump returns this water to the hot water generator.

Figure 13-10
Downfeed system

Review Questions (answers are on page 300)

1. What very common water system do you find in most buildings, even though it isn't required by most codes?

2. When a hot water system is installed, what does the code specifically require?

3. Why is it important to install safety equipment on hot water systems?

4. Since requirements for hot water distribution systems aren't specified in the code, who is responsible for designing and sizing the systems?

5. Who usually designs hot water supply systems for large commercial buildings?

6. Who usually designs hot water systems for small commercial and residential buildings?

7. What are the two principal objectives in designing a good hot water system?

8. What is the normal design range for hot water temperatures in most plumbing fixtures?

9. Whose responsibility is it to preset the water heater thermostat?

10. What three fuels are used in most direct water-heating units?

11. Why are electric water heaters the most common type used in homes today?

12. What is important about the location for gas- or oil-fueled water heaters?

13. What must be installed on water heaters designed to burn gas or oil?

14. In what type of building is it most efficient to have circulating hot water?

15. What conservation measure is advisable when installing hot water pipes in large buildings with circulating lines?

16. Though not needed in a cold water system, what must a hot water system design allow for?

17. How much expansion variance should be allowed for each 100 feet of piping?

18. In the average home, how long is the peak draw period for hot water assumed to be?

19. When sizing hot water storage tank capacity, what percentage of the tank's hot water supply is assumed available during the one-hour period of peak draw?

20. For a peak draw period of one hour, how many gallons of hot water should a 40-gallon water heater provide?

21. For a peak draw period of three hours, how many gallons of hot water per hour must a 40-gallon water heater provide?

22. For a storage tank heater, what determines the amount of hot water that's available during the peak demand periods?

23. What is the recommended storage capacity for a gas water heater installed in a two-bedroom house?

24. What is the recommended storage capacity for an oil-burning water heater installed in a three-bedroom house?

25. What is the recommended storage capacity for an electric water heater installed in a four-bedroom house?

26. What is the determining factor in choosing the storage capacity for any water heater, regardless of fuel type?

27. When might it be wise to split the hot water system and install two water heaters?

28. What size packaged high recovery rate heater will generally be adequate for up to 12 living units?

29. What does the code require all equipment used for the heating or storage of hot water be equipped with?

30. What purpose does the pressure relief valve on a hot water heater serve?

31. What should remain visible when you install a water heater?

32. In locating a water heater, you must ensure that it's accessible for servicing or replacement without doing what?

33. What type temperature relief valve must you install on a domestic hot water heater?

34. What type of relief valve is commonly used today?

35. Where, on a hot water heater pipe, is the placement of a shut-off or check valve prohibited?

36. Because of possible contamination, what part of a plumbing system does the code prohibit the hot water relief valve drip pipe from connecting to?

37. Name three fixtures above which a water heater relief valve drip pipe can never terminate.

38. Where should a water heater relief valve drip pipe terminate?

39. What type of end is prohibited on a water heater relief valve drip pipe?

40. If hot water storage tanks are placed above the roof of a building, where may the relief valve pipe discharge?

41. How is a water heater relief valve drip pipe sized?

42. What size relief valve drip pipe should serve a water heater with a 200,000 Btu rating?

43. What's the minimum size cold water supply pipe that you can connect to any water heater?

44. Under what conditions is a water heater drain pan required?

45. What's the required depth of a water heater drain pan?

46. What's the minimum size drain pan outlet?

47. How much clearance must there be between the drain pan sides and the water heater?

48. In a vertical installation, what's the maximum number of water heater drain pans that you can connect to a 1-inch riser?

49. In what location must a heat trap be installed on a water heater?

50. What's the purpose of a heat trap on a water heater?

Installing Water Systems

The plumbing code regulates the materials, size and installation methods for water piping. The basic requirement is for an adequate supply of potable water to all fixtures so they flush properly and stay clean and sanitary. The code also establishes safeguards to prevent pollution of the water supply due to backflow or cross connection. We'll discuss them in this chapter, as well as other requirements, limitations and restrictions.

Materials for Water Systems

All materials used in water service pipe and water distribution pipe, tubing and fittings must comply with the standards listed in Figure 14-1. Consider the kind of water in your area before selecting the material and size for the water supply system. Many communities have water that corrodes or leaves deposits on the interior walls of some kinds of pipe. And some soils and fills can corrode the exterior of certain pipe.

There's a wide variety of water piping materials that will meet code requirements. Some you can use only in underground installations, and others above ground only. Some materials are acceptable both above and below ground, for both hot and cold water use. Figure 14-2 shows the common water system materials and where you can use them.

There are two kinds of pipe that you can't use in a potable water system:

▮ Any pipe that can leave toxic substances in the water supply

▮ Any piping that has been used for other than a potable water supply system

For example, in a potable water supply system you can't use pipe or fittings that were once used in a gas system.

Water Service Supply Pipe

For a water service supply, you can use cast iron water pipe, cast iron threaded pipe, wrought iron pipe or steel pipe. Wrought iron and steel pipe and fittings must be galvanized inside and out. Other materials you can use for water service supply pipe under most codes are brass or copper pipe, Type K, L, or M copper tubing, and pressure-rated plastic pipes including ABS, PVC, PE, and PB pipe and tubing.

All plastic pipe and fittings must carry permanent identification markings from the ASTM (American Society for Testing Materials) or another national standard of acceptance (Figure 14-1). They also need a minimum working pressure of 160 psi.

Water Pipe in a Building

Water piping which is permanently inaccessible in a building, such as pipe installed under floor slabs, must be one of the following: cast iron water pipe, wrought iron pipe, steel pipe, brass, or Type K, L, or M copper tubing. Wrought iron and steel pipe and fittings must be galvanized, and you can use only the appropriate approved fittings. Chlorinated polyvinyl chloride (CPVC), polybutylene (PB), or cross-linked

Materials	ANSI	ASTM	AWWA	IAPMO
Nonmetallic pipe and fittings				
ABS plastic pipe, SDR		D2282		
ABS plastic pipe, Schedule 40, 80		D1527		
Asbestos-cement distribution pipe 4"- 16"	C400-80		R86	
Asbestos-cement pressure pipe				IS 15-82
CPVC plastic pipe, Schedule 40, 80		F441		
CPVC plastic pipe, SDR		F442		
CPVC plastic pipe and fittings		D2846		
Fiberglass pressure pipe			C905	
PB plastic pipe, SDR		D3000		
PB plastic pipe, SIDR		D2662		
PB plastic tubing		D2666		
PB plastic distribution systems		D3309		
PB plastic pressure pipe and tubing			C902	
PE plastic pipe, Schedule 40		D2104		
PE plastic pipe, Schedule 40, 80		D2447		
PE plastic pipe, SDR		D3035		
PE plastic pipe, SIDR		D2239		
PE plastic tube		D2737		
PE plastic pressure pipe and tubing			C901	
PVC plastic pressure pipe			C900	
PVC plastic pipe, Schedule 40, 80, 120		D1785		
PVC plastic pipe, SDR		D2241		
Ferrous pipe and fittings				
Cast iron fittings (threaded)		A126		
Cast iron pipe (threaded)	A40.5			
Cast iron water pipe		A377		
Ductile iron pipe, cement-lined	C151	A377	A21.51	
Stainless steel water pipe, Grade H		A268		
Steel pipe black		A53		
Steel pipe galvanized		A53		
Brass pipe		B43		
Copper pipe		B42		
Copper tube, Type K, L and M		B88		

Notes Although this standard may be referenced in some code tables for some of the pipe, tube or fittings, it is not always acceptable under every code. More complete standards may be referenced in some code tables. Check local code standard tables. See abbreviation for Plumbing-Related Organizations, Figure 8-1.

Figure 14-1
Plumbing material standards for potable water systems

Material	Water service pipe	Underground within buildings	Aboveground within buildings	Cold water piping	Hot water piping
ABS pressure rated plastic pipe	X	X		X	
Asbestos-cement pressure pipe	X				
Brass pipe	X	X	X	X	X
Cast iron water pipe	X	X			
Copper water tube, Type K, L, M	X	X	X	X	X
CPVC plastic water pipe	X	X	X	X	X
Ductile iron pipe, cement lined	X				
Fiberglass pressure pipe	X				
Galvanized steel pipe	X	X	X	X	X
PB plastic water distribution sys.	X	X	X	X	X
PE plastic pipe, Schedule 40, 80	X	X	X	X	
PVC plastic pipe, Schedule 40, 80, 120	X	X	X	X	
Stainless steel water pipe, Grade H			X	X	X

Notes

▌ Potable water piping shall have a minimum 160 psi at 73° F.

▌ Hot water piping shall be within the limits of its listed standard and the manufacturer's recommendations.

▌ Other codes may list materials that don't appear in above chart.

▌ Some of the pipe, tube or fittings referenced above are not acceptable by some codes. For example: (1) asbestos-cement pressure pipe is referenced only by the *Uniform Plumbing Code.* (2) Stainless steel water pipe is referenced only by the *Standard Plumbing Code.* (3) Some codes don't permit the installation of M copper underground within buildings.

▌ Always check local code requirements.

Figure 14-2
Common potable water system materials and locations where they can be used

polyethylene (PE) plastic pipe or tubing must be installed with approved fittings or bends. Your code will tell you what's approved for your area.

For water piping installed above first-floor slabs, you can use brass, copper Type K, L or M, cast iron, wrought iron, block tin or steel. Wrought iron and steel pipe and fittings must be galvanized inside and out. Use only approved fittings for the type of pipe you install.

It's now common to use approved plastic pipe and fittings below and above ground within buildings. The first plastic pipe and fittings approved by most codes was CPVC. It's a high-temperature vinyl plastic designed for use in both hot and cold water systems, available only in rigid lengths. It can withstand pressures up to 100 psi and temperatures up to 180 degrees F.

Most codes now accept, without limitation, polybutylene pipe, tubing and fittings. PB pipe and fitting resins are manufactured from 85 percent butene-1 in both rigid piping and flexible coils. It can withstand pressures up to 100 psi and temperatures up to 180 degrees F. In the area of creep resistance (expansion), polybutylene has a unique advantage over other approved plastic and metallic piping and fittings. It's flexible and doesn't require any provision for expansion or stress like rigid CPVC plastic or metal pipes do. Also, it can be joined by compression fittings, flare connections, crimp-type fittings and heat fusion. And it's not damaged by freezing temperatures.

Plastic pipe and fittings have a distinct advantage over metallic pipe and fittings; they resist corrosion, scale, sediment buildup, and aren't affected by soil conditions or electrolysis.

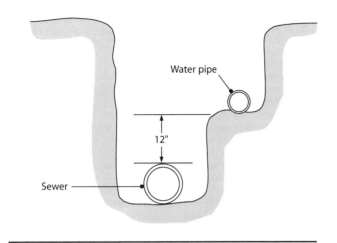

Figure 14-3
Water service supply pipe in sewer trench

Installing Water Service Supply Pipe

You can install the water service supply pipe in the same trench as the building sewer. See Figure 14-3. If you use a single trench for both the sewer and the water service, the installation must meet these conditions:

▮ Place the water service supply pipe on a solid shelf excavated at one side of the common trench and above the sanitary sewer line.

▮ The bottom of the water service supply pipe must be at least 12 inches (some codes require 10 inches) above the top of the sewer line.

▮ Keep the joints in the water service supply pipe to a minimum. Use full-length 21-foot pipe and 50-foot to 100-foot coils.

When installing metallic pipe on filled corrosive soil or where hydrogen sulfide gas is known to be present, protect the pipe with one or two coats of asphaltum paint or other approved coating. And protect the fittings by painting, wrapping with an approved material or applying other approved coatings.

Install water service supply piping in open trenches, laid on a firm bed of earth for its entire length. Make sure it's securely supported to prevent sagging, misalignment and breaking.

Cast iron water pipe, cast iron threaded water pipe, wrought iron pipe and galvanized steel pipe are preferred for their strength, durability and resistance to trench loads. They're especially good for outside use in water service supply piping. Because they resist trench loads, the depth of placement isn't critical except as protection against freezing. Avoid backfilling with large boulders, rocks, cinder fill or other materials that might physically damage the pipe, or encourage corrosion.

Most codes don't establish the depth for brass or copper pipe, or Type K, L, or M copper tubing. Naturally, the pipe must be deep enough to avoid freezing. These materials are soft and easily damaged. Select the backfill material with care. Use fine, rather than coarse, backfill that doesn't contain anything that promotes corrosion of the pipe exterior.

ABS, PVC, PE and PB plastic pipes are considered fragile. When installing them in open trenches, follow the installation methods recommended for plastic pipe used in building sewers. See Figure 8-16 back in Chapter 8.

Building codes set a minimum separation distance for water service supply pipe installed in a separate open trench. Look at Figure 14-4. Check your local code, but most require these separations:

▮ 5 feet from the sewer line

▮ 10 feet from any septic tank or drainfield

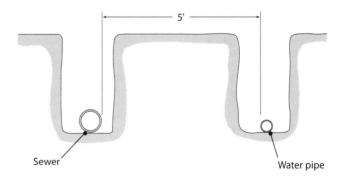

Figure 14-4
Water service supply pipe in separate trench

In climates where buried pipe is subject to freezing, the trench with the water pipe must be below the frost line. Thoroughly insulate pipe that enters a building above ground or in areas not protected from the cold to prevent freezing.

Where freezing isn't a problem, bury water pipe deep enough to avoid damage from edgers or other sharp tools. And bury the pipe deep enough so that it's not heated by the sun on warm days.

Water pipe passing under a building foundation must have a clearance of 2 inches between the top of the pipe and the foundation or footing. Refer back to Figure 8-17. Water supply pipe that passes through a basement wall or other cast-in-place concrete needs a sleeve to provide 1/2-inch clearance around the entire pipe. This prevents damage or breakage due to the building settling or the normal expansion and contraction of the pipe. This also protects the pipe from the corrosive effects of concrete.

When you connect a lawn sprinkler system to the potable water supply, install an approved backflow preventer on the discharge side of each valve. The backflow preventer must be at least 6 inches above the highest sprinkler head, and not less than 6 inches above the surrounding ground. This should eliminate the possibility of a cross connection. See Figure 14-5.

Each building must have a separate water control valve, independent of the water meter valve, installed in the water pipe. The control valve must be accessibly located at or near the foundation line outside the building, either above ground or in a separate approved box with a cover. Look back to Figure 12-1 in Chapter 12. Some codes require two control valves:

1) An accessible control valve near the curb

2) A control valve with a drip valve near where the water supply pipe enters the building. The drip valve is needed in cold climates to drain off water to prevent the pipes from freezing and bursting.

Water service supply piping must be electrically isolated from all other pipe, conduit, soil pipe, building steel and steel reinforcing. When the pipe comes in contact with other metals, you must wrap it with approved materials. The only exception is where an electric ground is required by the code.

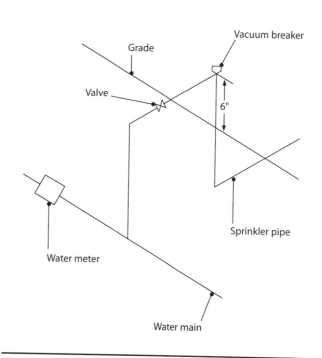

Figure 14-5
Placement of backflow preventer

No private water supply, such as a well, can be interconnected with the water service of any public water supply.

If a swimming pool water supply is connected to the potable water service pipe, provide a positive air gap to prevent a cross connection. There's more on this in Chapter 17, *Swimming Pools and Spas.*

Installing Water Distribution Piping

Most code sections that cover water distribution piping make good sense if you take the time to understand them. Unfortunately, it's not always easy for you or your tradesmen to take that time when you're trying to get a job done on schedule and within budget. When you're rushed, you can overlook important points. The purpose of the code is to ensure that the owners and occupants of a building have an effective, trouble-free plumbing system. If you comply with the code, you can be fairly sure that the owner is getting good value for each dollar he spends on plumbing.

Make sure underground water distribution pipe installed under a slab inside a building is firmly supported for its entire length on well-compacted fill. Lack of support can cause misalignment and sagging. That can lead to traps or depressions in the pipe that collect sediment or mineral deposits. Some mineral deposits solidify over a period of time, reducing the flow of water and causing premature failure of the system. Use fine rather than coarse backfill in the trench so compaction doesn't damage the pipe.

If at all possible, install the entire system so it will drain dry. This can be important if you have to make repairs or replace lost air in air chambers.

When you install the pipe, remember that the building will settle and the pipe will expand and contract during use. Make sure this movement doesn't damage the pipe. Provide sleeves for water supply piping that passes through cast-in-place concrete slabs or walls to provide $1/2$-inch clearance around the entire pipe. Be sure to caulk the sleeve with an approved flexible sealing compound so crawling insects can't get in.

You must protect water pipe installed under a concrete slab from corrosion. You also have to protect pipe in any location inside the walls of a building constructed on filled ground if hydrogen sulfide gas or other corrosive substances are known to be present. You can do that by painting the pipe and fittings with two coats of asphaltum paint or other approved coating. It's better still to install water pipe overhead in the attic if you're working on filled ground. In fact, some codes require it.

In climate zones where water pipe is subject to freezing, avoid installing pipe in crawl spaces or other unheated areas unless it's protected with adequate insulation. And in restaurants, packing houses and other commercial buildings with walk-in cold storage facilities, don't install the water pipe through the cold storage room.

The code says that water distributing pipe must be electrically isolated from all ferrous pipes, electrical conduit, building steel and reinforcing steel. This means that the pipe can't come into physical contact with any good conductor of electricity.

Here are some additional restrictions on the installation of water distribution piping:

- The code usually prohibits water supply piping under elevators or in elevator hoistways.

- Hot and cold water pipe must not contact each other when installed underground or within partitions. Contact or close installation can transfer the heat from the hot water pipe to the cold water pipe.

- Domestic cold water piping installed above a roof or within 10 inches of the roof must be adequately insulated with approved materials.

- Each separate apartment or store in a building must have its own independent control valve or individual fixture control valve controlling all the fixtures in that apartment or store.

- Each water closet and urinal supply must have an independent control valve installed above the floor.

- Hotels, office buildings, hospitals, clinics, places of public assembly and manufacturing plants require either separate fixture control valves or a single control valve for each group of fixtures in a single room.

- Residential buildings with more than two units and a single water service must have a separate control valve for each hose bibb (sill cock). This allows you to make repairs without interrupting the water supply to the residential units.

- No more than two fixtures can be connected to a $1/2$-inch cold water supply pipe.

- Water pipe installations must be protected from water hammer with properly located air chambers or other approved devices. Air chambers must be installed so they drain by gravity without disconnecting the fixture water supply. An air chamber constructed of pipe must be at least 12 inches long and one pipe size larger than the pipe it serves.

- Where water pressure within a building is in excess of 80 psi, install an approved water pressure regulator with a strainer to reduce the pressure in the building water distribution system to 80 psi or less.

Pipe Supports

Secure water piping inside walls and partitions with pipe straps or another approved method to prevent pipes from moving as they contract or expand. Water supply pipes installed above ground or in a vertical position must be securely supported by the building structure.

Horizontal water pipe requires the same type of support and separation as in drainage, waste and vent applications. See Figure 8-22. Here are the maximum distances between supports for horizontal water supply piping above ground:

▮ Screwed pipe must be supported approximately every 12 feet.

▮ Copper tubing 1¹/₂ inches or smaller must be supported at approximately 6-foot intervals.

▮ Copper tubing 2 inches and larger must be supported at approximately 10-foot intervals.

▮ CPVC, PB and PE plastic pipe must be supported every 4 feet.

Vertical water piping also requires the same type of support and separation as in drainage, waste and vent applications. Look back to Figure 8-23. Vertical water pipe requires the following support:

▮ Screwed pipe carrying cold water must be properly supported at the base and at every other story height.

▮ Screwed pipe carrying hot water must be properly supported to provide for expansion.

▮ Copper tubing 1¹/₄ inches and smaller that carries cold water must be supported at intervals of approximately 4 feet.

▮ Copper tubing 1¹/₂ inches and larger that carries cold water must be supported at each story height.

▮ Copper tubing carrying hot water must be properly supported to provide for expansion.

▮ CPVC, PB and PE plastic pipe carrying cold water must be supported at each story height.

▮ CPVC plastic pipe carrying hot water must be properly supported for expansion.

▮ PB plastic pipe carrying hot water needs no special provision for expansion.

Give good support to the base of water pipe risers, particularly in high-rise construction where the pipe is expected to last for the life of the building. Don't let the riser weight fall on the smaller water pipe branches. This would crack the joints of the smaller pipes.

Avoiding Cross Connection

The code requires the potable water supply outlet to terminate above the overflow rim of each fixture. This provides an air gap which prevents siphoning of the fixture contents back into the water outlet or faucet.

All water outlets equipped for hose connections (other than clothes washers) must have approved backflow preventers. Hose connections are common on restaurant faucets and service sinks.

Codes prohibit the use of fixtures with water supplied below the rim. But some special fixtures can't function properly without a below-rim supply — water-cooled compressors and degreasers, for example. For those fixtures, install a backflow preventer in the water supply pipe and individual backflow preventers in each individual special fixture.

Connect copper tubing to threaded pipe or fittings with a special brass or copper converter fitting. Unions in the water supply system must have metal-to-metal joints and ground seats.

Remember the importance of air chambers in water supply pipes. A system with water shock or hammer won't pass the final inspection. After the water distribution system is installed, it must be tested, inspected and proved to be tight. It'll be tested to at least the maximum working pressure it's designed for.

Threaded Pipe, Fittings and Valves

All standard pipe and fitting threads must conform to the standards adopted by the American Standards Association. The standard pipe thread has a taper of ³/₄ inch per foot. The last column in Figure 14-6 shows

Pipe size (in)	Threads per inch	Approximate length of threads (in)	Approximate thread engagement (in)
1/4	18	5/8	3/8
3/8	18	5/8	3/8
1/2	14	13/16	1/2
3/4	14	13/16	1/2
1	11 1/2	1	9/16
1 1/4	11 1/2	1	5/8
1 1/2	11 1/2	1	5/8
2	11 1/2	1 1/16	11/16
2 1/2	8	1 9/16	15/16
3	8	1 5/8	1
4	8	1 3/4	11/16

Figure 14-6
Standard pipe threads

Pipe size (in)	Pipe wrench size (in)
1/8, 1/4, 3/8	6 or 8
1/2, 3/4	10
1	14
1 1/4, 1 1/2	18
2	24
2 1/2, 3	36
4	48
* Larger sizes usually use chain tongs or special equipment.	

Figure 14-7
Pipe wrench sizes

the number of threads that must be engaged on screwed pipe. Use the right size wrench when joining valves or fittings on threaded pipe. A wrench too small for the job will require unnecessary effort for your hands, arms and back. A wrench too large will force the fitting too far onto the threaded pipe. This can result in a bad joint or a cracked fitting. The wrenches listed in Figure 14-7 should be adequate for the pipe sizes given.

Cutting, Reaming and Threading Pipe

Use heavy-duty pipe cutters to cut iron and steel pipe. The most common pipe cutter is the single wheel cutter. Begin each cut by *lightly* rotating the cutter completely around the pipe. This will give a "bite" or groove for the cutter wheel to follow. After each turn of the cutter wheel, tighten the handle slightly. Tightening the handle too rapidly will break the cutter wheel or spring the cutter frame, ruining the cutter. Occasionally put thread-cutting oil on the cutter wheel and rollers.

The pipe cutter wheel leaves a bur on the inside of the pipe (Figure 14-8). Mineral deposits can collect at the bur and cause premature failure of the line. Use a pipe reamer to remove the bur on threaded pipe up to 2 inches. On larger pipe, use a coarse half-round file.

Modern pipe threaders make plumbing installation much less exhausting than it used to be. Threaders cut the required standard threads and then disengage the dies.

Use plenty of thread-cutting oil on the threader dies when threading pipe to prevent overheating. Occasionally check to be sure you're cutting clean threads. Chipped or torn threads mean that the pipe at this location is too soft, too hard, has impurities in it, or that the dies are worn and need replacing. Always cut new threads when this happens. *Don't use pipe that has bad threads.*

Only the plumber who makes the fitting knows whether the bur was removed and whether the pipe has good threads. The contractor paying his salary doesn't know, the plumbing inspector checking the job doesn't know, and the people relying on his workmanship don't know. You and your crew demonstrate your integrity, good workmanship, and personal pride on every joint on every job.

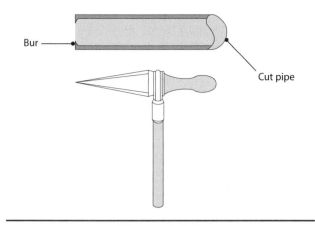

Figure 14-8
Pipe reamer

When you're satisfied that you've made a good joint, apply pipe compound (known as *pipe dope*) and screw the fitting snugly but firmly onto the threaded pipe. When installation is complete, open the house valve to test for leaks.

Measuring Offsets

When you're laying out and dimensioning piping arrangements, offsets are a problem anywhere you don't use 90-degree elbows. It takes careful calculation to find the exact distance between centers of the fittings in offsets.

Figure 14-9 shows a common pipe offset between parallel runs of pipe. The center-to-center offset (C) is 10 inches, the distance between the centers of pipes A and B. Assume that you're connecting pipes A and B

with 45-degree elbows. You need to find the length of pipe required for E. Use Figure 14-10. (The letters refer to the letters on Figure 14-9.) On the line for the 45-degree offset, you'll find the figure 1.4142 in the last column. Multiply distance C (assume 10 inches for this example) by 1.4142. The length of pipe E is 14.14 inches.

Cutting, Reaming and Joining Copper Tubing

You can cut copper tubing, both rigid and flexible, with a tubing cutter or a hacksaw. It's easier to get a clean square cut with a tubing cutter. Copper tubing cutters are similar to pipe cutters, though much smaller. Apply oil to the cutter wheel and rollers sparingly, as excessive oil will cause trouble when you're preparing the copper tubing for soldering. Be careful to remove the bur the cutter wheel leaves in the tubing. Use reamer attached to the tubing cutter frame for copper tubing up to 2 inches in diameter. For larger tubing, use a file to remove the bur.

The most common joint for copper pipe is the sweat joint. Clean the inside of the copper fittings with a wire brush designed for this purpose, or with emery cloth. Also polish the pipe end bright with emery cloth. Use cleaned parts as soon as possible and don't handle them with dirty or sweaty fingers. The bright surface oxidizes very quickly.

Apply a thin coat of noncorrosive soldering flux to the inside of the cleaned fitting and the outside of the cleaned pipe. That joins the fitting and pipe, ready for soldering.

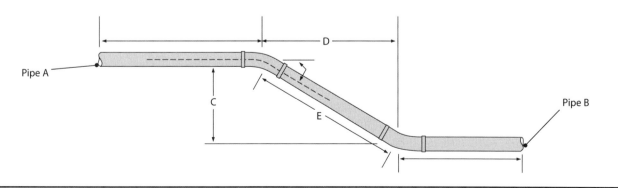

Figure 14-9
Measuring offsets

Degree of offset	When C = 1, D =	When D = 1, C =	When C = 1, E =
60	0.5773	1.7320	1.1547
45	1.0000	1.0000	1.4142
30	1.7320	0.5773	2.0000
22½	2.4140	0.4142	2.6131
11¼	5.0270	0.1989	5.1258
5⅝	10.1680	0.0983	10.2170

Figure 14-10

Finding the length of pipe to connect two parallel runs

Soldering joins the two metallic surfaces with a fusible alloy (solder) that has a melting point lower than that of the pipe to be joined. All solder and fluxes must meet approved standards. In a potable water supply system, you're prohibited from using solders and fluxes with a lead content exceeding 0.2 percent.

The soldering torch is fired from a tank of gas that's easy to move from one area to another. The tank and gas weigh only about 20 pounds when full but provide enough gas for three days of continuous soldering. The tank's regulator controls the flow of gas through a 6-foot-long rubber hose, and another valve controls the flame at the burner tip. Select a tip appropriate for the size of copper tubing you're joining. Too large a tip will overheat and burn the copper tubing and fitting, preventing the solder from joining the two surfaces. Too small a tip will heat the pipe and fitting unevenly and won't draw the solder into the joint. In either case, the result is a bad joint. That means that sooner or later the joint will leak. If you're lucky, it'll happen when it's first tested. If you're not, it'll be months later, after the piping's enclosed by the building walls.

It's more difficult to work with a system that's been pressurized because there's usually still some water present. The heat from the torch will turn the moisture in the tubing to steam. This can make pinholes in the newly-applied solder before it hardens. You can dry the pipe long enough to make a good joint if you apply heat to several inches on each side of the fitting and work quickly.

If you can't keep water away from the joint while you're working, stuff a piece of plain white bread (not the crust) as far into the pipe as possible in the direction the water's coming from. The bread will absorb the water, giving you time to make a good joint. When water pressure is returned to the system, the bread flushes out easily.

If the joint is properly cleaned and heated, surface tension will spread the solder to all parts of the joined surfaces. This results in a sound joint that will last as long as the tubing.

Soldering is an art in itself. It looks easy but requires care and precision. Apprentice plumbers should solder only under the supervision of an experienced journeyman. It takes a lot of practice in soldering before you have the knowledge and skill necessary to solder copper water systems properly.

Copper tubing is measured the same way as threaded pipe. The only exception is that you can heat and bend the smaller copper tubing when an offset is needed. In many minor offsets, you don't need fittings if you're using copper tubing up to 1-inch diameter.

Occasionally you'll have to make a flare joint on flexible copper tube to connect some fixtures and appliances. For example, you might install a connector for a gas range with a flare fitting. First, cut the tubing to the desired length. Then slip the flare nut that will make the connection onto the cut tubing. Use a flaring tool to flare the tubing ends so you get a perfect fit. Slide the nuts to each end of the flared pipe and gently bend or shape the tubing by hand so the ends fit together. Then firmly screw the flare nuts by hand on the male thread of each fitting. Tighten with the proper type and size wrench until the fitting is snug. Finally, test it for leaks.

Cutting, Reaming and Joining Plastic Pipe

Today you can use pressure-rated plastic pipe and fittings in all phases of the water distribution system. Plastic pipes are lightweight, easily handled by one person, yet rigid once cemented into place.

You can cut plastic pipe with a hacksaw blade or a special cutter similar to the copper tubing cutter. A square end cut is essential with plastic pipe. If you're using a hacksaw, use a fine-tooth blade and a miter box. A freehand cut is an invitation for trouble.

Remove the bur with a reamer if you're using a pipe cutter. A hacksaw doesn't leave a bur, but the interior and exterior of the pipe are left with a rough ridge. Remove this roughness with a knife or file.

You can measure offsets for rigid plastic pipe and fittings the same way as for threaded pipe. Refer back to Figures 14-9 and 14-10.

You'll always use special plastic cement to join rigid plastic pipe and fittings, though you may hear it referred to as a "welded" joint. First, check the fittings and pipe. Don't use any pipe or fittings that have gouges, deep abrasions or cracks. After cutting the pipe to the proper length, check the dry fit before cementing. The pipe must enter the fitting socket smoothly, but must not be so loose that the two surfaces don't make good contact.

Use liquid cleaner (developed for plastic pipe and fittings) or fine sandpaper to remove impurities and gloss from the surfaces to be joined. Use only the cement recommended by the manufacturer of the pipe and fittings you're using. Don't use cement that won't pour from the can or that has a rusty or dark brown color.

Plastic cement sets very quickly once it's applied to plastic surfaces, so you can only do one joint at a time. You have less than one minute to do all this:

▌ Brush a light coating of cement on the plastic pipe with the brush supplied with the can.

▌ Brush a thin coat in the fitting socket and quickly brush the plastic pipe a second time.

▌ Push the pipe fully into the fitting and then give the pipe a quarter turn. If a direction is required (a tee facing up, etc.), adjust the pipe at once. This is your last chance to make corrections.

▌ Wipe excess cement from the fitting and close the cement can immediately to prevent drying.

▌ Hold the pipe and fitting together for approximately 15 seconds until the cement sets.

Since a cemented joint looks like a dry-fit joint, check all joints to make sure that they have been cemented.

Wait at least half an hour after cementing the last joint before testing the system. If time isn't critical, it's best to let the joints harden overnight. Check for leaks

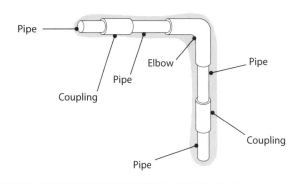

Figure 14-11
Replacing a bad joint

after turning on the house valve and releasing trapped air from the ends of each branch.

If a leak occurs in a rigid plastic system, cut out the bad joint and replace it with a new fitting. Remove enough pipe to allow room for two couplings and the fittings. Figure 14-11 shows how much pipe to cut out if there's a leak in a joint at the elbow. In this case it takes three fittings, two short lengths of pipe and six joints to make the repair. Do a first class job when cementing plastic pipe and fittings. *Plastic pipe fittings can't be reused if they don't work the first time.*

Use approved male-threaded plastic adapters if you have to connect plastic pipe or valves. And use only the thread compound recommended by the manufacturer.

PB and PE are the newest plastic pipes approved for use in water distribution systems in many states and by local codes. Their use for both hot and cold water piping is unrestricted.

You can cut the piping into desired lengths with the same tools and methods for other plastic pipe. You can join it with compression fittings, crimped fittings, or special heat fusion methods. It can also be flared (like soft copper) and joined together with flare fittings. When installation is complete, there's no waiting for joints to cure as in other plastic systems. You can turn on the water immediately to check out the system.

When the water system is complete, open a hose bibb on each section to flush out sand, pieces of pipe shavings and other impurities that may have collected inside the pipe during installation. This will save time when fixtures are set and prevents the damage sand and grit can do to washers.

Review Questions (answers are on page 302)

1. What three things does the plumbing code regulate for water piping?

2. Why does the plumbing code require that there be an adequate supply of potable water to all fixtures?

3. How do code-established safeguards protect our water supply?

4. Name two reasons why you should consider the kind of water and soil in your area before you select and size the material for water supply pipes.

5. What pipes are specifically prohibited in a potable water supply system?

6. All plastic pipes and fittings must display identification/acceptance from what standard?

7. What's the minimum working pressure required by code for plastic water service piping?

8. What plastic pipe and fittings were first approved for a building water distribution system?

9. What's the highest water temperature that CPVC plastic pipe can withstand?

10. What unique advantage does polybutylene plastic pipe have over other approved plastic and metallic pipe and fittings?

11. Polybutylene (PB) pipe, unlike other plastic pipe, is not damaged by what kind of temperatures?

12. List three advantages that plastic pipe and fittings have over metallic pipe and fittings.

13. Where must you place a water service pipe when installing it in the same trench as the building sewer?

14. How must you protect metallic water service pipe when you install it on filled corrosive soil?

15. Why should water service supply piping be securely supported?

16. Name two materials that are considered superior for outside water service supply piping.

17. Why shouldn't you use large boulders, rocks or cinder fill to backfill a metallic water service supply trench?

18. Why must you treat ABS, PVC, PE and PB plastic pipe with special care?

19. What is the minimum separation distance required between a water service supply pipe and a sewer line when they are installed in separate open trenches?

20. How deep should the trench for buried water service supply pipe be in climates where the pipe is subject to freezing?

21. How much clearance must you have between the bottom of a building foundation or footing and the top of the water service supply pipe?

22. What must you install on the discharge side of each valve when connecting a lawn sprinkler system to a potable water service supply pipe?

23. What must you install in each building's water service supply pipe, independent of the water meter valve?

24. From what must you electrically isolate water service supply piping?

25. What type of water supply does the code prohibit from interconnecting with a public water supply?

26. When you connect a swimming pool water supply to the potable water service pipe, what must you provide to prevent cross connection?

27. When mineral deposits in the water solidify in the distribution pipes over a period of time, what happens to the system?

28. How much clearance should you provide around the circumference of a pipe when it passes through cast-in-place concrete?

29. What do you need to do to protect water pipe installed in crawl spaces or other unheated areas if you work in a climate that's subject to freezing temperatures?

30. Why should you not allow hot and cold water pipes to come into contact with each other in underground or partition installations?

31. What does the code require that each water closet supply pipe have?

32. How many residential plumbing fixtures can you connect to a $1/2$-inch cold water supply pipe?

33. How do you protect water pipe installations from water hammer?

34. What protective device must you install when the water pressure within a building is more than 80 psi?

35. At what intervals must you support horizontal screwed water pipe?

36. What is the maximum distance allowed between supports for horizontal CPVC plastic pipe?

37. What's the maximum distance allowed between supports for vertical $1^1/_4$-inch copper tubing carrying cold water?

38. Why does the code require that potable water supply outlets terminate above the overflow rim of a fixture?

39. What must you install on all water outlets (except for clothes washers) equipped for hose connections?

40. In the water supply system, what kind of joints must unions have?

41. What standards must all pipe and fitting threads meet?

42. What can result when you use a wrench that's too large for a fitting?

43. What wrench size is recommended to tighten a fitting on a 2-inch pipe?

44. What type of heavy-duty steel pipe cutter is the most common?

45. When a pipe cutter wheel leaves a bur on the inside of a pipe up to 2 inches, what tool should you use to remove it?

46. Who's the only person who knows for certain whether an installed pipe has good threads?

47. When you're laying out and dimensioning piping arrangements, which offsets aren't difficult to calculate?

48. Name the two tools usually used to cut copper tubing.

49. What's the most common joint for copper pipe?

50. In a potable water supply system, what's the maximum allowable lead content for solders and fluxes?

51. Why does properly cleaning and heating a copper joint help make a sound solder joint?

52. When cutting plastic pipe, how can you be sure of getting a square end?

53. What's the common term for a cemented plastic pipe joint?

54. What should you use to remove impurities and gloss from the surfaces of plastic pipe and fittings before joining them?

55. In a plastic water system, why can you make only one joint at a time?

56. After you cement the last joint, how long should you wait before testing a plastic water system?

Private Water Supply Wells

According to the Water Quality Association, nearly one home in five isn't connected to a public water supply. Many suburban homes and practically all rural homes use water from private wells. The rapid spread of urban communities around large metropolitan areas has often outstripped the capacity of public water distribution systems. Many smaller towns and cities have lower population densities, which make the cost of public water systems prohibitive.

Domestic wells are regulated by either the local health department or the Department of Environmental Protection (DEP). These authorities set guidelines on how deep to dig a well, and how far it must be separated from sources of contamination. Then they inspect each well to make sure the owner has complied with the regulations. You need a permit for any drilled or driven well, regardless of whether the water is intended for domestic or irrigation purposes.

Well Water

Well water is ground water that has filtered down through the soil to the water table level. Rain water soaks into the ground and moves slowly down to the underground water reservoir. It's called meteoric water and makes up most of the estimated two million cubic miles of ground water in the upper crust of the earth.

The underground water table may be only a few feet or hundreds of feet below the surface. When you drive or drill a well, the bottom of the well casing must extend into the dry weather water table. Otherwise, during prolonged droughts, the water table may fall below the level of the well. See Figure 15-1.

Since it's been filtered through sand and rock, well water is usually cool, low in harmful bacteria and high in dissolved minerals. It's generally classified as *hard water* because of that high mineral content. Many people have to acquire a taste for this untreated water because it has a distinctly different taste from city water. It doesn't have the chlorine or other chemicals that city dwellers are accustomed to. Hard water also requires more soap to make a lather.

The minerals in untreated well water often stain the surface of plumbing fixtures a dark reddish brown. These stains are virtually impossible to remove. Minerals can also cause a buildup of scale in water heater storage tanks and water distribution pipes. This can lead to premature failure. You can greatly reduce scale buildup and improve the taste and smell of the water by installing a water softener on the building water service line. See Figure 15-2.

If you install a water softener on an existing system, install a full bypass. Otherwise the pressure drop through the softener will reduce the quantity of water available to operate the fixtures. On new installations, you don't need a full bypass. But you must consider the pressure drop through the softener when sizing the water pipe system. The local authority will have to approve your backwash disposal method.

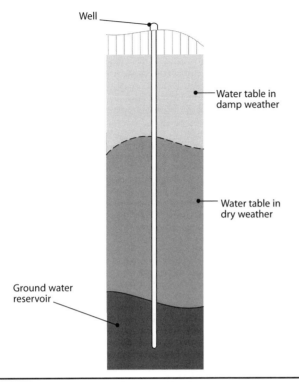

Figure 15-1
Well extended to dry weather water table

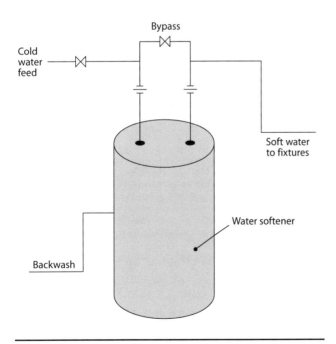

Figure 15-2
*Water softener installed in a building
water service line*

Well Installation

Generally, professional well drillers install wells, suction lines, pumps, and water pressure tanks. They're certified and licensed as a specialty contractor, permitted to do only this type of work. Although plumbing contractors are also licensed to do this work, they rarely do. If your contract includes a complete well supply system, you'll usually subcontract that part of the job to a professional well driller. But there are a couple of good reasons why you should be familiar with well systems. First, many of the people you deal with will assume you know something about wells. Second (and maybe more important), plumber's examinations include questions on well systems.

Local authorities will determine the depth of potable water supply wells, based on the depth of the water table. Even when the underground water table is within a few feet of the ground surface, the local authority may require a 30-foot minimum depth. The authority may also require a domestic well to be separated from a source of contamination (such as a septic tank, drainfield, soakage pit or discharge well) by 100 feet. Some codes may require more, some less.

Dug or Driven Wells

Wells that are dug or driven are classified as *shallow wells*. They're used where the water table is within 22 feet of the ground surface. The well must penetrate deep enough below the existing water table to assure a dependable supply of water, even in very dry seasons. In some parts of the country, plumbers install these shallow wells.

There are two types of driven wells, those with an open end casing and those with a casing equipped with a well point.

Open End Casings

Use an *open end casing* for domestic and irrigation wells in areas when the water table is close to the ground surface and located in a good rock formation. In some areas, rock or corrosion will clog the protective screening of a well point in a short time. When this happens, the well won't draw water. The open end well casing is preferable under these conditions.

With an open end casing, drive the pipe to the desired depth and then flush loose soil and rock from the driven section. You can flush with a smaller pipe with a sharpened point at one end for chipping into the rock. Attach a garden hose to the other end. Insert the smaller pipe into the open end casing and use water pressure to flush loose debris out of the casing. The pressure forms a water collecting pocket in the rock at the lower end of the well casing during the flushing.

When you're sure that a good well has been installed, connect a three-horsepower gasoline-driven centrifugal pump to the top of the well casing. Pump water out of the well until the water is free of rocks and sand.

Well Points

A well point is best in areas where the water table is in loose shale or sand. There are two types of well points. One, used in sand, has a screen or fine perforations. The other type, for gravel or loose rock formations, has larger openings. Attach the well point to the well casing and drive it to the desired depth. See Figure 15-3.

Drilled Wells

Drilled wells are classified as *deep wells* and may penetrate hundreds of feet into the earth. The water from deep wells may be better because there's less chance of contamination in a well that deep. And the water level in these wells is little affected by seasonal rainfall or dry years.

You're expected to know the installation methods and local code requirements for well drilling even though you may never have to do this work. Be aware of the following requirements:

■ Unless specifically approved by the local authority, you can't locate a well within any building or under the roof or projection of any building or structure.

■ The well casing must be continuous, of new pipe, and must terminate in a suitable aquifer. Well casing pipe 6 inches or less in diameter must be of galvanized steel.

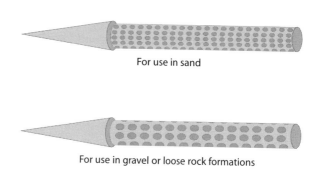

For use in sand

For use in gravel or loose rock formations

Figure 15-3
Well points

■ Pour a concrete collar a minimum of 4 inches thick and 36 inches wide around the top of the well casing. The concrete pad must be placed immediately below the tee and suction line. This collar must slope away from the casing to prevent surface water from carrying pollutants down the well casing to the water reservoir. See Figure 15-4.

■ Install a tee the same size as the well casing at the top of the well. This provides access to the casing for inspections, adding disinfecting agents, measuring well depth and testing the static water level. See Figure 15-4.

Suction Lines

The suction line must be large enough to provide the water volume and pressure required to operate the plumbing fixtures in the building. The suction line or water service pipe from the well to the pump can't be smaller than 1 inch. It must be installed with a pitch toward the well. If the well requires a suction line longer than 40 feet, increase the suction pipe to the next pipe size shown in Figure 15-5. This table is primarily for residences or small commercial buildings with flush tanks.

As an example, consider a residence with 30 fixture units and a suction line 65 feet long. The table shows a suction pipe size of 1^1/4 inches. Since the suction line is longer than 40 feet, use the next pipe size — 1^1/2 inches.

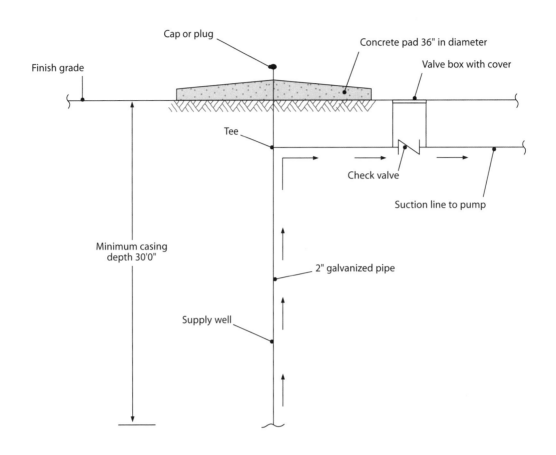

Figure 15-4
Supply well detail (potable water) isometrically illustrated

Fixture units	Supply required (GPH)	Diameter of suction pipe	Diameter pressure pipe	Diameter service pipe	Size of tank	Horsepower	Well size
23	720	1	$3/4$	$3/4$	42	$1/2$	$1\frac{1}{2}$
30	900	$1\frac{1}{4}$	1	1	82	$3/4$	2
40	1200	$1\frac{1}{2}$	1	1	120	$3/4$	2

Figure 15-5
Tank and pump size requirements

Install a soft seat check valve rated at 200 pounds water test as close as practical to the well. You can use either spring-loaded or flapper-type brass check valves up to a 2-inch diameter. The suction line must have a union or slip coupling installed just before the pump.

Materials and installation methods for the suction line are usually the same as for the water service piping described in Chapter 14.

Pressure Tanks

The two types of pressure tanks most commonly used today are the hydropneumatic tank and the diaphragm pressure tank. Let's take a closer look at each.

The Hydropneumatic Tank

The *hydropneumatic tank*, with its pressure switch, has been the dominant well water supply system since the 1920s. See Figure 15-6. While the tank itself has given good service, an unattended system isn't reliable. It's essential that air in hydropneumatic tanks be maintained automatically with air chargers and air volume controls. The air pressure and volume controls need regular service.

A hydropneumatic tank uses compressed air to control water pressure. Compressed air acts like a huge spring to drive water out of the tank to any needed location. When the pressure drops in the system, say to 20 pounds, the pump starts up. This forces more water into the tank and compresses the air in the tank to the set pressure of approximately 40 pounds.

Unfortunately, some of the air in the tank is gradually absorbed by the water passing through the tank. If the lost air isn't replaced, eventually there won't be enough air available to pressurize the tank. Water itself is nearly incompressible. If the tank loses air, it causes the pump to cycle on and off too often.

Manually draining the water from the tank will cause it to refill with air at atmospheric pressure. Then as the pump fills the tank with water, it traps a normal supply of air for proper operation of the water supply system. Figure 15-6 shows a typical pressure tank.

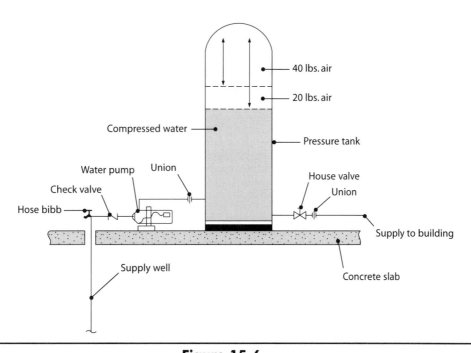

Figure 15-6
Typical pressure tank (hydropneumatic) installation

Startup cycle

Diaphragm is pressed against the bottom of the chamber.

Fill cycle

Water is pumped into the reservoir, which forces the diaphragm upward into the air chamber.

Hold cycle

Pump cutoff pressure is attained, diaphragm reaches its upper-most position. Reservoir is now filled to its rated capacity.

Delivery cycle

Pump remains shut off while air pressure in top chamber forces diaphragm downward, delivering water to system.

Corrosion-resistant base

Base rotates for easy alignment to pipe connection.

Figure 15-7
How a diaphragm tank works

Hydropneumatic tanks should be large enough to prevent excessive cycling of the pump. When sizing the tank, it's important to consider these points:

1) The draw-off capacity of the tank between the "cutout" and "cut-in" limits. (The tank size should provide a draw-off of 6 gallons of water while maintaining an operating range of 20 to 40 psi water pressure.)

2) The pump capacity must be large enough to quickly refill the tank.

3) Tanks are sized by the fixture units of a building. That's the best indicator of the amount of water used. For other sizes of tanks and specific requirements, see Figure 15-5. The minimum size hydropneumatic tank for a single-family residence is 42 gallons.

The Diaphragm Tank

The diaphragm tank, also known as a *breather*, has been around since the 1950s. The tank shell is constructed of lightweight drawn steel, with an epoxy finish and a bonded plastic lining for extra corrosion resistance. It has a replaceable air charge valve at the top of the tank for safe and easy operation, and a strong, flexible butyl diaphragm with a locking retainer ring for positive separation of air and water. The diaphragm tank is approved by most codes. See Figure 15-7.

Because a diaphragm tank is precharged with air, it will always occupy less space for similar amounts of pressurized water than a plain hydropneumatic galvanized steel tank. This kind of tank is suitable for most small systems. It gives consistent service and doesn't require air to be injected into the tank. The tank volume for a home may be as small as 14 gallons or as large as 119 gallons, depending on the water need.

Your sizing procedure is the same as for a hydropneumatic tank, based on the water-use habits of the occupants. It's usually wise to allow a little extra capacity because it adds very little to the tank cost. It also prevents too-frequent pump starts (excessive cycling).

The water piping from the pump to the pressure tank can't be smaller than the discharge outlet of the pump. Install a gate valve, with the handle removed, between the pump and the tank if the tank's capacity is more than 42 gallons.

Install a minimum $3/4$-inch gate valve on the discharge side of the tank. This serves as the house valve to control water in the building.

Be sure to set the tank level in an enclosed area, or on a concrete pad built for the pump and tank. Place them so there's reasonable access for repair or replacement. Interior water piping materials and installation methods are the same as described in Chapter 14 for water distribution pipes.

Review Questions (answers are on page 305)

1. Approximately what percentage of homes in the U.S. depend on private wells as their source of water?

2. Name one of the agencies that has the authority to approve and inspect domestic wells.

3. What document is required before you can drill or drive a well?

4. How far down must the bottom of the well casing extend when you drive or drill a well?

5. How is well water generally classified?

6. Why is well water considered hard water?

7. Why does well water have a distinctly different taste from "city water"?

8. What problems may the minerals in untreated well water create?

9. What can you do to improve the taste and smell of well water?

10. What tradesmen are most likely to install wells, suction lines, pumps and water pressure pumps?

11. Under what classification are professional well drillers certified and licensed?

12. Even though plumbing contractors very seldom install domestic wells, why should you learn about well systems?

13. What minimum depth will local authorities usually require for a potable water supply well?

14. How are wells that are dug or driven classified?

15. What are the two types of driven wells?

16. Where are open end well casings commonly used?

17. Once you're sure you've installed a good well, what must you do before it's ready for use?

18. Where would you use well casings equipped with a well point?

19. Where would you use a well point with a screen or with fine perforations?

20. Where would you use a well point with large openings?

21. How are wells that are drilled classified?

22. Why is water from deep wells more desirable than water from shallow wells?

23. Where may you not locate a well without specific approval from your local authority?

24. Where should the well casing in a drilled well terminate?

25. Why is a 36-inch-wide sloping concrete collar required around the top of the well casing?

26. What must you install in the top of the well to provide access for inspections, measure well depth, test the static water level, and allow disinfecting agents to be added?

27. What's the minimum size for a suction line from the well to the pump?

28. Where should you install the check valve on a well suction pipe?

29. Name one of the two types of check valves commonly used on suction lines.

30. What must you install on the well suction line just before the pump?

31. What two types of pressure tanks are commonly used today?

32. Which of the two commonly-used pressure tanks has been the dominant well water supply system since the 1920s?

33. What's the psi operating range for a hydropneumatic tank?

34. What happens when there's not enough air to pressurize the tank in a hydropneumatic system?

35. What's the minimum size hydropneumatic tank needed for a single-family residence?

36. How long has the diaphragm tank been approved for use in domestic well water systems?

37. What problem can you usually prevent by adding a little extra capacity to a diaphragm tank?

38. What's the purpose of installing a gate valve on the discharge side of the tank?

39. What's the minimum size gate valve required?

40. What are two important considerations in locating the tank and equipment?

Fire Protection

Every plumber needs to know the code requirements for installing fire protection systems — but they're not easy to learn. The sections about fire protection equipment are scattered throughout the code. In this chapter I'll bring together the various code requirements and explain what the code language means.

Standpipe Systems

Many larger buildings must have emergency fire hose connections to give firemen immediate access to an adequate supply of water in case of a fire. These hose connections are called *standpipes*. Standpipe systems can connect to a public water main if it provides the quantities and pressures required by the code. Generally, they use a public water supply if the street water main is at least 4 inches in diameter and located within 150 feet of the nearest point of the building. If the public water main can't provide adequate quantities and pressures, the system will have to use an alternate method, like a fire pressure pump (Figure 16-1) or a gravity tank.

Standpipe Requirements

A fire standpipe system is required in buildings over 50 feet high. That means that every seven-story building must have standpipes, assuming there are 8 feet between floors. And where standpipes are required, the system must be pressurized (*wet*) with a primary water supply constantly or automatically available at each hose outlet.

When a building with required standpipes is under construction, the standpipes must be constantly available for fire department use during construction. There must be a fire department connection on the outside of the building at the street level and at each floor up to the highest completed floor. Arrange the standpipe locations so they're protected from mechanical and fire damage. Throughout construction, the standpipes must be maintained with a water source.

Here's how to determine the number of standpipes and hose stations: All parts of all floors must be within 30 feet of the nozzle end of the hose when a hose up to 100 feet long is connected to the standpipe. That makes all areas accessible to a stream of water. Hoses over 100 feet long are prohibited.

Standpipes must be located within an enclosed stairway. If the stairway isn't enclosed, the standpipe must be within 10 feet of the floor landing of an open stairway. You can't locate the valve or hose connections behind a door.

Sometimes it takes more than one standpipe and hose station to provide protection on each floor. If there's only one stairway, locate the extra standpipes and hose stations in hallways or other accessible locations approved by the authority.

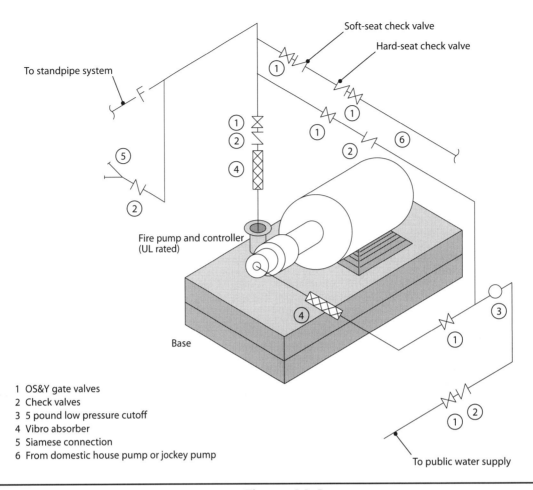

Figure 16-1
Fire pump detail

There are special requirements for buildings designed for theatrical, operatic or similar performances. Some codes require a 2¹/₂-inch diameter standpipe on each side of the stage, with a hose not over 75 feet long at each standpipe hose station. The *Uniform Fire Code*, however, requires automatic sprinklers.

Fire Flow On-Site Well Systems

Where adequate public water isn't available for a required standpipe system, an on-site well system may be the best solution. The system must meet these requirements:

■ The well's diameter and depth must be adequate, and straight enough to receive the pump.

■ The well must be cased to a proper depth and sealed to prevent loose or foreign material from entering.

■ All wells must be cased and sized for a flow of 500 gallons per minute (gpm).

■ All casings must have a wall thickness of at least ³/₈ inch.

■ The well must be properly developed so it's free of sand or loose gravel.

■ The well must be test-pumped at 150 percent of the capacity of the pump to be installed for two hours after it's free of sand.

■ The drawdown in the well can't exceed 4 feet during pumping at 150 percent of pump capacity. See Figure 16-2.

In figure:
Soft-seat check valve
Hard-seat check valve
To standpipe system
Fire pump and controller (UL rated)
Base
To public water supply

1 OS&Y gate valves
2 Check valves
3 5 pound low pressure cutoff
4 Vibro absorber
5 Siamese connection
6 From domestic house pump or jockey pump

- The flow rate must be 500 gpm at 20 psi where the pump discharges.

- The hookup must meet the requirements of the fire department with jurisdiction. Each fire department connection on the discharge side of the pump must have a shutoff valve with a diameter not smaller than the size of the discharge opening.

- The required fire department connections, rated by gpm, must be national standard threads. These connections need a flow of 500 gpm with one $4^1/_2$-inch connection and one $2^1/_2$-inch connection. See Figure 16-3.

- All on-site systems must be tested before final approval by the fire department with jurisdiction.

- All on-site systems must be located a minimum of 50 feet from the buildings being protected whenever physically possible. The fire department must have access to the on-site system on a roadway suitable for fire equipment. The fire department connections can't be more than 8 feet from the roadway.

- Direct connections between on-site fire protection systems and the potable water supply aren't permitted.

Even where a standpipe system isn't required, certain occupancies may need a standard fire well if a public water supply isn't available. See Figure 16-4. The well installation is nearly the same as the on-site well system, with two exceptions: No pump is required and the hose connection can be a single $4^1/_2$-inch American National Standard hose connection.

Fire Hydrant Locations

Fire hydrants are usually required in commercial, industrial and residential areas to protect existing and future buildings. Fire hydrants must be provided along required fire apparatus access roads and adjacent public streets.

The average spacing between fire hydrants may be as little as 200 feet but not more than 500 feet. The spacing is determined by the type of buildings, the square footage of those buildings, and the fire flow gpm required by the local fire authority.

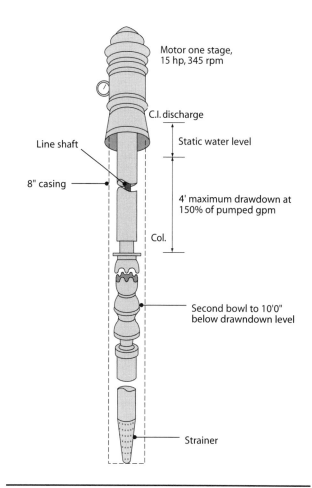

Figure 16-2
On-site fire flow well (sectional view)

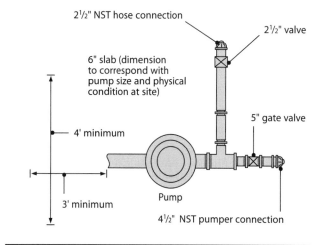

Figure 16-3
On-site fire flow well (plan view)

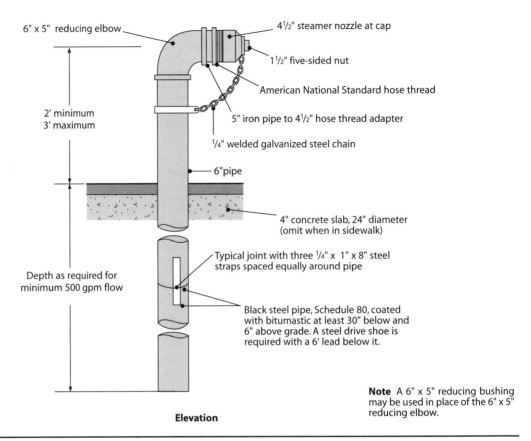

6" x 5" reducing elbow

4¹/₂" steamer nozzle at cap

1¹/₂" five-sided nut

American National Standard hose thread

2' minimum
3' maximum

5" iron pipe to 4¹/₂" hose thread adapter

¹/₄" welded galvanized steel chain

6"pipe

4" concrete slab, 24" diameter
(omit when in sidewalk)

Typical joint with three ¹/₄" x 1" x 8" steel
straps spaced equally around pipe

Depth as required for
minimum 500 gpm flow

Black steel pipe, Schedule 80, coated
with bitumastic at least 30" below and
6" above grade. A steel drive shoe is
required with a 6' lead below it.

Note A 6" x 5" reducing bushing
may be used in place of the 6" x 5"
reducing elbow.

Elevation

Figure 16-4
Standard fire well detail

Hydrants must have two 2¹/₂-inch connections with national standard threads compatible with the local fire department equipment. The fire department must approve the location of all fire hydrants. Figure 16-5 shows a typical hydrant installation.

Standpipe Materials and Installation Methods

All underground fire lines and fitting materials must be approved by the local authority. Aboveground fire lines in the exterior wall of a building (usually the standpipe) must be black steel pipe, hot-dipped galvanized steel pipe or copper pipe. These lines, the fittings and all connections must be able to withstand 175 psi.

Installation methods are the same for underground fire lines as for water service pipe of the same materials with this exception: Each change of direction in underground fire lines must have poured concrete thrust blocks resting on undisturbed soil (Figure 16-5). The thrust blocks prevent the pipe and fittings from blowing apart under the 200 psi pressure test required for all underground fire lines.

Buildings up to 100 feet high need standpipes at least 4 inches in diameter. Buildings higher than 100 feet need standpipes at least 6 inches in diameter. The standpipes can't be longer than 275 feet. If the building is over 275 feet high, it'll have to be zoned for multiple standpipes.

Each zone needs a separate pump if pumps are required. Where pumps at the same level supply two or more zones, supply each zone by separate lines no smaller than the risers they serve. When it's necessary to pump the water supply from a lower zone to a higher zone, the risers can't be smaller than the largest riser they serve.

If the system can't supply all of the higher zones with a residual pressure of 65 psi, you have to provide auxiliary means acceptable to the authorities.

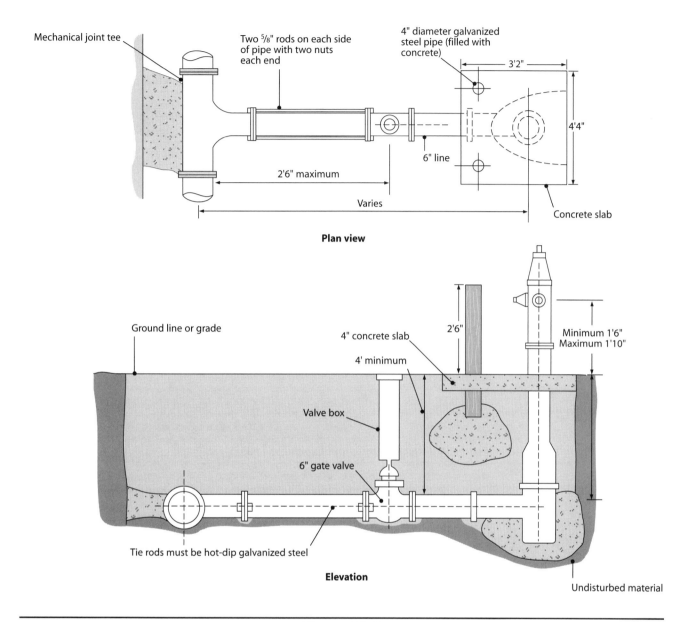

Figure 16-5
Underground fire line thrust blocks

Standpipes in buildings 50 feet or higher must extend above the roof a minimum of 30 inches. This extension must be with the same diameter pipe as the rest of the standpipe. Install a UL-approved duplex roof manifold with 2¹/₂-inch fire department connections on each standpipe.

Standpipes in stairway enclosures must have 2¹/₂-inch valves adapted for fire department hose connections with 2¹/₂-inch national standard threads. There must be accessible outlets on each floor at the stair enclosure, including the basement if there is one. Don't place hose outlets within 10 feet of the standpipe or hose station, and not lower than 5 feet or higher than 6 feet above the finished floor.

Each hose station must be within 10 feet of the standpipe. The connection between the hose station and the standpipe is usually 2¹/₂-inch pipe. Never locate hose stations in a stair enclosure. See Figure 16-6 for a typical layout for standpipes and fire hose cabinets.

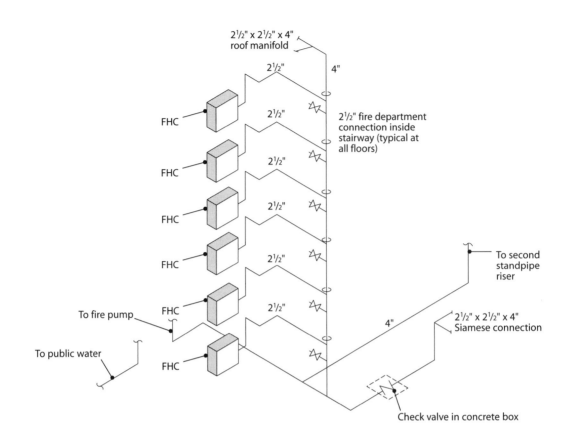

2½" x 2½" x 4"
roof manifold

2½"   4"

FHC   2½"

2½" fire department
connection inside
stairway (typical at
all floors)

FHC   2½"

FHC   2½"

FHC   2½"

To second
standpipe
riser

To fire pump   FHC   2½"

2½" x 2½" x 4"
Siamese connection

4"

To public water   FHC

Check valve in concrete box

Figure 16-6
Standpipe layout and location of fire hose cabinets

Install an approved wall-mounted hose reel, cabinet or rack for each hose station. The station must be located so it's accessible at all times. Each hose must be able to withstand 100 psi working pressure and have an adjustable nozzle that can turn the water on and off. The nozzle should also adjust from a fog spray to a strong stream of water. Where the pressure may exceed 100 psi, install pressure regulatory devices to control the pressure to the hose.

When more than one standpipe is required to serve a building, each standpipe must be interconnected at its base. Provide a Siamese (duplex) fire department connection for each of the first two required standpipe risers. If more than two standpipes are required, the Siamese connections must be remotely located. Each Siamese connection must have the same diameter as the largest standpipe. For example, a building that needs a 4-inch standpipe must have a 4-inch Siamese connection.

Install an approved UL-listed check valve between the Siamese connection and the fire standpipe system. Each Siamese connection must have a 2½-inch national standard thread fire department hose connection.

Each Siamese connection must be installed on the street side of the building at least 1 foot but no more than 3 feet above grade. The Siamese connection and its related piping can't project over public property (such as sidewalks) more than 2 inches. Attach a permanent sign identifying the standpipe to the exterior of the building near the fire department Siamese connection. It should say *standpipe* in letters 1 inch high.

The water supply for standpipes must be adequate to maintain 65 psi residual pressure at the highest outlet, with a required flow of 500 gpm. If more than one standpipe serves a building, the required flow must be 750 gpm. In most cases, you need fire pumps to supply this additional pressure and water flow. Refer back to Figure 16-1.

The fire pumps and controllers required to supply the required flow must be UL listed. The pump controllers may use limited service motors of 30 hp or less. Fire pumps must operate automatically with compatible controls. Pumps must either be supplied with a separate electric service or connected to a standby generator through a separate automatic transfer switch.

Refer back to Figure 16-1 to understand these requirements for fire pumps:

▮ There must be a 15 psi minimum pressure on the roof in a standpipe system. There are two ways to maintain this pressure. Use a jockey pump actuated by a pressure switch (a secondary pump to boost the pressure) or connect to a suitable domestic system through two 170 psi check valves. One of these valves must have a soft seat and one must have a hard seat.

▮ Provide a full-size bypass with an approved gate and check valve.

▮ The fire pump drives must be flexibly coupled.

Combined and Automatic Sprinkler Systems

In recent years, fire protection systems on new construction may use a *combined system* or an *approved automatic sprinkler system*. Plumbing contractors with local or state certification can install a fire protection system consisting of standpipes and fire hoses. However, many states now require that only plumbing contractors certified by the state fire marshall can install fire protection systems with automatic sprinklers.

Let's review code requirements for both combined and automatic sprinkler systems.

Combined Systems

In a combined system, water-supplied risers serve both the 2¹/₂-inch outlets for the fire department and the automatic sprinklers. The minimum size for risers for a combined system is 6 inches in diameter.

The water supply you need for a combined system depends largely on the occupancy class of the building. There are three classes:

▮ *Class I:* Combustibility is considered moderate. Class I occupancies include bakeries, distilleries, canneries and laundries, to name a few. The minimum water supply must provide 500 gpm for a period of at least 30 minutes. Where there are two or more standpipes, the minimum water supply to the first standpipe must be 500 gpm, and 250 gpm for each additional standpipe for a period of at least 30 minutes. The total supply can't exceed 2,500 gpm for a minimum period of 30 minutes. It must maintain a residual 65 psi pressure at the most remote outlet.

▮ *Class II:* Combustibility is considered low. Class II occupancies include churches, hospitals museums and the like. The minimum water supply has to provide 100 gpm for a period of at least 30 minutes. The water supply must maintain a residual pressure of 65 psi at the highest outlet with 100 gpm of water flowing.

▮ *Class III:* Combustibility is considered high. Feed mills, paper process plants and tire manufacturing plants are a few examples of high combustibility. The minimum water supply requirements for Class III occupancies are the same as for Class I occupancies.

Automatic Sprinkler Systems

Where standpipe risers are required, they can supply the automatic sprinkler system. If standpipe risers aren't required, sprinkler-fed risers sized by a professional engineer can supply the automatic sprinklers. Locate 2¹/₂-inch fire department connections on each floor, similar to those in the standpipe and hose system. See Figure 16-7.

Here are the other requirements for automatic sprinkler systems:

▮ Where more than one standpipe riser is required, risers must be looped at the lowest floor. Make the loop lines the same size as the risers. For example, if the riser pipe is 4 inches, the loop piping must also be 4 inches.

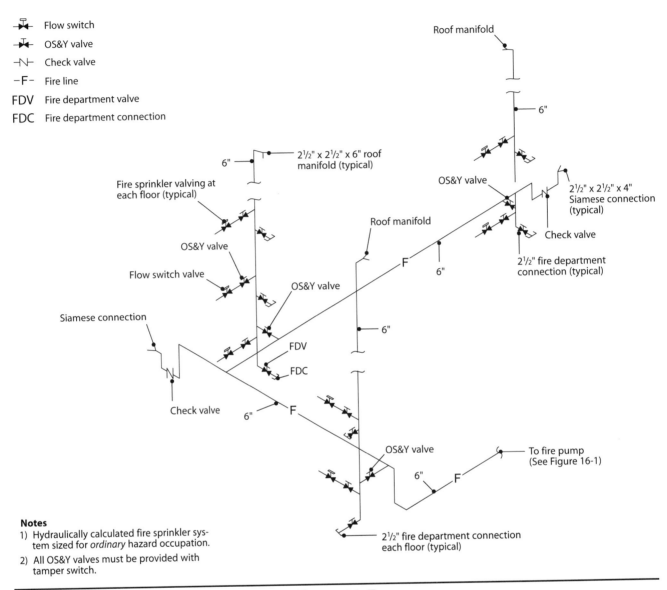

Flow switch
OS&Y valve
Check valve
–F– Fire line
FDV Fire department valve
FDC Fire department connection

Roof manifold

6"

OS&Y valve

2½" x 2½" x 4"
Siamese connection
(typical)

Check valve

2½" fire department
connection (typical)

2½" x 2½" x 6" roof
manifold (typical)

6"

Fire sprinkler valving at
each floor (typical)

Roof manifold

OS&Y valve

6"

Flow switch valve

OS&Y valve

6"

Siamese connection

FDV

FDC

Check valve

6"

F

F

OS&Y valve

To fire pump
(See Figure 16-1)

6"

F

2½" fire department connection
each floor (typical)

Notes
1) Hydraulically calculated fire sprinkler system sized for *ordinary* hazard occupation.
2) All OS&Y valves must be provided with tamper switch.

Figure 16-7
Automatic sprinkler-feed risers with sprinkler branch lines

▪ Take branch lines off the riser at the floors they serve.

▪ Where more than one riser is required, take off branch lines on alternate floors from different risers.

▪ When there's more than one branch line on a floor, take each branch off from a separate riser, where practicable.

▪ Branch lines can't supply areas exceeding 4,000 square feet in apartment or hotel occupancies.

▪ Locate a post-indicator valve and check valve on the water supply line located outside the building.

▪ Place an outside screw and yoke (OS&Y) valve at the bottom of each riser.

▪ When a loop system is used, install other OS&Y valves to isolate each riser from the loop.

▪ Each branch line has to have an OS&Y valve with a tamper indicator and flow switch. These should be monitored at the central control station. See Figure 16-8.

■ Install water flow devices to actuate a local alarm on each floor, when water flow is detected on that floor.

■ You can use copper piping, but copper connections must be soldered using solder with a thermal strength of not less than 95 percent tin and 5 percent antimony.

■ Water supply lines aren't required to be graded.

■ Where two or more fire pumps are used, each pump must operate independently.

■ Fire pumps usually must have electric motors. Engine-driven fire pumps are acceptable only if approved in advance by local authorities.

■ You can omit 1¹/₂-inch fire hose and cabinets, provided each standpipe on each floor is equipped with a 2¹/₂-inch hose valve, a 2¹/₂-inch by 1¹/₂-inch reducer and an attachment chain for fire department use. See Figure 16-8.

■ Where a secondary supply of water is required, the water has to be available automatically when the principal supply fails. Look back to the on-site fire flow wells in Figures 16-2 and 16-3.

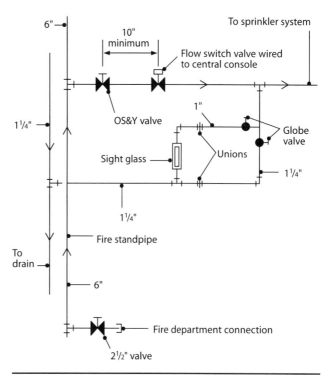

Figure 16-8
Typical fire sprinkler valving at each floor

Review Questions (answers are on page 306)

1. What's the minimum size public water main that you can use to supply a fire standpipe system?

2. When a public water main can't provide the required quantity and pressure for a fire standpipe system, what alternative methods can you use?

3. At what building height is a fire standpipe system required?

4. What special requirement ensures that there's a constant water supply available for a fire standpipe system?

5. Where must you locate fire department connections for buildings under construction that require standpipes?

6. What two types of possible damage to standpipes must you consider when determining fire standpipe locations?

7. What's the maximum distance the nozzle end of a 100-foot fire hose can be from any part of a building floor?

8. Where must fire standpipes be located if the stairways are enclosed?

9. Where must fire standpipes be located if the stairways are *not* enclosed?

10. Where may additional fire standpipes be located when stairways are not available?

11. What special provisions are required for fire protection in buildings designed for theatrical performances?

12. What alternate water system will the code accept if there's no adequate public water supply and a fire standpipe system is required?

13. What's the maximum drawdown for an on-site fire well system when pumping at 150 percent pump capacity?

14. How many gallons per minute at 20 psi must an on-site fire well system be sized for?

15. What is the requirement for fire department hose connections?

16. What must you do before you can receive final approval from the fire department having jurisdiction over an on-site well system?

17. What water supply system may *not* be directly connected with an on-site fire well system?

18. What's one of the major differences between an on-site standard fire well and a regular on-site well system?

19. What water supply device is usually required for fire protection in commercial, industrial and residential areas?

20. Who must approve the location of all fire hydrants?

21. Who must approve underground fire line and fitting materials?

22. What pressure must aboveground fire lines and fittings be able to withstand?

23. What must you provide at each change of direction in underground fire lines?

24. How much pressure must underground fire lines be proven able to withstand?

25. What's the minimum size requirement for fire standpipes in buildings up to 100 feet high?

26. What's the minimum size requirement for fire standpipes in buildings over 100 feet high?

27. What's the maximum length allowed for a building standpipe?

28. How far above the roof are fire standpipes required to extend on buildings that are 50 feet or higher?

29. What fire department connection are you required to install on each standpipe extension above the roof?

30. What size valve is required for the fire department hose connections on standpipes located in stairway enclosures?

31. What's the maximum distance allowed from the standpipe or hose station to a hose outlet?

32. What pipe size is generally used to connect a hose station to the fire standpipe?

33. Where must a fire hose cabinet be located?

34. How much working pressure must each fire hose be able to withstand?

35. What must you install if the pressure exceeds 100 psi at the fire hose outlet?

36. What certification must a fire pump have if it's required to supply a 500 gpm flow rate?

37. What type of electric service is required for fire pumps with a 500 gpm flow rate?

38. What type of equipment can you use to maintain the 15 psi minimum pressure required on the roof in a fire standpipe system?

39. Who is qualified to install fire protection systems consisting of standpipes and fire hoses?

40. In some states, what type of fire protection system can only be installed by plumbing contractors who hold state fire marshall certification?

41. What's the minimum size for risers in a combined fire system?

42. How is the water supply need for a combined fire system determined?

43. What's the combustibility level of a Class I building?

44. What's the combustibility level of a Class II building?

45. What's the combustibility level of a Class III building?

46. What must you do if more than one fire standpipe riser is required in an installation?

47. Where must you take off each fire standpipe riser branch line?

48. What two fittings should you install on the water supply line for automatic sprinkler systems?

49. How must each pump operate if you use two or more fire pumps in an automatic sprinkler installation?

50. Under what conditions can you use an engine-driven fire pump?

Swimming Pools and Spas

Swimming pools and spas are big business for some plumbers. There are hundreds of thousands of swimming pools and spas in the United States, and most, by far, are privately owned.

Specialty swimming pool contractors with a certificate of competency usually install the necessary piping and equipment in new pools and spas. These specialists also repair and maintain existing pools and spas.

You should know that your plumbing contractor's license also qualifies you to perform all work usually done by a swimming pool contractor. Some plumbing contractors act as subcontractors on pool work. In most parts of the country, anyone who wants to be licensed as a plumber must be familiar with pools and spas to pass the journeyman's and master's examination.

Swimming Pools

According to the code, a swimming pool is any structure that's suitable for swimming or recreational bathing that's over 24 inches deep. It can be permanent or nonpermanent, in the ground or aboveground. A private pool is located at a single-family residence, available only to the family and their guests. A public pool is used collectively by a number of persons for swimming or bathing, whether a fee is charged or not. There are more definitions for pools and spas in the glossary at the back of the book.

The most common type of mechanical system plumbed into swimming pools today is a recirculating system. It's equipped with a pump to recirculate water from the pool through a filter system. The recirculating piping (also known as *return piping* or *pool inlet piping*) connects to the discharge side of the pump. It returns water to the pool after filtering.

Some private pools have an automatic feeding device that adds chlorine or fluorine. But many private pool owners add their own chemicals, or use a professional pool company to maintain the quality of pool water.

Even when it's heavily used, a recirculating swimming pool uses a minimum of water. The owner just adds fresh water as needed when it's lost by evaporation, splashing or backwashing. Good filtration equipment assures the water is clean, free of organic matter, and safe from harmful bacteria.

Water Supply

Water for the pool can come from the public water system. Many private pool owners use a garden hose attached to a hose bibb to fill the pool. If so, there should be a vacuum breaker on the hose bibb to prevent cross-connection. Public pools and some private pools have a direct connection to the public water system. If there's a direct water supply, the fill spout must have an air gap above the overflow rim of the pool. See Figure 17-1.

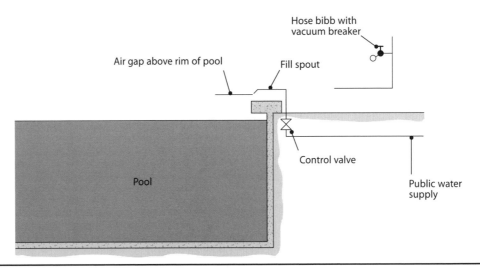

Figure 17-1
Water supply layout

If a well supplies the water for a swimming pool, the water must be clean and meet the bacterial requirements for a domestic water supply. If it's not reasonably free of objectionable minerals, the filtration system must remove them. The color of well water is also a concern. The iron content can't exceed 0.3 parts per million *before* filtration. If the raw well water doesn't meet these specifications, it's got to be treated before entering the pool.

Waste Water Disposal

Swimming pools must have some means of disposing of backwash water, and of being emptied. Pools equipped with pressure diatomite filters need piping to carry backwash waste to a settling basin before final disposal. Any of the following waste water disposal methods is allowed, as long as it's approved by the authority with jurisdiction:

▌ It can empty into a public or privately-owned sewage system.

▌ It can flow into a disposal well.

▌ It can drain into an open waterway, bay or ocean, where this is permitted by the Director of Environmental Protection (DEP), or by the local health department, depending upon which agency has jurisdiction.

▌ It can empty into an adequately-sized drainfield, soakage pit or drainage trench.

▌ It can flow through a sprinkler system used for irrigation purposes. The waste must be confined to the property from which it originates. It can't flow on or across any adjoining property, public or private. In this instance, only the pool water may empty through a sprinkler system. The *backwash* water from the pool, containing hair and other bits of debris, *must not* be discharged through a sprinkler system or it will clog the system.

▌ Pool waste and backwash water may be puddled on private property provided the disposal area is big enough and properly graded to retain the waste water within the confines of the property. The pool of standing water can't remain for more than one hour after discharge. The disposal area must be a minimum of 50 feet from any supply well. Figure 17-2 gives percolation rates for soakage pits.

Minimum Equipment for Swimming Pools

Figures 17-3 and 17-4 show a piping diagram and a plan view of a typical swimming pool. They show the equipment that's required for every pool.

Pool capacity (gal)	Diameter S & G filter (in)	Soil percolation rates (minutes/in)											
		1		2		3		4		5		6	
		SF	Gal	SF	Gal	SF	Gal	SF	Gal	SF	Gal	SF	Gal
17,000	24	53.5	2,000	96	3,590	130	4,860	158	5,910	182	6,800	202	7,560
17,000 to 26,000	30	83	3,100	149	5,560	200	7,550	247	9,240	280	10,500	315	11,780
26,000 to 38,000	36	120	4,490	215	8,050	292	10,910	358	13,400	408	15,290	452	16,900
38,000 to 52,000	42	163	6,100	293	10,970	400	14,980	485	18,150	555	20,800	618	23,100

Note Effective depth of soakage pit is 5'0". SF refers to the area of the bottom of the soakage pit.

Figure 17-2
Minimum area and volume of soakage pits for swimming pools

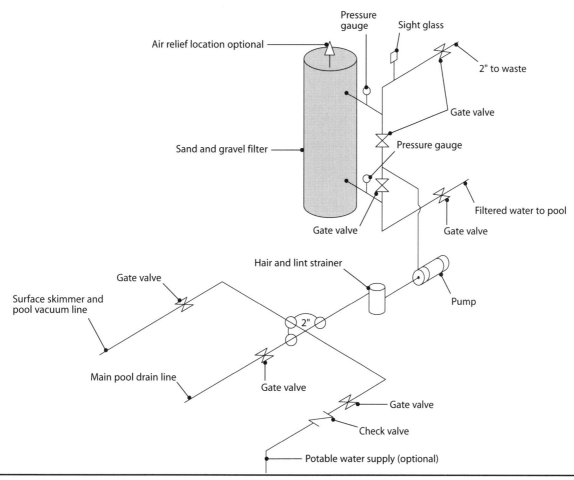

Figure 17-3
Pool equipment piping diagram

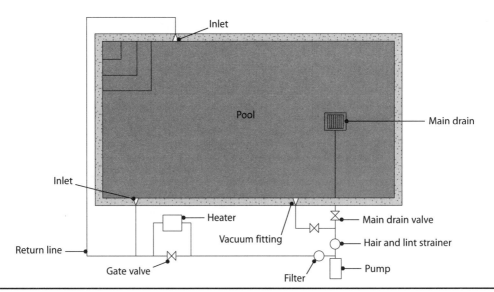

Figure 17-4
Plan view of swimming pool

Size and space the recirculation inlet or inlets to produce uniform circulation of the incoming water throughout the pool. One inlet is required for each 350 square feet of pool water surface or fraction thereof, with a minimum of two inlets located at least 10 feet apart. Size the entire recirculation system so velocities don't exceed 10 feet per second (fps) at the design flow.

The suction line can't exceed 5 fps at the design flow during the filtration period. When the main drain is used for a return, it's considered an inlet, but you have to size it as a suction line. Install a hair and lint strainer in the suction line ahead of the pump. The strainer must have an easily-removable screen with a free area 5 times the cross sectional area of the suction pipe.

Provide a vacuum fitting that's $1\frac{1}{2}$ inches in diameter on all pools. Install it a maximum of 10 inches below the water line and in an accessible location. Some skimmers have the vacuum line built in. If it doesn't, connect the vacuum fitting to the piping that's connected to the pump's suction side.

Install a valve on the main drain (outlet, or suction) line, accessibly located outside the walls of the pool (see the gate valve in Figure 17-4). Residential pools must have a minimum turnover rate of once every 12 hours of operation.

Filters

Every pool needs filtration equipment. Most pools use a sand filter or a diatomite filter. They may use another type if tests show it's as efficient as a sand and gravel filter. Here are the requirements for *sand filters*:

- Pressure sand filters must have a filtration rate not over 5 gpm per square foot of filter area and a backwash minimum rate of 12 gpm per square foot of filter area.

- Sand filters must hold a minimum of 19 inches of suitable grades of screened, sharp silica sand properly supported on a graded silica gravel bed.

- They need sufficient free-board above the surface of the sand and below the overflow troughs or pipes to permit 50 percent expansion of the sand during the backwash cycle. There should be no loss of sand.

- The inflow and effluent lines must have pressure gauges.

- The backwash line must have a sight glass installed so backwash water can be visibly checked for clarity.

- Tanks larger than 24 inches must have an access hole measuring a minimum of 11 inches by 15 inches.

Diatomite filters must meet these requirements:

- They may be either the vacuum or pressure type.

- They must have a filtration rate of no more than 2 gpm per square foot of effective filter area.

- The installer must be able to introduce a filter aid into the filter tank to evenly precoat the filter septum or element before it's placed in operation.

- The filter piping must be designed and installed so the filter aid recirculates or discharges through the waste pipe during the precoating operation, and doesn't return to the pool waters.

- There must be a way to remove the caked diatomite, either by backwash or disassembly.

- The filter elements must be easy to remove.

- Install pressure or vacuum gauges on these filters to measure the differential across the filter. That shows when the filter needs to be cleaned.

- There has to be an air relief device at the highest point on each pressure filter tank.

- If discharge from the backwash line isn't visible, there must be a sight glass in the line to check the clarity of backwash water.

Surface Skimming for Pools

Provide at least one skimming device for each 1,000 square feet of pool surface or fraction thereof. Skimmers must be built into the wall of the pool and meet the following requirements:

- The rate of flow through each skimmer should be at least 25 gpm.

- The skimmer weirs must adjust automatically to variations in water level over a range of 3 inches. Skimmers must be at least 5 inches wide.

- Place a basket with a minimum of 75 cubic inches where it can be removed easily for cleaning.

Pumps must be able to filter and backwash the pool water at the pressure and rate for the filter and piping system you're using. All valves, pumps, filters and other installed equipment must be readily accessible for operation, maintenance and inspection.

Pool Piping

Figure 17-5 shows the types of pipe you can use in pool installations, and where you can use them.

Thermoplastic pipe must be continuously marked on opposite sides with the size, type, schedule, and the U.S. Commercial Standard and National Sanitation Foundation seal of approval. All fittings for ABS or PVC plastic pipe must be Schedule 40. Polyethylene pipe fittings must be the insert type connected with stainless steel clamps. All fittings used in gutter lines (usually in public pools) must be the drainage type.

If you're using thermoplastic pipe and fittings, the trenches and backfill must be free of rock. In general, you install and support pool piping the same as water piping of the same material. (See Chapter 14.)

Don't install short radius 90-degree piping elbow fittings on pool or spa suction piping below grade. And suction piping for pools and spas must be a minimum of 2 inches in diameter.

Dielectric fittings are required if you're installing dissimilar metals in pool and filter piping.

The entire pool pressure piping system, including the main drain, must be water-tested at 40 psi and proved tight before the installation is concealed.

Type of pipe	Permitted use
Copper, Type K or L	All lines
Galvanized steel, standard weight	All lines
Wrought iron, standard weight	All lines
Brass pipe or tubing	All lines
Cast iron, service weight	Gutter lines only
Stainless steel, AISI, Type 300 series	All lines
Monel	All lines
Polyethylene pipe	Pressure lines only
ABS and PVC Schedule 40	All lines

Figure 17-5
Swimming pool materials and permitted installations

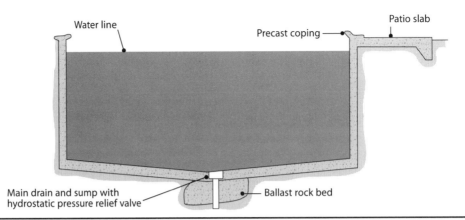

Figure 17-6
Swimming pool main drain

The Main Drain

Every pool has a main drain to empty the pool. It must be at the lowest point of the pool so the pool will drain dry for cleaning, painting and general repairs. See Figure 17-6.

Unfortunately, the suction of water through the main drain in pools and spas seems to hold some kind of fascination for children. There have been numerous injuries and drownings when children sat on the main drain and were held under water by its strong suction. That's why most codes no longer accept the older, flat main drain previously used in pools and spas. All new pools and spas must meet the following requirements:

▮ They must have an antivortex cover with at least 6 square inches of open unobstructed area, securely fastened over the main outlet so it can't be removed without using tools, or

▮ They must have an open grate drain with a minimum open unobstructed area of 85 square inches, and

▮ Some states now require two main drains in lieu of one.

Pool Heaters

Gas-fired swimming pool heaters and swimming pool boilers must comply with AGA and ASME standards. Oil-burning equipment must be approved by Underwriters Laboratory or another nationally-recognized testing agency.

All pool-heating equipment must be at least 70 percent thermally efficient across the unit. Water heaters and boilers need either a thermostatic or high-temperature control with a maximum temperature differential of 15 degrees F (or some other acceptable overheat protection device). The temperature of the heated water can't exceed 105 degrees F.

Install pool water heating equipment the same way you install domestic water heaters. If you install the heater in a pit, provide drainage for its protection.

Spas

The growing popularity of spas has forced local authorities to compile and adopt codes to provide adequate protection for the users. The code defines a residential spa as one that's permanent or nonpermanent, and used by not more than two families and their guests. Spas are also defined by their capacity in gallons — a maximum of 3,250 gallons of water.

Spas and pools have many similar requirements, but here we'll focus on the ones that apply only to spas.

Spas more than $3^{1}/_{2}$ feet deep must have adequate and suitable handholds around 60 percent of their perimeter area. Some approved handholds include suitable slip-resistant coping, ledges, flanges or decks located along the immediate top edge of the spa.

Ladders, steps, or seat ledges are also acceptable. They can't have any protrusions, extensions, means of entanglement or other obstructions that can entrap or injure bathers.

Spa materials and installation methods are the same as for swimming pools and domestic water systems of the same materials. (See Chapter 14.)

Spa Filter Requirements

Spa filters are similar to swimming pool filters. They're almost always designed by the manufacturer to meet these code requirements:

- Filters must maintain spa water under anticipated operating conditions, and the filtration surfaces must be easily inspected and serviced to restore them to the design capacity.

- Any filter or separation tank with an automatic internal air release as its principal means of air release must have, as part of its design, a way to provide a slow and safe release of pressure.

- Filters must meet the safety performance standards of the National Sanitation Foundation, or other approved testing agency.

Pumps for Spas

Every spa needs a pump to circulate the spa water. The pump must meet these conditions for filtering and cleaning of the water:

- There must be a hair and lint strainer (equipped with an easily removable screen) to filter out

such things as solids, debris, hair, and lint, installed before the circulation pump.

- The design and construction of the pump and component parts must provide safe operation.

- Pumps must be mounted on a solid formed base, elevating the bottom of the motor at least 4 inches above the surrounding area.

Air Induction Systems

To prevent electrical shock hazards, be sure your air induction system does not allow any water backup. Also, the placement of air intake sources must minimize the possible introduction of deck water, dirt or other pollutants into the spa.

Surface Skimming Devices

Spas have the same requirements as pools when it comes to skimming devices, with these exceptions:

1) Spas must have at least one skimming device.

2) Spa skimmers must have a vacuum break.

Spa Heaters

The heater must include a thermostatic control that allows a maximum temperature of 105 degrees F. There must be a consumer use label posted on, or near, the spa, which includes the maximum spa water temperature limit.

Review Questions (answers are on page 309)

1. What type of certification does a contractor need in order to specialize in the installation of piping and equipment for swimming pools and spas?

2. Why should you be knowledgeable about swimming pools and spas if specialists do most of the work?

3. How does the code define a swimming pool?

4. How does code define a private swimming pool?

5. How does code define a public swimming pool?

6. What is the most common mechanical system plumbed into swimming pools today?

7. What basic equipment is required for a recirculating-type swimming pool?

8. What other terms may be used to identify recirculating piping?

9. What chemicals do you use to maintain the quality of swimming pool water?

10. Name two of the three ways that water is lost from a swimming pool.

11. What is the purpose of having a good swimming pool filtration system?

12. What must a homeowner install on the hose bibb to prevent cross-connection if he uses a garden hose to fill his swimming pool or spa?

13. What is required to prevent cross-connection if a swimming pool or spa has a direct connection to the public water supply?

14. What are two of the several approved methods for disposing of swimming pool water?

15. Where should the main drain for a swimming pool be located?

16. What is the minimum number of inlets required for a swimming pool?

17. What other function is the main swimming pool drain considered to have when it's used for a return?

18. How must you size the main swimming pool drain when it's used for a return?

19. What's the minimum diameter size for a swimming pool vacuum fitting?

20. To what part of a swimming pool pump do you connect a vacuum fitting?

21. What's the filtration rate for pressure sand filters?

22. What must you provide on the inflow and effluent lines for a swimming pool sand filter?

23. What must you provide on a swimming pool backwash line?

24. What two types of diatomite filters are used for filtering swimming pool water?

25. What's the filtration rate for a diatomite swimming pool filter?

26. What provisions must be made for removing caked diatomite from the swimming pool filter?

27. What must the design and installation of a diatomite filter permit?

28. What must you install on each swimming pool pressure filter tank at its high point?

29. How many square feet of swimming pool surface can one surface skimming device accommodate?

30. What's the required rate of flow through a swimming pool skimming device?

31. Where can you use copper, Type K or L pipe in a swimming pool installation?

32. Where can you use cast iron, service weight pipe in a swimming pool installation?

33. What weight fittings can you use with ABS or PVC plastic pipe in swimming pool installations?

34. What type fittings must you use in a swimming pool gutter line?

35. Where may you not install short radius 90-degree elbow fittings on swimming pool or spa piping?

36. What's the minimum size suction piping required for swimming pools and spas?

37. What kinds of fittings are required in swimming pool filter piping if you're installing dissimilar metals?

38. What pressure is required to water-test a swimming pool pressure piping system, including the main drain?

39. Why are the older, flat-type main drains for swimming pools no longer considered acceptable by most codes today?

40. What are the two code-accepted main drain covers used for swimming pools and spas today?

41. With what standards must gas-fired swimming pool heaters and swimming pool boilers comply?

42. Name a national testing agency that can approve oil-burning equipment for swimming pools.

43. What's the maximum temperature acceptable for heated swimming pool water?

44. How does the code define a residential spa?

45. What's the maximum gallon capacity acceptable for a spa to be considered a spa and not a pool?

46. What's the maximum temperature of a spa heater?

Harnessing Solar Energy

The increasing cost of fossil fuels and the uncertainty of the foreign oil supply have spurred new interest in the search for alternate sources of energy. The investment in research to harness the sun for heating and cooling seems to be paying off.

It could, in fact, result in solar energy becoming the standard for most homes. Even now, solar energy is a competitive source for heating water for all domestic purposes, including swimming pools. If you've ever paid the gas or electric bill for heating a swimming pool, you'll understand why more and more people are turning to solar energy for their pools. As solar energy systems become more practical and more widespread, you'll need to know how the professional plumber fits into the picture.

Plumbing codes are constantly being modified to protect consumers and contractors against untested or inferior products. That's why some local codes are adopting the new *Uniform Solar Energy Code* (written in 1997) to cover solar energy systems. You can get a copy by contacting:

> *IAPMO Publication Order Desk*
> *20001 Walnut Drive South*
> *Walnut, CA 91789-2825*
>
> *Phone (909) 595-8449*
> *Fax (909) 594-3690*

To install, repair or alter any solar energy system, you must first get a plumbing permit. When you're planning a solar energy system, the building department will probably require two sets of plans, one to file with the building department and one to be placed at the job site for inspection purposes. They usually require that the plans be done by a registered professional engineer, and show structural calculations, mounting frames and anchorage details. The plumbing drawings should show the entire solar system and specify only approved plumbing items and UL-approved electrical components.

When you've completed your installation, everything has to be tested and proved tight under a water-, fluid- or air-pressure test. The system must be able to withstand 125 psi for at least 15 minutes, without leaking. You won't want to call for an inspection until you're sure the system holds.

Domestic Hot Water Heating

In Florida and other sunny areas, solar energy has been used to heat water since the beginning of the 20th century. Since the systems are tested and proven, the cost of adding them in either existing buildings or new construction is relatively modest. Using a solar system, a family of four could save more than $200 a year on their energy costs.

How the Solar Hot Water System Works

Have you noticed how water trapped in an ordinary garden hose gets extremely hot on a sunny day? That's how *solar energy* converts to *heat energy*. A solar system has three major parts: solar collector, circulation system and storage tank. If you're working with a pumped system, you'll have a fourth component, a *control center*. It consists of a circulating pump operated by an automatic controller with thermostat sensors, complete with drain, cutoff switch, check and pressure-temperature valves.

Some systems may have a fifth component — *a backup heat source* to ensure availability of hot water even during periods of peak water use and low solar radiation. This requires a built-in backup heater element in the solar storage tank.

The heat collector converts radiant solar energy into useful heat energy (Btu) as it heats water passing through copper pipes or other approved materials. The circulation system then sends the heated water into the storage tank. It's essential that this tank be properly sized and highly insulated. The water flows either through thermosiphon action or with the help of a pump. Both types are discussed and illustrated later in this chapter.

Let's take a closer look at each component of a solar system, beginning with the solar collector.

The Solar Collector

The heating element (source of energy) for a solar water heater is the solar collector. See Figures 18-1 and 18-2. The flat plate solar heat collector is the most practical and least expensive for residential use, producing temperatures up to 200 degrees F. Its heat deck consists of a metal plate, and tubing. The plate absorbs heat and transfers it to the liquid in the tubing.

Heat deck materials are usually copper, steel, and in some cases, aluminum. Thermally, these materials are the same. But *both the tubing and the collector plate should be of the same metal* so they expand and contract at the same rates. And take note of this: Codes generally won't permit potable water to flow through aluminum tubing, except in a closed system.

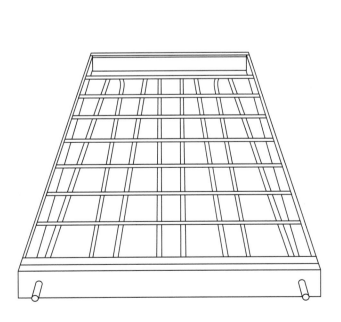

Figure 18-1
Flat plate collector

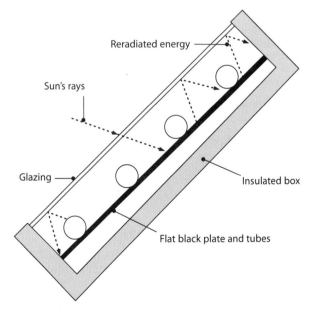

Figure 18-2
Collector cross section

Be sure the solar collector box is well insulated. This accomplishes two purposes: It shields the heat deck plate from the weather and it reduces heat loss. Paint the heat deck plate and tubing flat black to maximize absorption of the sun's energy. Never use light colors, which reflect the sun's rays.

The transparent cover on the collector box permits the sun's rays to strike the metal collector plate *and* reduces the loss of radiated heat back into the atmosphere. The best cover is glass with a low iron content so it's as transparent as possible to incoming rays. This glass admits solar radiation but is opaque to the long-wave energy trapped inside the collector box. The trapped heat is transferred to the fluid in the tubing.

A clear plastic sheet is better than no cover at all, but plastic has a couple of drawbacks. It tends to transmit both incoming and outgoing energy. And most plastics deteriorate quite rapidly under heat and moisture exposure. If a more heat-resistant plastic ever becomes available, it would be ideal because it's less breakable than glass. But most experts agree that glass is preferable for the time being.

In cold climates you'll need to use a collector with a double layer of glass. This is the only efficient way to prevent heat loss by convection when cold air strikes the transparent surface.

As a professional plumber you won't be expected to build solar collectors. But you'll want to have a good grasp of how they work so you can answer your customers' questions.

There are many companies that manufacture solar system components. It's your responsibility as the installer and permit holder to check for the approvals required by your local plumbing code *before you begin installation*.

Here are some of the requirements for solar heat collectors:

- Install only solar heat collectors approved by your local authority.

- Always use exterior-quality materials to secure frames and braces to solar heat collectors on a roof.

- Anchor a solar heat collector to a roof or other structure so it will resist dead loads, live loads, snow loads, wind loads and seismic loads.

- Make sure your installation doesn't impair proper roof drainage.

- Make sure all joints are watertight around pipes, ducts, bolts, or anything which penetrates the roof.

- If you install a solar collector panel that's not an integral part of the roof, mount it at least 3 inches above the roof surface.

- If you install a solar collector panel at ground level, it must be at least 6 inches above the ground to meet code requirements.

- Install all solar collector panels and related piping to drain dry.

- Take care that solar collector boxes have drainage holes at the low point to drain rain, condensation or other liquids that might collect there.

- Use only tempered glass with a low iron content.

Sizing Solar Heat Collectors

In most cases, you'll size a solar system to provide one day's supply of hot water. According to government figures, *each* of the first two people in a family uses 20 gallons of hot water (40 gallons for two people) per day. Each additional person should use 15 gallons per day. Here's how it works for a family of four:

20 + 20 + 15 + 15 = 70 gallons of hot water per day

You base the size of a solar heat collector on the number of bedrooms in a residence. For a one- or two-bedroom residence, never use a collector smaller than 4 feet by 12 feet (or 48 square feet of collector surface). A collector this size should heat approximately 80 gallons of water daily — usually plenty for a family of four.

For each additional bedroom, allow another 24 square feet of collector surface. Here's a word of caution. If you're working in a northern state, you may need a larger collector surface per bedroom. Check your local code requirements.

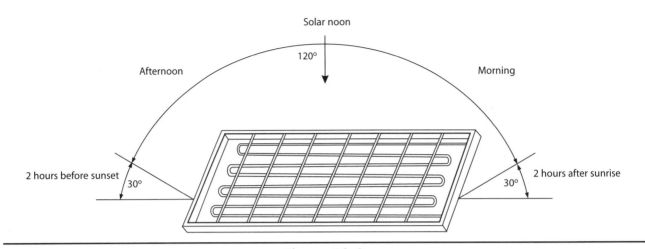

Figure 18-3
The optimum heat collection period

Locating the Solar Heat Collector

Only about 30 to 65 percent of the solar energy that strikes the glass surface of a collector actually heats the water circulating through the tubing. The rest is lost back into the atmosphere through the glass plate.

It's important to have the most efficient collector possible. Of course, the most efficient solar collector would be perpendicular to the sun's rays every moment of the day, every season of the year. There are motorized collectors that turn and tilt to follow the sun's path, but they're far too expensive for most homeowners.

The best compromise is to tilt the flat plate collector in the general direction of the sun's path across the sky at an angle to the particular latitude, plus 10 degrees. Try to make sure the collector faces south. If that's not possible, collectors facing southeast or southwest work about 75 percent as well as those facing due south.

Locate the collector wherever it's most convenient and most attractive, as long as it's in full sun during the major sunlight hours — from two hours after sunrise to two hours before sunset. If the collector is shaded only early or late in the day, when the sun is low on the horizon, it won't appreciably affect the solar system. At those times most of the sun's energy is reflected back into the atmosphere anyway. See Figure 18-3.

Mounting the Solar Heat Collector

In many areas, the plumber is responsible for mounting solar heat collectors. You'll find this especially true in new construction, where collector units are built into the roof as an integral part of the house. Built-in collectors look better because they're designed to blend with the exterior. They resist wind loads better because they're usually flush, or nearly flush, with the roof surface. For new construction, coordinate the installation with the roofer to minimize leaks.

Roofing contractors make several recommendations for mounting collectors on existing roofs. The collector should have a structural frame that's securely bolted to the roof rafters. Install flashing and a rain collar around the pipes and around the collector. Then fit the collector into the frame and anchor it.

Roof-mounted collectors invite roof leaks. Be sure to seal every hole drilled through the roof membrane. To prevent leaks, bore holes smaller than your securing bolts and caulk around each one very carefully.

A pitch pan is a practical solution on built-up roofs. If you use a pitch pan, seal it carefully before you secure it to the roof and fill it with pitch (hot tar). Otherwise, the pitch will heat up under the summer sun and create a pocket. Then water can get into the pocket and stay there. This moisture will eventually rust away the top of the securing bolt and cause a leak.

A pitch pan is ideal around pipes that penetrate the roof. Advise the owner to check the roof annually to be sure expansion and contraction haven't opened fissures where bolts and pipes penetrate the roof membrane.

Install a bearing plate under the cliff angle (the support that secures the heat collector to the roof) to prevent the collector from rocking. Screw the bolt all the way through the sheeting and well into the rafters. If you fasten lag bolts only to the sheeting, they'll pull out or loosen when the wind load causes a collector to vibrate. That's an instant leak.

On roofs with asphalt shingles, place a layer of plastic cement on top of the shingles, but beneath the cliff angle. Apply another layer of plastic cement to the top of the cliff angle. Then fasten the cliff angle securely to the roof rafters with lag bolts.

Mount the collector as close as possible to the storage tank to reduce heat losses and friction in the pipes. Make sure all piping is well insulated. In northern states, shield the transparent cover to protect it from winds that would otherwise cool the surface.

If you're considering installing a collector as an awning or a fixed overhang, be sure to get local authority approval before you begin working. This kind of installation may not be acceptable under your code.

Collectors can be mounted on the ground almost anywhere in full sun. They must be securely anchored, as in Figure 18-4. Unfortunately, any ground installation is much more vulnerable to accidental breakage and vandalism than a roof installation.

In climates subject to freezing temperatures, all the tubes should be installed so that they will drain dry. Draining the pipes is the best and least expensive way to keep water from bursting the pipes during a freeze.

Water Circulation

The most common way to move hot water from the collector piping to the storage tank is a circulating pump. But you can accomplish the same thing with a natural thermosiphon circulation system. This system requires no external energy source, no pumps, no controls, and no moving parts.

Here's how it works. The intensity of the sun controls the rate of movement of hot water from the collector to the storage tank, and of cold water from the tank to the collector. A thermosiphon results when hot water (which is lighter) rises to the storage tank and replaces the heavier cold water, which is drawn into the collector. See Figures 18-4 and 18-5.

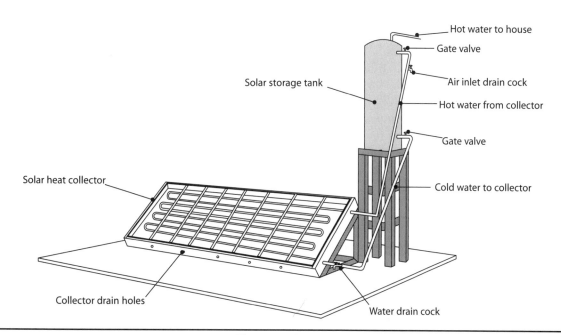

Figure 18-4
Ground-mounted thermosiphon solar water heating system

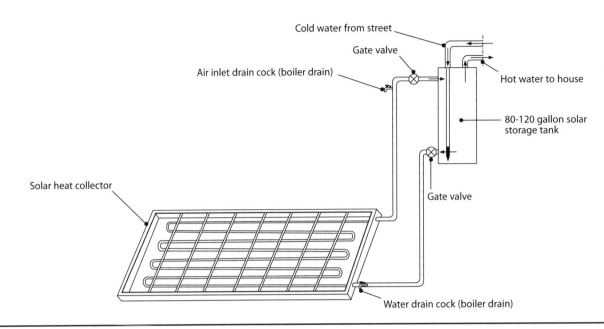

Figure 18-5
Thermosiphon solar water heating system

The thermosiphon won't work properly unless the storage tank bottom is located at least 2 feet higher than the top of the solar collector. That means you'll probably have to put a large, heavy storage tank on a roof or in an attic. This presents problems of weight, construction and appearance. A leak in an attic-mounted tank can cause considerable water damage inside the home. Solar energy codes usually require that you install a drain pan beneath any hot water storage tank located above the first floor. The drain pan should be at least as large as the base of the water storage tank, have a turn up of 2 to 4 inches, and be equipped with a 3/4-inch drain pipe. The pan will catch any dripping water and the drain pipe will convey it safely to the exterior of the building.

In a thermosiphon system you must use a minimum 3/4-inch inside diameter pipe in both the collector and the circulation system. This reduces flow resistance. Be sure the connecting pipe or tubing has a continuous fall with no sags that might allow the formation of an air pocket. An air pocket will stop the circulation.

A pumped system uses the same basic components as the thermosiphon system, but adds a circulating pump to force hot water from the heat collector to the large storage tank. See Figure 18-6.

An obvious advantage of a pumped system is that you can place the storage tank in any convenient place. You can avoid the problems which go with roof- or attic-mounted tanks. You may also use 1/2-inch copper tubing in a pumped system. The pump must be controlled so that it circulates water through the heat collector only when water in the tank is cooler than the water leaving the collector. The added expense of installing the pump and controls may be offset by the lower cost of installing the heavy storage tank at ground level.

Closed Solar Heating Systems

In a closed solar energy collection system, a fluid like antifreeze, instead of water, is heated in the collector. A closed system has a built-in heat exchanger, either within the storage tank body or on the outside of the storage tank, as shown in Figure 18-7. As the fluid circulates through the solar collector, it transfers its heat to water in the storage tank through the heat exchanger.

A closed system does, however, have some disadvantages. It's less efficient, it's more complicated, and it's more expensive. Its one advantage is that in cold

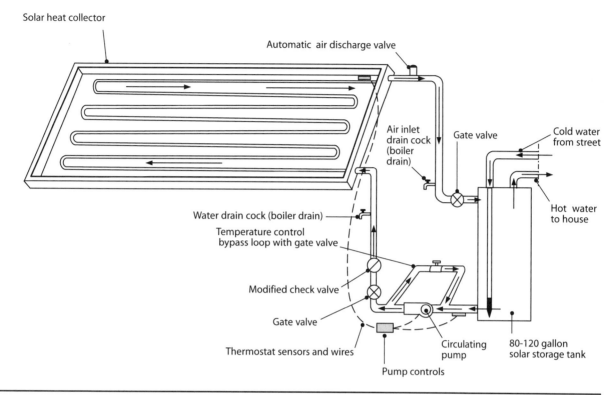

Figure 18-6
Pumped solar water heating system

climates it doesn't have to be drained, and because the system contains antifreeze instead of water, it won't freeze. Obviously, freezing can destroy the solar heat collector.

If your installation is in an area with freezing temperatures, it may be best to install a closed system. Whether you use the thermosiphon or pumped system, make sure it's protected from the cold and that it can be drained dry in winter.

Materials and Installation

The growing interest in solar energy for heating water for pools and domestic use has created a problem for many building code administrators. They're often faced with new installation problems not covered in some codes. If your particular code doesn't address this topic specifically, it will have some general provisions as to materials, types and installation methods. Be sure you follow them.

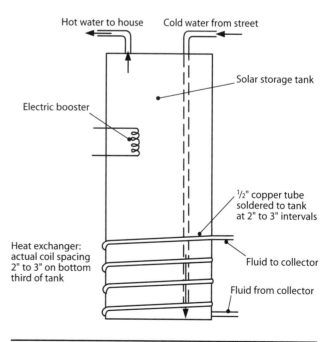

Figure 18-7
One type of heat exchanger

The Uniform Solar Energy Code

The *Uniform Solar Energy Code* is one of the few codes that addresses solar energy. It says that pipes and fittings used for conveying fluids within a solar system must comply with code standards for a potable water system. Look back to Chapter 14 for potable water systems.

Galvanized steel or Type K or L copper pipe and fittings are the most common materials for a solar circulation system. Also approved are cast iron and malleable iron pipe and fittings. These must be galvanized up to and including 2 inches in size, the largest size addressed by the *Uniform Plumbing Code*. Solar systems requiring larger pipe sizes would be unusual, and would need to be approved by your local authority.

There are two kinds of pipes you *can't* use for conveying heated fluids in a solar system:

■ You can't use plastic pipe because it's prohibited where temperatures could exceed 180 degrees F.

■ You can't use aluminum tubing in a potable water system. It's permitted in a closed solar heating, *if first approved by the local authority*.

The piping in built-in heat exchangers in the body of the storage tank must be at least Type L copper tubing. The heat exchanger can't have seams, joints, fittings or valves. It must be constructed of double-wall material designed to prevent leaks which could create a cross-connection with the potable water supply.

Here are some additional requirements:

■ Remember to allow for expansion, contraction and normal pipe movement when installing piping.

■ To minimize heat loss, insulate all piping which carries heated water, fluids or gases from a solar collector or heat exchanger to a storage tank. Use insulation that limits the maximum heat loss to 5 Btu per hour per linear foot of pipe.

■ Make all threaded, soldered and flare joints in a solar system the same as for domestic water piping (Chapter 14).

■ Make sure all piping in a closed solar heating system is isolated so that gases, fluids or other substances can't enter any portion of the potable water system.

■ Use only approved plumbing items and UL-approved electrical components throughout the solar heating system.

■ Your entire solar system must be tested and proved tight under a water-, fluid- or air-pressure test. It must be able to withstand 125 psi for 15 minutes without leaking.

All fittings in a solar piping system should be of the same material as the pipe. When you can't avoid using dissimilar materials, make sure they're electrically isolated with properly-installed approved fittings.

There's an exception to the "no dissimilar material rule" for valves or similar devices. Valves installed in a solar piping system up to 2 inches in diameter must be of brass or other approved materials. The fully-opened valve must occupy 80 percent of the cross-sectional area of the nominal size of the pipe to which the valve is connected. You'll have to install control valves so they can isolate the solar system from the potable water supply. As always, be sure they're readily accessible.

Install an approved pressure regulator where excessive water pressure is likely to occur. Since a solar energy system is an integral part of the building's water supply system, you can use the regulator in the water service pipe for that purpose. If you do, you won't need to install a second regulator in the solar supply pipe.

A combination temperature and pressure relief valve is required for each pressure-type water storage tank. It's important to install the temperature sensing element in the hottest water — that's in the top one-eighth of the tank. See Figure 13-5 back in Chapter 13. Please note that the temperature setting can't exceed 210 degrees F.

Also, the pressure setting can't exceed 150 percent of the maximum designed operating pressure of the solar system. Always check the relief valve setting to make sure it doesn't exceed the manufacturer's recommendations.

Relief valves inside a building must have a full-size discharge line of galvanized steel or hard drawn copper pipe and fittings. Your line should discharge to the outside and turn down to within 6 inches of grade. Make certain that no part of the drain pipe is trapped and that the end isn't threaded. Of course, the line

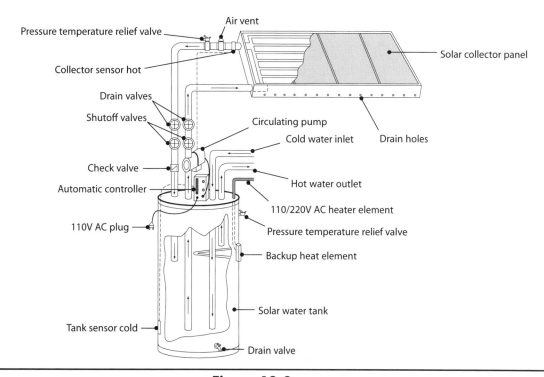

Figure 18-8
Typical pumped solar hot water system (solar collector, water tank and control center)

must be securely strapped to the building structure. A discharge line can terminate at other locations if you first get approval from your controlling authority.

Some authorities may require you to locate the pressure temperature relief valve at the highest point of the solar piping system, near the automatic air discharge valve (air vent). See Figure 18-8. When a relief valve is located on the roof, codes don't normally require a separate discharge line to the ground. They'll usually accept the discharge of the solar system's relief valve onto the roof.

You must install automatic air discharge valves at all high points of a solar piping system. See Figures 18-6 and 18-8.

Solar Storage Tanks

All solar storage tank plans must be submitted for approval, unless they're listed by an approved listing agency. Your plans must show dimensions, reinforcing, structural calculations and any other pertinent data that the building department requires.

In a conventional water heater (electric, gas or oil), the energy source works continuously to produce hot water. The storage tank can be rather small, often 30 or 42 gallons. The tank in a solar water heating system must be large enough to keep water warm through cloudy days and the hours of darkness. The usual recommendation for a family of four is an 80-gallon tank.

The guide in Figure 18-9 may be useful *if your work is in sunny areas* from Florida to California. But note that it gives the *minimum* recommended storage capacity for solar tanks. Since larger tanks are available at very little additional cost, it's worth it to have the extra hot water storage for times when demand is unusually high. The minimum recommended size for the average family of four (two bedrooms) is 70 gallons. Suggest an 80-gallon tank instead. It's always better to have more hot water than not enough.

If there's not enough space for a larger tank, install a solar storage tank with a backup heater element. See Figure 18-8.

Let's assume that you designed a system sized for three bedrooms and six people. You installed at least a 100-gallon storage tank. But eventually three or four of these people move away. The reduced water usage

Number of bedrooms	Number of people	Solar tank capacity (gal)
1	2 (20 + 20)	40
2	4 (20 + 20 + 15 + 15)	70
3	6 (20 + 20 + 15 + 15 + 15 + 15)	100
4	8 (20 + 20 + 15 + 15 + 15 + 15 + 15 + 15)	130

Figure 18-9
Sizing solar storage tanks

may cause water in the storage tank to get too hot. If this becomes a problem, you may need to install a tempering valve in the hot water pipe leading from the tank to the fixtures. This valve will temper the hot water flow so it combines with cold water and arrives at the fixture at a set temperature, usually about 140 degrees F.

Here are other requirements for solar storage tanks:

▮ They must be of sound, durable materials, watertight, not subject to excessive corrosion or decay, designed to withstand all anticipated loads and pressures, installed level, and on a solid bed. The required test pressure for tanks is two times the working pressure from utility mains (with or without a pressure-reducing valve) but never less than 300 psi.

▮ Use insulation that limits heat loss to no more than 2 percent of the stored energy in a 12-hour period.

▮ Install an adequate and accessible drain valve. See Figure 18-8.

▮ Make sure the tank and any devices attached to it are accessible for repair or replacement.

▮ Storage tanks must have a permanent label showing the maximum allowable working pressure and the hydrostatic test pressure designed into the tank. Install it so all markings are accessible to the inspector.

With prior approval from the controlling authority, you can locate a tank below the ground if its design and construction will resist trench load and corrosive soil effects. Obviously, you can't cover or conceal any part of the tank until it's inspected and approved.

Troubleshooting Guides for Solar Systems

If you're ever asked to check a malfunctioning solar system, it'll probably be a problem listed in the troubleshooting guide in Figure 18-10. For each problem, it offers a likely solution.

If the trouble's in the controller, follow these steps, in the order they're listed, to find the problem:

1) Short the two leads to the collector sensor. The controller will cause the circulating pump to run.

2) Short two leads to the solar tank sensor. The circulating pump should stop running.

3) If the solar controller checks out, do the same at each sensor. This should determine if there are any shorts or breaks in sensor wires.

4) Check to see that the sensors are properly wired and the wire nuts are sealed with silicone. Corrosion at the connection points can cause the solar controller to operate improperly.

5) If the solar controller and sensor wires check out, install new sensors, starting with the tank sensor.

6) If you've checked all of these and the problem appears to be with the solar controller itself, return it to the distributor or manufacturer. Install a new controller.

Problem	Likely solution
Water leaking from roof	Probably a leaking pressure temperature relief valve. Replace with a new pressure relief valve.
A solar system circulating pump runs all night	Probably caused by a faulty controller. Check the controller troubleshooting guide in the text.
A solar system circulating pump won't run	Make sure pump is plugged into controller. If it is, then check for faulty controller.
A solar system controller light is on but pump won't run	Make sure pump is plugged into controller. If it is, then check pump impeller. May be jammed with debris. Clean out impeller.
A solar system circulating pump is running but there's no hot water in tank	Probably air trapped in pump collector. Check automatic air discharge valve on collector. (Valve cap should be loose to allow trapped air to escape.) If automatic air discharge valve seems to operate properly, then purge system by manually opening pressure relief valve.
Solar system appears to be working, yet very little hot water available in the mornings	May have a leaking lower check valve. (It's best to check for this after sundown or before sunrise.) Feel along collector line for several feet from tank. If one line is hot to the touch, check valve may be leaking. Replace or clean check valve. (Clarification of problem: Hot water in tank rises to collector, having no sun to heat water. Cools water and drains it back into tank through leaking check valve. This dilutes temperature of stored hot water. If check valve proves okay, see if tank has backup heater element. Setting may be too low or element may be faulty. Set temperature to 120° Fahrenheit.)
Solenoid valve hums	Solar system performance is not affected but noise can be annoying. Tighten nut at solenoid drain connection.
Poor general performance	Most likely cause is inferior insulation or poorly-installed insulation. Check all pipes, fittings and tank. Make sure all insulation joints are tight. Glue or tape as necessary to seal properly.
Solar circulating pump runs periodically at night	System may be thermosiphoning. This is natural circulation due to temperature differential. May be caused by a leaking lower valve. Clean check valve or replace with new one.
Solar collector freezes	May be caused by flat horizontally-mounted collector whose pitch doesn't ensure drainage. Tilt collector so that output end is higher. Should work.

Figure 18-10
Troubleshooting guide for solar water heating systems

Review Questions (answers are on page 311)

1. How is solar energy used in homes today?

2. Why should the professional plumber know about installing solar energy units?

3. What's the first requirement for installing, repairing or altering a solar energy system?

4. In most cases, who must prepare the plans for a solar water heating system?

5. What must be included in the plumbing drawings for a solar water heating system?

6. How much pressure should a solar energy system be able to withstand when tested?

7. Why is solar energy for heating domestic water not considered new?

8. Approximately how much could a family of four save by using solar heated water for domestic purposes?

9. What are the three major components of a solar water heating system?

10. If you're dealing with a pumped solar system, what's the fourth component you'll need?

11. What type of solar heat collector is most practical for residential use?

12. How high are the water temperatures produced by solar water heating systems?

13. What three materials are acceptable for the heat deck of a solar heat collector?

14. What's the thermal difference between using copper, aluminum or steel as heat deck materials?

15. Why must both the tubing and the collector plate be of the same metal?

16. What two purposes do you accomplish by ensuring that the solar heat collector box is well insulated?

17. Why should you use glass with a low iron content in a solar heat collector?

18. What provision should you make to prevent heat loss in a solar collector in a cold climate?

19. What quality of materials should you use for the frames and braces needed to secure solar heat collectors to roof structures?

20. How high above the roof surface should you mount solar collector panels that aren't an integral part of the roof?

21. What must a solar collector panel box have at its lowest point?

22. According to U.S. government figures, how many gallons of hot water per day will each of the first two people in a family use?

23. According to U.S. government figures, how many gallons of hot water per day should you allow for each additional family member after the first two?

24. How many gallons of hot water per day will a 4- by 12-foot solar heat collector provide?

25. What's the minimum size solar storage tank recommended for a family of four?

26. What percentage of the solar energy that strikes the glass surface of a collector actually heats the water circulating through the tubing?

27. To be the most efficient, which direction should a flat solar heat collector face?

28. Why should a solar heat collector be mounted as close as possible to the storage tank?

29. Before you install a solar heat collector as an awning or as a fixed overhang on a residence, what must you get?

30. How does a natural thermosiphon solar water heating system work?

31. When a solar storage tank is attic-mounted, what do most codes require you to install?

32. What's the minimum size piping that you can use in a thermosiphon circulation system?

33. In a pumped solar system, where can you locate the hot water storage tank?

34. What size copper tubing is permissible for use with a pumped circulation system?

35. What fluid(s) can you use in a closed solar energy collection system?

36. In a closed solar energy collection system, how is heat transferred to the water in the storage tank?

37. What standards does the *Uniform Solar Energy Code* require for pipe and fittings used within a solar system?

38. What are the two most common materials used for pipe and fittings in a solar circulation system?

39. Why can't you use plastic pipe in a solar circulation system?

40. Why should you insulate the piping that carries heated water, fluids or gases from the solar collector to the storage tank?

41. What material is required for valves up to 2 inches in diameter installed in a solar piping system?

42. Where must you install control valves in a solar system?

43. What's the required location of the combination temperature and pressure relief valve in a solar hot water storage tank?

44. If authorities require a second relief valve, besides the mandated relief valve on the storage tank, where should you place it?

45. Where must you install automatic air discharge valves?

46. When water usage in a household is reduced, resulting in hot water being stored for longer periods of time, how can you prevent the water in the solar storage tank from becoming dangerously hot?

47. What's the minimum working pressure for a solar storage tank?

48. When the solar system circulating pump is running but the water in the tank isn't hot, what's most likely causing the problem?

49. What is the probable cause for a solar collector freezing?

50. If a solar controller is malfunctioning and you've ruled out any shorts or breaks in the sensor wires, what may be the likely cause of the malfunction?

Gas Systems

Every substance known to man is in one of three states — a liquid, a solid, or a gas. Although gases are lighter than the other two, they do have weight. Gas is made of constantly-moving atoms, with neither a fixed shape nor a fixed volume. When these atoms are forced into a container, they'll take on the container's shape but occupy only about one thousandth of the container's interior space. The spaces between these atoms are empty. That's why gas can be forced through very small spaces.

Gas particles liquefy when they're cooled below their boiling point. When they reach this temperature, the gas particles are pulled together to form a liquid. This is the principle used to make liquid oxygen.

The first recorded discovery of natural gas was made by an ancient Greek shepherd. He discovered that a substance coming from the ground made him lightheaded and talkative, and caused his sheep to act strangely. The Greeks, believing these vapors were the breath of Apollo, built a temple at the site.

The Chinese are credited with being the first to use natural gas for industrial purposes, nearly 3,000 years ago. Using hollow bamboo, they piped natural gas for use in evaporating brine to make salt.

The first discovery of natural gas in America was in West Virginia in 1775. The first commercial use of natural gas was at Fredonia, New York, in 1821. By 1850 many cities and towns were using gas for lighting purposes. But the gas piped to these lighting outlets was yellow and gave off a poor light. This problem was solved when scientists learned to mix air with the gas to produce a better flame.

Today, gas is used in cooking, refrigeration, heating and many industrial applications. Texas and Louisiana produce about 70 percent of the natural gas found in the United States. Most of the rest comes from the Midwest and Rocky and Appalachian Mountain areas. A network of natural gas pipelines transports the gas all over the country.

Since natural gas is clean, dry, and has no odor, a gas leak could go undetected until it causes an explosion. Gas producers add a chemical scent to gas before it enters the pipelines. This odor warns anyone in the area of escaping gas before the concentration can reach a danger level.

Kinds of Gas

Natural gas (methane) contains chemical impurities, which are removed before the gas is piped to the consumer. They're valuable for uses other than as a fuel. The natural gas which millions of consumers use as fuel in homes and industries is known as *dry* or *sweet gas*. While natural gas isn't poisonous, it can cause suffocation in a closed space. It's also explosive under certain conditions.

Manufactured gas is produced chiefly from coal. It's generally added to other fuels to increase its heating capacity. Manufactured gas is also used by consumers as fuel in homes and industries. It can be poisonous, since it contains carbon monoxide, and it's explosive under certain conditions.

Liquefied petroleum gas is also known as *LP* or *bottled gas*. It's produced in plants that process natural gas. LP consists primarily of butane or propane, or a mixture of both. LP gas under moderate pressure becomes a liquid, which makes it easy to transport and store in special tanks. When an LP tank supplies a building's gas, the liquid returns to its original gaseous state when it drops to normal atmospheric pressure and temperature.

LP gas is heavier than air, colorless and nonpoisonous. Since it's easily containerized and transported, it's convenient to use as fuel for homes and businesses in remote areas.

Gas Piping and the Plumber

The plumber has different responsibilities when sizing and installing a building's gas supply system instead of a water supply system. For water service pipe, the plumber sizes and installs the pipe from the meter (located at the property line) to the building. For gas service pipe, on the other hand, the gas supplier sizes and installs the piping all the way to the gas meter. The plumber's only responsible for sizing and installing the gas supply piping within the building.

When it comes to sizing or installing gas supply systems, most plumbing codes will refer you to the local gas code. The *Uniform Plumbing Code* is an exception. Chapter 12 of the *UPC* includes specifications for gas supply systems.

Your local gas code is a separate book with a complexity all its own. Regardless of your experience level, complying with the gas code can be frustrating and discouraging. But even if you don't specialize as a gas fitter and installer, you must be familiar with all the key requirements of the gas code.

You'll find all the information relevant to the plumber included in your adopted national gas code or locally-written gas code. This chapter and the next will give you a working knowledge of the essential requirements. I've simplified the information and arranged it by the way it's used. But the information here doesn't replace the gas code. The code is always your final authority.

Sizing Gas Systems

You need two factors to size the gas main and branch lines of a building:

1) The maximum gas demand at each appliance outlet

2) The length of piping required to reach the most remote outlet

There are a few other variables, like pressure loss, specific gravity and diversity, but they're already accounted for in the code tables.

Gas appliance manufacturers always attach a metal plate in a visible location on each appliance. This data plate shows the Btu input rate (the maximum gas demand). *Btu* is the abbreviation for British thermal unit, the quantity of heat required to raise the temperature of 1 pound of water 1 degree F. While the appliances are rated in Btu, the code tables size gas piping in cubic feet per hour (cfh). So you'll have to convert each Btu input rating to cubic feet of gas before sizing the distribution piping.

You can assume that each cubic foot of natural gas releases 1,000 Btu per hour. Some gas has Btu ratings that vary slightly, but 1,000 Btu per cubic foot is generally a safe assumption. Assume you're sizing pipe for a gas range with a maximum demand of 68,000 Btu per hour. Divide the value in Btu by 1,000 to find the demand in cubic feet per hour:

$$68,000 \ Btu \div 1,000 = 68 \ cfh$$

Occasionally you may be asked to install a used appliance where the Btu rating is missing or illegible. Here's a rule of thumb for a safe installation: Make the appliance inlet pipe at least as large as the supply pipe serving the appliance. You could use a larger supply pipe without violating the code, but it won't make the appliance function any better. Just don't make the supply pipe smaller than the appliance's inlet pipe, and never smaller than 1/2 inch.

Figure 19-1 gives you a sizing method that's quick and easy. But a word of caution: Use it *only* to help you understand how to use the table in your local gas code. The low pressure gas table is the one most commonly used by plumbers. Low pressure gas is used in millions of home and business installations.

Length (ft)	Nominal iron pipe size (in)						
	½	¾	1	1¼	1½	2	2½
10	176	361	681	1,401	2,101	3,951	6,301
20	121	251	466	951	1,461	2,751	4,351
30	98	201	376	771	1,181	2,201	3,521
40	83	171	321	661	991	1,901	3,001
50	74	152	286	581	901	1,681	2,651
Residential 60	67	139	261	531	811	1,521	2,401
70	62	126	241	491	751	1,401	2,251
80	58	119	221	461	691	1,301	2,051
90	54	111	206	431	651	1,221	1,951
100	51	104	196	401	621	1,151	1,851
125	45	94	176	361	551	1,021	1,651
Commercial 150	41	85	161	326	501	951	1,501
175	38	78	146	301	461	851	1,371
200	36	73	136	281	431	801	1,281
Assume 1,000 Btu per cubic foot. More complete sizes will be found in your code.							

Figure 19-1

Maximum capacity of pipe in cubic feet of natural gas per hour

Commercial Gas Sizing Example

Figure 19-2 shows a typical gas piping arrangement for a restaurant. It's similar to what you might find on the journeyman or master plumber examination. We'll size the gas piping system using the data in Figure 19-1.

Assume the total developed length of gas piping in Figure 19-2 is 147 feet from the meter to outlet A. *The developed length is the only distance you use to calculate the size of any section of the gas piping.* Now find that distance in Figure 19-1. You have to use the next longer distance if column 1 doesn't include the exact length. So you'll use the length of 150 feet. (This line is shaded in Figure 19-1.) Read across to find the maximum volume that each pipe size will carry.

Now you're ready to size each section of pipe shown in Figure 19-2. Maximum gas demand of outlet A is 365,000 Btu per hour, or 365 cubic feet per hour assuming 1,000 Btu per cubic foot. Looking across the 150-foot line, the first figure that exceeds 365 is in column 6 (501 cfh). Look up to the top of the column — to find the size of pipe you should select for section A. You'd use a 1½-inch pipe.

Use the same procedure to size each remaining section. The maximum gas demand for section B is 26 cfh, so the correct pipe size is ½ inch. The maximum gas demand for section C includes the combined demand of sections A and B, a total of 391 cfh. Again, you'd use a 1½-inch pipe. The maximum gas demand for section D includes the combined demand of five appliances:

$$30,000 + 30,000 + 25,000 + 230,000 + 90,000 = 405,000\ Btu$$

The conversion to cubic feet yields 405 cfh. Here too, you'd use a 1½-inch pipe. The maximum gas demand for section E includes the combined cfh of sections A, B, C, and D, for a total of 796. You'll have to go to the 2-inch pipe column to find a capacity that exceeds 796.

The maximum gas demand for section F is 289 cfh. Note that the two hot top burners have a *combined* Btu rating of 94,000. The correct pipe size for section F is $1^1/_4$ inches (column 5). The maximum gas demand for section G includes the combined demand of sections A, B, C, D, E and F, a total of 1,085 cfh. We have to use figure 1,501 in column eight. The correct pipe size for section G is $2^1/_2$ inches.

The maximum gas demand for section H is 95 cfh and the correct pipe size is 1 inch. The maximum gas demand for section I includes the combined demand of sections A, B, C, D, E, F, G and H, a total of 1,180. Again we'll use column 8 since the combined total doesn't exceed 1,501. The pipe size for section I is $2^1/_2$ inches. The maximum gas demand for section J is 32 cfh and the correct pipe size is $^1/_2$ inch. The maximum demand for section K is the cumulative total for the entire restaurant — 1,212 cfh. The correct pipe size is $2^1/_2$ inches.

We've just sized the entire gas piping system for a commercial building. Review this section until you're sure you understand how to size each section in the system.

Residential Gas Sizing Example

Size the gas piping for a single-family residence just like you sized the restaurant. The total developed length of the gas piping for a residence is usually shorter and the Btu ratings of the appliances are lower. But the sizing procedure is the same.

Figure 19-3 shows a simple gas piping system similar to what you'll find in most single-family residences. Study this illustration and the explanation in the rest of this chapter until you can size each section of pipe correctly. When you master these illustrations, you can size the pipes in any type of low pressure gas system.

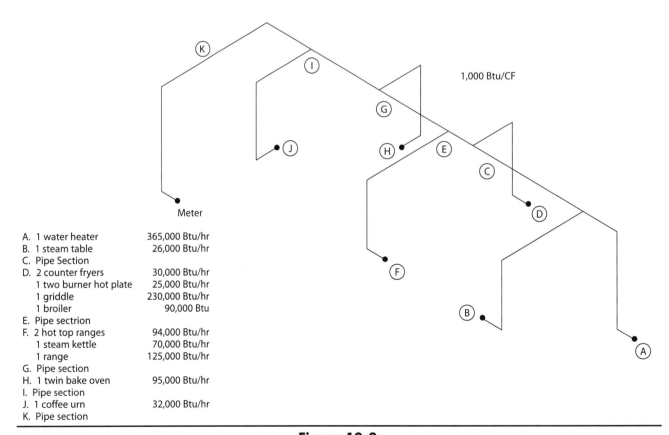

A. 1 water heater   365,000 Btu/hr
B. 1 steam table   26,000 Btu/hr
C. Pipe Section
D. 2 counter fryers   30,000 Btu/hr
   1 two burner hot plate   25,000 Btu/hr
   1 griddle   230,000 Btu/hr
   1 broiler   90,000 Btu
E. Pipe sectrion
F. 2 hot top ranges   94,000 Btu/hr
   1 steam kettle   70,000 Btu/hr
   1 range   125,000 Btu/hr
G. Pipe section
H. 1 twin bake oven   95,000 Btu/hr
I. Pipe section
J. 1 coffee urn   32,000 Btu/hr
K. Pipe section

1,000 Btu/CF

Meter

Figure 19-2
Natural gas installation for a commercial kitchen

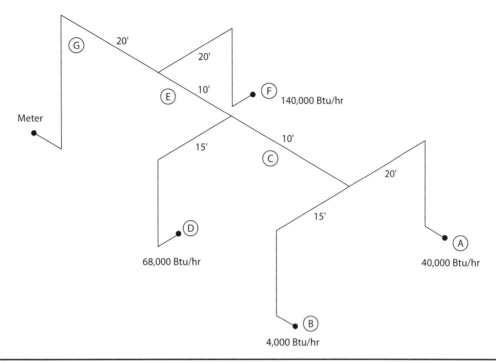

Figure 19-3
Natural gas installation for a residence

The developed length of the gas piping (measured from the meter to the most remote outlet (A in this case) is 60 feet. In Figure 19-1, find the number 60 in column 1. The pipe size at the top of each demand column is the correct pipe size to use for the volume given.

The maximum gas demand of outlet A is 40,000 Btu per hour, or 40 cfh. The correct pipe size is $^1/_2$ inch (column 2). The maximum demand of outlet B is 4,000 Btu, or 4 cfh. Again, use a $^1/_2$-inch pipe. The combined maximum demand for pipe section C is 44 cfh: Use a $^1/_2$-inch pipe. Maximum demand of outlet D is 68,000 Btu per hour, or 68 cfh. The correct pipe size is $^3/_4$ inch. The combined maximum demand for pipe section E is 112 cfh, for a pipe size of $^3/_4$ inch. The maximum demand of outlet F is 140,000 Btu per hour (140 cfh). Use a 1-inch pipe. The combined maximum gas demand for pipe section G is 252 cfh. That's the maximum gas demand for the entire residence. So the correct pipe size is 1 inch.

That's all you need to know to size gas supply systems. In the next chapter you'll learn how to install them.

Review Questions (answers are on page 313)

1. What are the three physical states of matter?

2. What does a gas consist of?

3. At what cooling point do gas particles liquefy?

4. Where was natural gas first discovered in America?

5. About 70 percent of the natural gas produced in the United States is found in which two states?

6. What is added to natural gas as a warning aid to help curb the danger of accidental explosions?

7. What three types of gases are used for fuel today?

8. By what other terms is natural gas (methane) known?

9. Although natural gas itself is not poisonous, how is it lethal?

10. What type of gas is chiefly produced from coal?

11. What poisonous substance is in manufactured gas?

12. What other terms are used for liquefied petroleum gas?

13. Liquefied petroleum gas consists primarily of what substances?

14. What physical change occurs in LP gas under moderate pressure?

15. What happens to liquefied petroleum under normal temperature and atmospheric pressure conditions?

16. Why is LP gas a convenient fuel to use in remote areas?

17. Who's responsible for sizing the gas service pipe to a building?

18. Who governs the sizing and installation methods for interior gas piping?

19. What two factors must you know before sizing any gas building main or branch lines?

20. What is the meaning of the abbreviation "Btu"?

21. How do you define one Btu?

22. How many Btu can you assume to be in each cubic foot of natural gas?

23. If you know the maximum Btu rating for an appliance, how do you convert that into cubic feet?

24. What pipe size should you use when connecting a gas supply pipe to an appliance with a missing Btu rating plate?

25. Regardless of circumstances, what's the minimum size gas supply pipe outlet that you can use?

Materials and Installation Methods for Gas Systems

Your local gas code regulates the materials and installation methods you can use for gas systems. Some materials are limited to certain installations, while others are completely prohibited.

When selecting the materials for gas supply pipes, tubing, or fittings, consider the characteristics of your particular gas supply and its effect on the pipes. For example, gases in certain areas are classified as corrosive. These gases contain an average of 0.3 grains of hydrogen sulfide per 100 cubic feet. If your community gas supply is corrosive, the code won't accept certain types of materials in common use for gas piping.

Materials

A wide variety of materials are acceptable for gas piping. Some are acceptable for underground installations, some for above ground only. Some materials you can use for both. But no matter which material you use, always use fittings of the same material.

Steel and Wrought Iron Pipe

Some piping materials are widely used because of their versatility. The following materials are acceptable for interior use as well as for exterior use underground (except under a concrete slab):

- galvanized steel pipe

- black steel pipe

- galvanized wrought iron pipe

These materials are also acceptable for use with corrosive gas. See Figure 20-1. Pipe up to 2 inches is generally threaded. Seal all threaded joints tight with an approved pipe compound. Larger size pipes can be welded or flanged.

Brass and Copper Pipe

Under most codes, you can use yellow brass and copper pipe (75 percent copper) for interior gas piping. Most codes allow you to install both brass and copper pipe outside underground, but not under a concrete slab. All codes agree you can't use brass and copper pipe if the gas is corrosive.

Use brass and copper seamless tubing, Type K or L, for interior gas piping. But first, make sure it's approved by the local code authority or the area gas supplier. Solder joints between fittings and tubing with a hard solder — generally a silver solder. Or you can braze the joints, usually with a filler of brass to join the metals. If the joint isn't concealed, you can also use approved gas flare fittings for joining copper tubing. This type of gas piping generally requires special installation, so you won't see it in most general construction work.

Plastic Pipe

Plastic pipe and fittings, when approved for use by the local authority, must conform with the ASTM (American Society for Testing and Materials) specifications.

Material	Exterior underground	Interior above ground
Galvanized steel pipe	X	X
Galvanized wrought iron pipe	X	X
Black steel pipe	X	X
PE plastic pipe	X	
Yellow brass pipe	X	X
Copper pipe (hard)	X	X
Copper tubing (soft)	X	X

Notes

▮ Metallic piping installed underground must have a minimum 12" of earth cover. Plastic piping must have a minimum 18" of earth cover.

▮ Other codes may list materials that don't appear in this chart.

▮ Yellow brass pipe and copper tubing can't be used where gas is corrosive.

▮ Always check local code requirements.

Figure 20-1
Common gas piping materials and where to use them

Corrosive gas doesn't affect plastic pipe and fittings, so you can use PE pipe for exterior underground use *except* under a concrete slab. You can't use PE pipe in interior installations, although some codes allow the use of PVC pipe.

Plastic pipe must meet these requirements:

▮ Plastic pipe, tubing and fittings can be joined with solvent cement, heat fusion, or compression couplings or flanges.

▮ If you use solvent cement or heat fusion joints, they have to produce gas-tight joints at least as strong as the tubing being joined.

▮ Don't make solvent cement or heat fusion joints between different kinds of plastics.

▮ Use only heat fusion or mechanical joints to join polyethylene (PE) pipe, tubing or fittings. You can't use solvent cement.

▮ If you make compression mechanical joints, the gasket material in the fitting has to be compatible with the plastic piping, and approved by the gas company.

▮ Connections between metallic and plastic piping can only be made outside the building, and underground.

▮ Plastic pipe or tubing can't be threaded.

Installing Gas Piping

In general, these are the rules you have to follow when installing gas piping:

▮ Don't install it in any air duct, clothes chute, chimney or vent, ventilating duct, dumbwaiter, or elevator shaft.

▮ Don't install it underground closer than 8 inches from a water pipe or a sewer line.

▮ Don't install it in the same ditch with water, sewer or drainage pipe, unless first approved by the local authority.

▮ You can install it in accessible above-ceiling spaces used as an air plenum, but you can't put the valves there.

Exterior Installations

When you use horizontal metallic gas piping, be sure to place it a minimum of 12 inches underground. If you use plastic gas piping, it has to be 18 inches underground. If you're installing metallic pipe in corrosive soils, protect it with an approved wrapping or one or two coats of asphaltum paint. Insulate the pipe where it enters a building above ground, in crawl spaces, or anywhere it isn't protected from the cold. Many gases contain moisture which can freeze and block the pipe.

Lay underground gas piping in open trenches on a firm bed of earth. In areas where freezing temperatures occur, the trench bottom should be below the frost line to prevent freezing and rupturing of the pipe. Make sure the pipe is securely supported to prevent sagging and excessive stress during backfill. Finally, use only fine materials for backfilling.

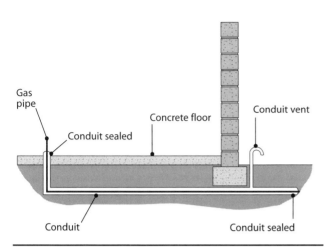

Figure 20-2
Installation beneath concrete floor

Installation Under a Slab

Occasionally you have to install gas piping underground under a building slab. When this is unavoidable, the code permits this type of installation only if the following conditions are met:

1) The entire length of gas piping up and through the floor must be encased in conduit of a material approved for installation beneath buildings, and not less than Schedule 40 pipe.

2) The termination of the conduit above the floor must be sealed to prevent the entrance of any gas into the building in the case of a leak.

3) The termination of the conduit outside the building must be tightly sealed to prevent water from entering the conduit.

4) A vent must be extended above the grade and secured to the conduit. This vent conveys any leaking gas to the outside of the building. See Figure 20-2.

Interior Installations

If you're installing gas equipment or appliances that are subject to vibration or require mobility, connect them with an approved flexible gas hose connector. The gas hose should be no longer than necessary

and no longer than 6 feet in any case. Only use approved gas hose connectors to connect the hose to the gas outlet pipe.

Installation in Concrete or Masonry

Gas appliances that aren't placed next to walls make it more difficult to conceal the gas piping. You'll have to install the gas pipe in an open channel in the concrete floor, then conceal it with a removable grill or cover that provides access to the piping. See Figure 20-3.

In rare cases where it's unavoidable, you can get approval from your local authority and the gas supplier to embed the pipe in concrete. You can use galvanized steel or wrought iron pipe, but you must meet the following conditions:

1) The concrete can't contain cinder aggregates or additives designed to set concrete more quickly than normal.

2) The pipe must be embedded directly in a portland cement concrete slab with a minimum of $1^{1}/_2$ inches of concrete on all sides.

3) The piping can't be in contact with any metallic materials.

4) You must protect the pipe from the corrosive effect of the concrete where it enters and exits the slab. Use an approved coating or sleeve.

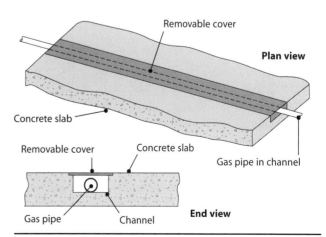

Figure 20-3
Concrete floor open-channel installation

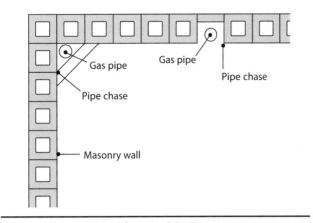

Figure 20-4
Vertical masonry wall chase — plan view

When gas piping must pass through masonry walls, protect the pipe against corrosion by sleeving or painting. Vertical masonry walls must provide adequate chases to protect the pipe. See Figure 20-4.

The horizontal and vertical supports for gas piping and tubing must meet the same requirements as cold water piping. Look back to Chapter 14.

Installation in Partitions with Wood or Metal Studs

When installing gas piping or tubing horizontally in *wood partition* walls, take these steps to protect the building structure and the pipe or tubing:

1) Install short runs of horizontal gas piping or tubing that don't require additional joints through a hole drilled in the center of the partition stud. Be sure the hole isn't large enough to weaken the stud.

2) Install longer runs of horizontal gas piping or tubing in notches cut deep enough to conceal the pipe or tubing. Don't cut deeper than one-third of the total width of the stud to avoid weakening the partition.

3) Protect soft tubing in a notched partition with a metal stud guard to avoid penetration by lath nails.

Metal stud partitions are replacing wood partitions in many new buildings. The metal studs are hollow rather than solid. Simply install the gas pipe or tubing through the manufactured openings in the center of the stud. Wrap the pipe or tubing with an approved material to prevent contact with the metal. Secure the pipe with tie wire.

Gas pipes or tubing installed in a concealed location can't have unions, right or left couplings, tubing fittings, running threads, bushings or swing joints. To make a new connection to an existing line in a concealed threaded gas piping system, use a ground joint union. The center nut is punched to prevent it from working loose under vibration. You can't make new connections on a concealed gas tubing installation, regardless of the tubing material used.

The procedure for cutting, threading and reaming of gas pipe is the same as described for water pipe in Chapter 14. Threads for gas pipe must conform with the standards adopted by the American Standards Association. See Figure 14-6 for the number and length of standard pipe threads. Avoid using pipe with chipped or torn threads.

Installing Drip Pipe and Shutoff Valves

Install gas mains so they can drain dry, with the pitch or grade toward the gas meter. And put in an appropriate drip pipe to catch any condensation that forms in the pipe. The drip pipe, usually assembled from a tee, nipple and cap, must be accessible for emptying. Don't make it smaller than the pipe or pipes it serves, or less than 18 inches long. Protect the drip pipe from freezing in colder climates. See Figure 20-5.

Gas branch pipes must connect to other horizontal pipes at the top or the side of the feeder pipe and never at the bottom. This will keep condensate from filling and obstructing the branch lines. See Figure 20-6.

Buildings with multiple tenants and a master meter must have a gas shutoff valve for each apartment. Locate each shutoff valve on the outside of the building in an accessible location near the meter. Since it is accessible, to avoid accidental or malicious tampering, the shutoff valve usually has a square nut head that requires a special tool.

Each gas appliance inside a building must have an accessible, manually-operated shutoff valve. This interior valve must have a lever handle that doesn't require a tool. Shutoff valves are available in two types for convenient use: a straight pattern or an angle pattern. Locate it as close as possible to the gas outlet pipe, not more than 6 feet from the appliance it serves.

Don't conceal the completed gas installation until it's been pressure tested. Cap each outlet and install a pressure gauge on one of the outlets. Pressure test at 10 to 20 pounds. The system must remain airtight (no loss of pressure) until after inspection. If a leak should occur during the test of the rough piping, test for leaks by running a small brush dipped in liquid soap around each joint. The leaking joint will blow bubbles.

Installing and Venting Gas Appliances

It's the plumber's job to connect the gas from the wall outlet to the appliance, with either a rigid pipe or an approved flexible connector. When the appliances are connected, the gas supplier will purge the lines of air, check for leaks at the joints that connect the appliance to the gas system, light the pilot lights, and finally, adjust each appliance.

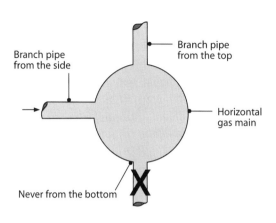

Figure 20-6
Branch pipe connection to horizontal pipe

Gas appliances such as water heaters and clothes dryers can be installed on the floor of a residential garage under these circumstances:

▮ The floor of the garage must be higher than the driveway or adjacent ground.

▮ The combustion chamber must be a minimum of 18 inches above the floor or adjacent ground.

If you install an appliance in a separate room off the garage, the walls, ceiling and door of the room must have a one-hour fire rating if the appliance has an input of 100,000 Btu or less. It must ventilate through permanent openings with a total free area of 1 square inch for each 1,000 Btu per hour of input rating (with a minimum of two 50-square-inch openings). One vent opening should be a minimum of 12 inches above the floor. The second opening should be a minimum of 12 inches below the ceiling. This provides enough air circulation for the combustion and dilution of the flue gases.

Always install gas appliances so there's access to the appliance for repairs and cleaning as well as the intended use. You can't locate water heaters in any living area that may be closed off, like bedrooms or bathrooms.

Install any appliances that must be vented as close to the vent pipe as possible. If a draft hood is required, the vent pipe can't be smaller than the opening of the draft hood. Provide sufficient clearance between gas

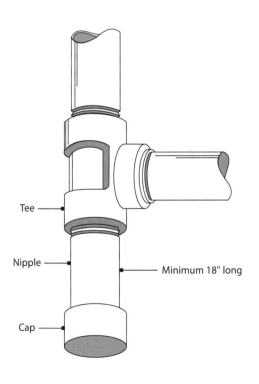

Figure 20-5
Drip pipe

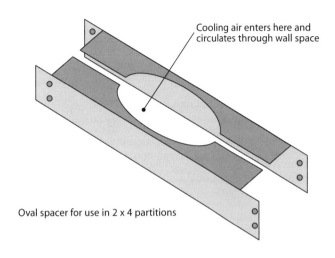

Cooling air enters here and circulates through wall space

Oval spacer for use in 2 x 4 partitions

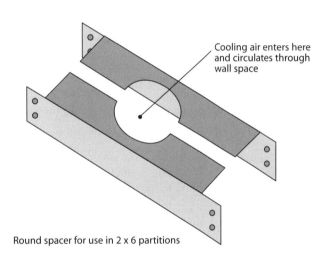

Cooling air enters here and circulates through wall space

Round spacer for use in 2 x 6 partitions

Figure 20-7
Vent pipe spacers

appliances and vents and combustible materials to avoid a fire hazard. Place gas water heaters with an insulated jacket at least 2 inches from any combustible material, or 1 inch from one-hour fire-rated materials.

There are two acceptable types of concealed vent pipe material. The most common is the *double-wall metal pipe* and fittings. The clearance distance recommended by the manufacturer is stamped in the metal. It's usually at least 1 inch from any combustible material. The second vent material is *asbestos cement flue pipe*. It should have a clearance of $1^1/2$ inches from any combustible material. Vent pipes in partitions built of combustible material must have an approved metal spacer that keeps the surface temperature below 160 degrees F. See Figure 20-7.

You can use *single-wall vent pipe* for exposed vent pipe installed in a room built of noncombustible material. Support all horizontal vent pipes to prevent sagging or misalignment. Use straps or hangers that are at least 20 gauge sheet metal. The horizontal vent section length can't exceed 75 percent of the vertical vent length.

Here are the requirements for locating the exit terminal of a vent pipe:

▪ Don't put it closer than 4 feet to any opening through which combustion products could enter the building.

▪ Don't put it closer than 2 feet from an adjacent building.

▪ Make sure it's at least 4 feet from any property line except a public way.

All vent pipes above the roof of a building must terminate in a UL-approved cap. Figure 20-8 shows several examples of acceptable vent pipe terminations outside a building.

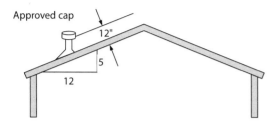

A Roof pitch 5'/12' (22½°) or less (includes flat roofs). Maintain minimum clearance of 12" as illustrated

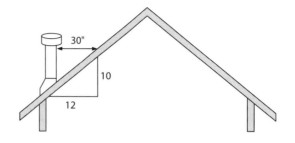

B Roof pitch 5½'/12' to 12'/12' (45°). Maintain 30" horizontal distance as illustrated.

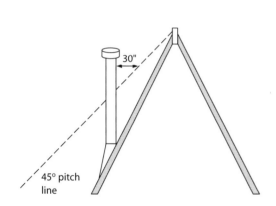

C Roof pitch greater than 12'/12' or vertical wall. Maintain 30" horizontal distance from 45° pitch (12'/12') line.

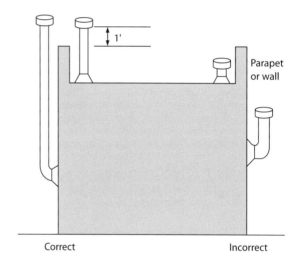

D Vent top should be located 1' above parapet or wall when within 30"

Figure 20-8
Acceptable pipe terminations

Review Questions (answers are on page 314)

1. What must you consider when selecting piping materials for a gas system?

2. What three piping materials are code-accepted for both underground and above ground gas installations?

3. Name two piping materials that are acceptable for use in a gas system where the gas is corrosive.

4. What percent of yellow brass pipe must be copper if you're using it in a gas installation?

5. Where are you not allowed to install brass and copper pipe in an underground installation?

6. Under what condition should you not use copper piping in a gas system?

7. If approved by your local code or gas supplier, which two weights of copper pipe and tubing can you use for interior gas piping?

8. When joints are necessary in a copper gas piping system, what type of solder must you use?

9. Under what conditions can you use approved gas flare fittings in a copper gas piping system?

10. When plastic pipe and fittings are approved for a gas system, what national organization's specifications must they conform to?

11. In what locations can you make connections between metallic and plastic piping?

12. Name three locations in a building where you should never install gas piping.

13. What's the minimum distance you must maintain between gas piping and a water pipe or a sewer line in an underground installation?

14. What's the minimum depth for placing underground horizontal metallic gas piping?

15. What's the minimum depth for placing underground horizontal plastic gas piping?

16. What's one method of protecting gas piping if you're installing it in corrosive soil?

17. How deep should you install gas piping in areas subject to freezing temperatures?

18. What type of backfill should you use when backfilling a trench containing gas piping?

19. Under what conditions can you install gas piping under a slab?

20. What type of connection do you use for gas equipment or appliances subject to vibration or requiring mobility?

21. How should you install gas piping to serve an appliance located in the center of a room?

22. What must you provide to protect gas piping in vertical masonry walls?

23. For what type of gas pipe installation may you drill a hole in the center of a partition stud?

24. Why shouldn't you notch a partition stud deeper than one-third its total width?

25. How should you protect soft tubing in a notched partition?

26. How should you secure gas piping installed in metal stud partitions?

27. When are bushings permitted in a concealed gas piping system?

28. What must you do to prevent a union in an existing concealed gas line from loosening?

29. When are you allowed to make a new connection on an existing concealed gas piping or tubing installation?

30. The procedure for preparing threads for gas piping is the same as for what other type of piping?

31. To what standards must the threads for gas piping conform?

32. What must you install to catch any condensation that may form in a gas main?

33. Why should gas branch pipes connect only at the top or side of a gas feeder pipe?

34. Why should a shutoff valve be installed near the gas meter?

35. How is accidental or malicious tampering with the outside gas shutoff valve avoided?

36. What type of shutoff valve is required for each gas appliance in a building?

37. What two types of shutoff valves are manufactured for appliances?

38. What's the maximum distance allowed from a shutoff valve to the appliance it serves?

39. What must you do before you can conceal a completed gas installation unit?

40. What's the safest way to check gas piping for leaks?

41. At what minimum height above the garage floor can you set the combustion chamber for a gas water heater?

42. If you install a gas appliance having a 100,000 Btu input or less in a separate room off the garage, exactly what number and size of ventilation opening(s) must you provide, and at what location in the room?

43. Where must you never install a gas water heater?

44. Where must you install gas appliances that require venting?

45. What size vent pipe is required for a 30-gallon gas water heater with a 4-inch draft hood?

46. What's the minimum separation required between a gas water heater with an insulated jacket and any combustible material?

47. What are the two acceptable types of concealed gas vent piping materials?

48. What must you provide for gas vent pipes installed in partitions constructed of combustible material?

49. What gauge metal straps or hangers should you use to support horizontal gas vent piping?

50. How must all gas vent pipes extending above a roof terminate?

21

Plumbing Fixtures

Over the past 100 years, plumbing organizations have developed well-recognized standards that control the quality and design for all plumbing fixtures in use today. As a plumbing professional, it's your responsibility to install only high-quality fixtures that conform in design to the standards in Figure 21-1. Never allow substandard fixtures that will detract from the quality of the work you do.

Fixtures must be acid resistant and free from defects and concealed fouling surfaces. All fixtures must be set level and in proper alignment with the adjacent walls. And fixtures constructed of pervious materials (such as baths or showers made of tile or marble) must have a waste outlet that can't retain water.

Installing Plumbing Fixtures

Never install fixtures in a room that doesn't have adequate light and ventilation. If there's no window for natural ventilation, you must install a fan and duct. Because inadequate lighting or ventilation promotes unsanitary conditions, most codes prohibit locating fixtures in areas without them.

Water Closets

Grout all water closets, wall-hung or floor-mounted, with white cement or any suitable material to provide a watertight seal at the joint with the wall or floor.

This prevents the accumulation of odor-causing materials, avoids other unsanitary conditions and keeps roaches and other insects away from these areas.

Water closets installed for public use (anyplace except a single-family residence or apartment building) must have an elongated bowl equipped with an open front seat. Seats for water closets must be made of smooth nonabsorbent materials and they must fit the water closet bowl. For example, you can't install a round front seat on an elongated bowl.

Wall-hung water closets should be rigidly supported with brass bolts on a concealed metal carrier. Make sure the fixture pipe connection doesn't carry any of the load.

Water closets with tanks designed to use ballcocks must refill after each flushing and then close tight when the tank is full. The flush valve operates manually but the flushing operation must be automatic after it's manually activated. The flushing device and the connection between the tank and the bowl should have enough flow capacity to allow the water to flush all surfaces of the bowl.

The tank has to have a refill tube reaching and turning down into the overflow tube. Water from this tube automatically restores the closet bowl water seal. Make sure there's an antisiphon valve built into the unit to prevent contamination of the potable water supply.

Plumbing fixtures and authorized approvals	
Bathtubs, plastic units	ANSI Z124.1
Ceramic plumbing fixtures, non-vitreous (Fireclay)	ANSI/ASME A112.19.9M
Drains for prefabricated and precast showers	JAPMO PS-4-90
Drinking water coolers	ANSI/UL 399-86
Enameled cast iron plumbing fixtures	ANSI/ASME A112.19.1M
Enameled steel plumbing fixtures	ANSI/ASME A112.19.4M
Floor drains	ANSI A112.21.1M
Lavatories, plastic	ANSI Z124.3
Roof drains	ANSI A112.21.2M
Shower receptors and shower stalls, plastic	ANSI Z124.2
Sinks, plastic	ANSI Z124.6
Stainless steel plumbing fixtures (residential use)	ANSI/ASME A112.19.3M
Vitreous china plumbing fixtures	ANSI/ASME A112.19.2M
Water closet bowls and tanks, plastic	ANSI Z124.4
Whirlpool bathtub appliances	ANSI/ASME A112.19.7M
Plumbing fixture trim	
Faucets and fixture fittings	ANSI/ASME A112.18.1M
Trim for water closet bowls, tanks and urinals	ANSI/ASME A112.6.1M
Water closet seats, plastic	ANSI Z124.5
Water hammer arrestors	ASSE 1010

Note For maximum allowable water usage for plumbing fixtures, see Figure 11-1.

Figure 21-1
Plumbing fixture standards

Each tank also has to have an overflow tube adequate to prevent tank overflow and remove excess water at the rate it enters the tank. Consider what would happen if the flush ball were securely in place on the flush valve seat and the ballcock locked in an open position. The flush valve seat must be a minimum of 1 inch above the rim of the bowl.

A water closet using a flushometer instead of a tank has to have a vacuum breaker located a minimum of 6 inches above the rim of the bowl. It must complete the normal flushing cycle automatically after being manually activated, and deliver enough water to flush all surfaces of the bowl. It should open fully, and close tight under the normal water working pressure. Each flushometer can serve only one water closet, and must be readily accessible for repair. The valve must have some way to regulate the water that flows with each flush.

The *Uniform Plumbing Code* requires all water closets, tank or flushometer, to have a maximum average consumption of 1.6 gallons of water per flush.

Urinals

Wall-hung urinals must be rigidly supported by a concealed metal carrier or other approved backing so no strain is transmitted to the pipe connection. Grout the joint between the urinal and the finished wall surfaces with white cement or other suitable material to provide a watertight seal.

When you install floor-mounted stall urinals, put them slightly below the finished floor to provide drainage. Then put beehive strainers in the waste opening. The trap size is 2 inches.

A urinal with a flushometer must complete the normal flushing cycle automatically after it's manually activated. It must deliver enough water to flush all surfaces of the urinal. The valve must open fully, and close tight at normal water pressure. The urinal must also have some means of regulating water flow. Only one urinal can be served by a single flushometer.

Where a single flushing tank serves several urinals, it must operate automatically once it's manually activated, and have enough capacity to cleanse all urinals simultaneously. For *Uniform Plumbing Code* compliance, urinals must have an average consumption of 1.0 gallons of water per flush.

Lavatories

Wall-hung lavatories for commercial use are supported by a concealed metal carrier. The carrier provides the lavatory with enough support so the fixture pipe connection and the finished wall carry no strain. That makes it unlikely that the lavatory can pull away from the wall. In residential installations, you'll generally support wall-hung lavatories with a metal bracket screwed securely to wood backing fastened to the bathroom partition studs.

The point of lavatory contact to finished wall surfaces must be sealed with white cement or other suitable material.

Cabinet-mounted lavatories are securely fastened to the countertop by special rim clips and are made watertight with a caulking compound or other adhesive. The weight is transferred to the cabinet top and places no strain on the fixture piping.

Other wall-mounted fixtures must be adequately supported and grouted for sanitary purposes. Even shower rods must have a suitable backing so they don't work loose from the wall.

Waste outlets for lavatories must be a minimum of $1^1/4$ inches outside diameter.

Where circular-type multiple wash sinks are used, each 18 inches of wash sink circumference is considered one lavatory (one fixture unit). Straight-line multiple wash sinks must have a separate set of faucet combinations no closer than 18 inches from center to center. Each faucet set is considered to be one lavatory (one fixture unit).

According to the *Uniform Plumbing Code*, lavatory faucets must have aerators and a maximum water flow rate of 2.5 gallons per minute.

Bathtubs and Showers

The minimum size waste and overflow for bathtubs is $1^1/2$ inches. There are several approved tub waste and overflows in use today. But some codes prohibit a trip waste because they're difficult to adjust to properly retain and discharge tub water. Any bathtub that's recessed into tile or other finished wall materials must have waterproof joints. The walls must be of smooth, noncorrosive and nonabsorbent waterproof materials to a height of 4 feet above the rim of the tub.

The waste outlet for a shower compartment floor must be at least 2 inches in diameter, located so water will drain from the shower floor without puddling. The strainer must have a diameter of at least $3^1/2$ inches, and be removable so the trap can be cleaned. Shower traps can't be smaller than the waste outlet pipe used in the shower compartment.

Shower compartments need a minimum floor area of 1,024 square inches and a minimum span between walls of 32 inches — adequate for use by adults. The floors must be smooth and sound. Institutional or gang showers used by more than one bather at a time have to be designed so waste water from one bather doesn't pass over areas occupied by other bathers.

Where shower pans are required, use pans of lead, copper or other approved materials. Here are the minimum weights:

▮ Lead pans — at least 4 pounds per square foot

▮ Copper pans — at least 12 ounces per square foot

▮ Nonmetallic — pan may be constructed (on site) of three layers of 15-pound asphalt impregnated roofing felt

A shower pan of lead or copper must be protected from corrosion where it joins concrete or mortar, with asphaltum paint inside and outside. Place a layer of 30-pound asphalt-saturated felt or a $1/2$-inch thick layer of sand under the pan. This protects the pan against rough surfaces and helps avoid accidental puncturing before it's protected by the finished floor material.

Cut the shower pan large enough to allow a turned-up edge on all sides at least 2 inches above the finished curb, or $3^1/2$ inches above the rough curb. Shower pans must be securely fastened to the shower strainer base at the invert of the weep holes. Use a clamping ring to make a watertight joint between the shower waste outlet stub and the pan.

Support shower pan material with an adequate backing that's secured to the partition studs. This prevents the pan sides from sagging before the interior of the shower compartment is in place to hold the pan rigid. If you have to puncture the shower pan material to secure it in place, make sure the penetration isn't lower than 1 inch from the top of the pan's turn-up. Figure 21-2 shows a correctly-installed pan.

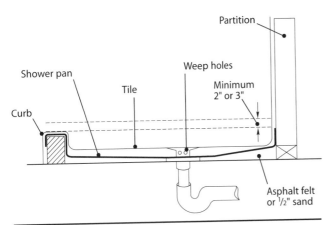

Figure 21-2
Shower pan installation detail

To test each shower pan, remove the shower strainer plate and plug waste outlet. Fill the pan with water. The pan must be full and ready for the tub and water pipe inspection. Otherwise, the plumbing contractor may have to pay for a reinspection.

You can omit shower pans in shower compartments built on a concrete slab on the ground floor. Pour the bottom, sides and curbs of the shower compartment at the same time as the floor slab. Pour a curb 1 inch higher than the existing slab around all sides of the shower compartment. This will usually keep the water level below the height of any surrounding wood plates or studs and help keep the compartment watertight. See Figure 21-3.

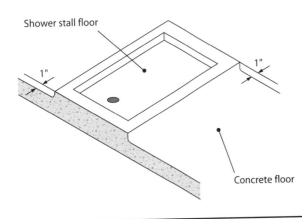

Figure 21-3
Shower compartment that doesn't require a pan

Shower pans aren't required in prefabricated shower stalls. But these stalls require individual approval by the plumbing inspector for watertightness. Walls of shower compartments must be waterproof, smooth, noncorrosive and nonabsorbent to 6 feet above the floor.

Sinks and Laundry Tubs

Waste for sinks and laundry tubs must be at least $1^{1}/_{2}$ inches in diameter. Tail pieces and continuous waste pipe must be at least $1^{1}/_{2}$ inches outside diameter. Each compartment in a laundry tub needs a waste outlet with a suitable stopper for retaining water.

Domestic sinks need a waste opening at least $3^{1}/_{2}$ inches in diameter if there's a waste disposer unit installed.

The *Uniform Plumbing Code* requires that kitchen sink faucets be designed with aerators and have a maximum water flow rate of 2.5 gallons per minute.

Fixture Overflows

Bathtubs and lavatories are two of the more common fixtures provided with overflows. But the code doesn't require overflows on lavatories, so they're often omitted. Integral overflow passageways provide an escape for excess water below the flood-level rim of the fixture, as well as secondary protection against self-siphonage of a fixture trap.

When a plumbing fixture does have an overflow, the waste pipe must serve two purposes. It has to prevent water from rising into the overflow when the stopper is closed, as well as stop water from staying in the overflow when the drain is open for emptying.

Connect the overflow pipe or passageway from a fixture on the inlet side of the fixture trap. This prevents sewer gases and odors from entering the room through the overflow. The code prohibits connecting the overflow of a fixture to any other part of a drainage system. Figure 21-4 shows one prohibited connection.

Fixtures must have durable strainers or stoppers, but they can't prevent rapid drainage of the fixture. There's an exception for fixtures with integral traps, such as water closets, some bidets and some wall-hung urinals. A strainer can't be smaller than the fixture waste outlet it serves and should be easy to remove for cleaning if it's not fixed (manufactured as part of the fixture and not removable).

Food Waste Disposers

A few codes permit one trap and one waste line for a food waste disposer. But when you install a two-compartment sink in a residence, you've got to make sure it's separately trapped and separately wasted to the stack or vented branch. Use a hi-lo fitting 2 inches in diameter with two $1^1/_2$-inch double vertical tappings not more than 6 inches apart. See Figure 21-5.

This double trapping and wasting has two advantages:

1) It permits the use of one compartment of the sink even when there's a stoppage in the other compartment.

2) It prevents the waste disposer from pumping garbage into the other compartment of the sink.

You can install a food waste disposer in the two-compartment sink in an existing home or apartment even if there's not a second waste opening available. The waste can flow through a single $1^1/_2$-inch trap if you use a special directional tee or wye, as shown in Figure 21-6.

Commercial food waste grinders must waste directly into the sanitary drainage system, never through a grease interceptor. The waste pipe from a commercial food waste grinder must be equal in size to the discharge opening of the machine, and at least 2 inches. The grinder has to be individually trapped and vented just like any other fixture.

Dishwashers

Waste from a domestic dishwasher with a pump discharge must rise to a height equal to the height of the underside of the dishwasher top. It should connect to the sink waste with a dishwasher wye branch as shown in Figure 21-7.

Some codes require you to make this connection with a directional tee installed in the sink tail piece (Figure 21-8). Other codes require a waste air gap

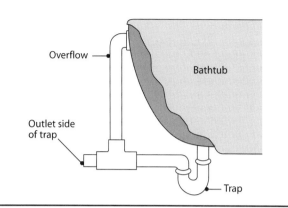

Figure 21-4
Prohibited overflow connection

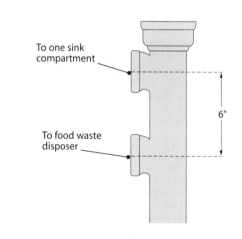

Figure 21-5
Hi-lo fitting

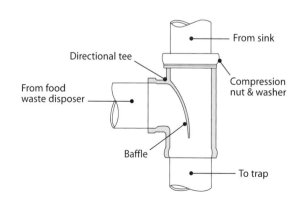

Figure 21-6
Directional tee

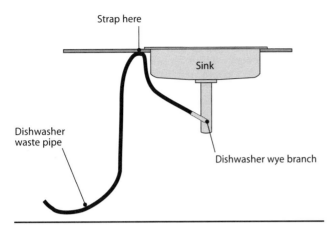

Figure 21-7
Dishwasher connection through wye branch

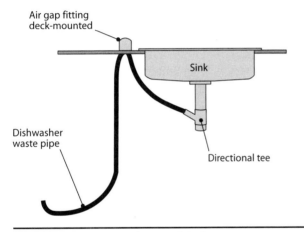

Figure 21-8
*Dishwasher connection through directional tee
with air gap fitting*

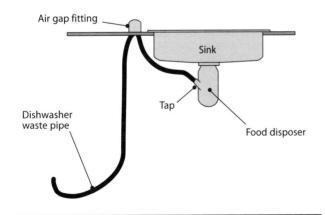

Figure 21-9
*Dishwasher connection through food disposer
with air gap fitting*

fitting, either deck-mounted to the sink or cabinet top, or wall-mounted. See Figures 21-8 and 21-9. You can't locate dishwashers more than 5 feet from the sink waste connection.

If there's a food disposer unit installed in the sink, the waste from the dishwasher must connect to the tap in the body of food disposer (Figure 21-9).

Floor Flanges

Securely fasten all floor fixtures to an approved floor flange with brass bolts or screws. Flanges for floor fixtures have to be set on top of the finished floor, not recessed flush with it. See Figure 21-10.

Plumbing fixtures with a flanged connection between the fixture and the drainage pipe must be set in setting compound or have an approved gasket or washer. Graphite-impregnated asbestos and felt are approved gasket materials.

The flanges must be of the same material, or compatible with, the materials in a drainage system. Here are the requirements for each kind of system:

▌ If you use lead stubs to secure the fixture to the drainage system, solder a brass or hard lead flange securely to the stub.

▌ If you use copper stubs in a copper drainage system, solder a brass flange securely to the copper stub.

▌ In a cast iron drainage system, cast iron stubs are acceptable under some codes. Then you can use a cast iron flange with a lead and oakum joint to secure the floor flange to the cast iron stub.

▌ In a plastic drainage system, cement-weld the plastic floor flange to the plastic stub.

Floor Drains

Floor drains are considered fixtures. The traps of floor drains must have a permanent water seal fed from an approved source of water, or an automatic priming device designed and installed for that purpose. This permanent water supply has to retain the trap's seal and prevent evaporation from drying out the trap. A dry trap lets sewer gases enter the building.

You can connect condensate drain waste from air conditioning units to a floor drain, but it's not enough to supply a permanent water seal. A single drinking fountain waste can discharge to a floor drain if it's not in a restroom. This is usually considered an adequate water supply to protect the trap seal.

The discharge from a garbage can washer can't discharge through a trap serving any other device or fixture. Connect the waste pipe directly into the greasy waste line, discharging through a grease interceptor. The receptacle (floor drain) that receives the waste from the garbage can washer must have a basket to collect solids $\frac{1}{2}$ inch or larger in size. It's essential that the basket be easily removable for cleaning.

The hot and cold water connection must be properly valved and have an approved vacuum fitting to prevent cross-connection. See Figure 21-11.

Floor drains serving indirect waste pipes from food or drink storage rooms or appliances can't be located in toilet rooms or in any inaccessible or unventilated closet or store room. You can't install any type of plumbing fixture in a room containing air handling machinery.

Special fixtures such as baptisteries, ornamental pools, aquariums, ornamental fountains, developing tanks or sinks and similar fixtures that have a waste and water connection should have the water supply protected from back-siphonage with an approved vacuum breaker.

Plumbing Fixture Clearances

Install all plumbing fixtures with spacing that permits easy access for cleaning and repair, as well as the intended use. Figure 21-12 illustrates the minimum clearances required in most codes.

▌ *Water closets* must have a minimum spacing of 30 inches center-to-center when set in battery installations. They must be set a minimum of 15 inches from the center of the bowl to any finished wall or partition, and a minimum of 12 inches from the center of the bowl to the outside edge of a tub apron. Finally, they must have a minimum clearance of 21 inches from the front of the bowl to any finished wall, door or other plumbing fixture.

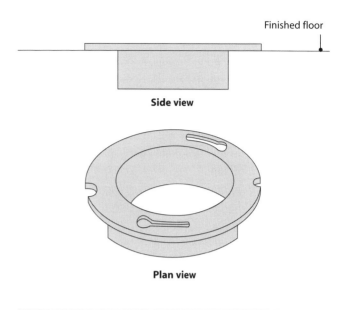

Side view

Plan view

Figure 21-10
Closet floor flange

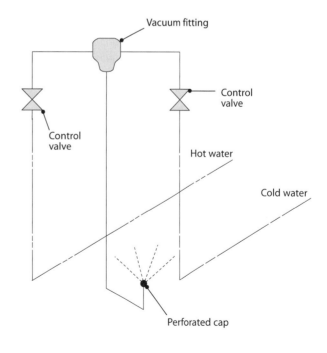

Figure 21-11
Garbage can washer piping diagram

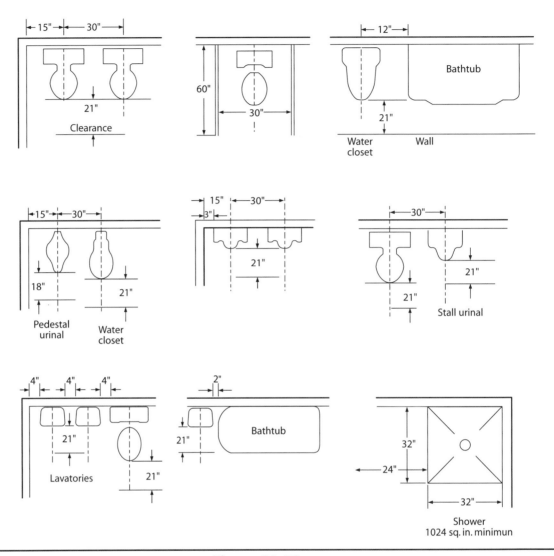

Figure 21-12
Minimum fixture clearances

- *Urinals*, whether pedestal, stall or wall-hung, require a minimum 30-inch center-to-center spacing when set in a battery installation. They must have a minimum 15-inch clearance from the center of the urinal to any finished wall. Pedestal urinals must have a minimum clearance of 18 inches from the front of the urinal to any finished wall, door or other plumbing fixture. Stall and wall-hung urinals must have a minimum clearance of 21 inches from the front of the urinal to any finished wall, door or other plumbing fixture.

- *Lavatories* are manufactured in various designs and widths, so center-to-center measurements don't apply. The minimum clearance is measured from the edge of the lavatory to the nearest obstruction. Lavatories must have a minimum clearance of 4 inches from the edge to any finished wall, and 2 inches from the edge to the edge of a bathtub. Lavatories in battery installation must have a minimum clearance of 4 inches from lavatory edge to lavatory edge. Finally, they must have a clearance of 21 inches from the front of the lavatory to any finished wall, door or other plumbing fixture.

- *Shower* compartment or stalls must have a minimum clearance of 24 inches from the opening to any finished wall, door or other plumbing fixture for easy entry and exit.

Fixtures for the Handicapped

Federal and state laws both now require public buildings and privately owned multistory buildings to have toilet facilities available for the physically handicapped. The law states that all *required* restrooms in a public building must provide facilities for the handicapped. In large multistory public buildings, you must provide facilities for each sex on each floor and mark halls and restroom doors with readily visible signs showing the location of the restrooms. Single-family residences and buildings that are considered hazardous (where handicapped persons are not likely to be employed) are exempt from this requirement.

In each *required* restroom with one or more water closets, one toilet must comply with standards created by the President's Committee on Employment of the Physically Handicapped and by the American National Standards Institute (ANSI). You'll seldom find these standards in the plumbing code or in other reference books.

1) Toilet rooms must have at least one toilet stall that's at least 3 feet wide and 4 feet 8 inches deep. It has to have handrails on each side, parallel to the floor and 33 inches high. Handrails should be 1^1/$_2$ inches in outside diameter, with a 1^1/$_2$-inch clearance between the rail and wall. Handrails must be securely fastened at each end and at the center. If there's a door, it has to be at least 32 inches wide and swing out. The water closet must have a seat that's 17 to 19 inches from the floor.

2) Toilet rooms must have at least one lavatory with a narrow apron which, when mounted at standard height, is accessible by people in wheelchairs. It may be mounted higher than standard height, when the particular design demands, as long as it's accessible to wheelchairs. The drain pipes and hot water pipes must be insulated to prevent burns to someone in a wheelchair who has no feeling in his or her limbs.

3) Toilet rooms for men must have at least one floor-mounted urinal that's level with the main floor of the toilet room, *or* a wall-mounted urinal with a basin opening no higher than 19 inches from the toilet room floor.

4) There must be an appropriate number of water coolers or fountains accessible to and usable by the physically handicapped. They can be hand-operated or hand- and foot-operated, with up-front spouts and controls. Wall-mounted, hand-operated coolers can serve the able bodied and the physically handicapped equally well when mounted 36 inches from the floor. Fully-recessed water fountains aren't acceptable for use by the handicapped, but they can be set in an alcove if it's at least 32 inches wide.

Minimum Fixture Requirements

Whenever the code requires plumbing fixtures, it regulates the minimum number and type of fixtures. This includes all premises intended for human occupancy or use.

The minimum number and type of fixtures set by most plumbing codes is based on the type of occupancy and the anticipated number of people who'll use the facilities. Codes vary considerably in how they arrive at the number of fixtures needed. This book will help you understand and interpret your particular code's method of computing the required toilet facilities. But refer to your local code for the exact requirements.

Residences

The fixtures required vary according to the number of dwelling units in the building.

▮ *Single-family residences:* Minimum requirements are one kitchen sink, one water closet, one lavatory and one bathtub or shower unit, and provision for a clothes washing machine. Hot water is optional in some codes and mandatory in others.

▮ *Duplex residential units:* One kitchen sink, one water closet, one lavatory, and one bathtub or shower unit are mandatory. There must also be provision for a clothes washing machine for each unit, or for one machine if it's available to all residents. Hot water is optional in some codes and mandatory in others.

■ *Apartment units:* Each unit must have at least one kitchen sink, one water closet, one lavatory, and one bathtub or shower unit. Provide for a clothes washing machine for each unit unless there's a centrally-located laundry room with the correct ratio of machines according to the particular code. The *Uniform Plumbing Code* requires two laundry trays or two automatic washer standpipes, or a combination, for each 10 apartment units. For example, an apartment building with 15 units would double the minimum requirements, so you'd need four laundry trays or four automatic washer standpipes, or a combination. The *Standard Plumbing Code* requires only one automatic washer standpipe for each 12 apartments. For that 15-unit apartment building, you'd only need two washer standpipes. Never forget to check your local code for their requirements.

When there's a central washing facility for residents in a complex of several buildings, most codes establish a maximum distance to the most remote unit. The distance from the entrance of the most distant apartment building can't exceed 400 feet.

Hot water generating facilities are generally installed in all buildings. Again, hot water is optional in some codes and mandatory in others.

Places of Employment

The number of toilet fixtures in manufacturing plants, heavy industry, warehouses and similar establishments is based on the number of employees. Your local authority can change the percentage ratio and type of fixtures required for males and females. They may alter the requirements in Figure 21-13 if you can provide data which shows that some other fixture ratio is more appropriate.

Let's consider one example. Assume that toilet facilities are needed for a medium-sized manufacturing plant employing 100 persons. Some building codes require a ratio of 50 percent male and 50 percent

Uniform Plumbing Code						
Males				Females		
No. of males	Water closets	Urinals[1]	Lavatories	No. of females	Water closets	Lavatories
1 - 10	1		1	1 - 10	1	1
11 - 25	2		3	11 - 25	2	3
26 - 50	3		5	26 - 50	3	5
51 - 75	4		8	51 - 75	4	8
76 -100	5		10	76 -100	5	10

Notes
■ Over 100, add 1 water closet for each additional 30 persons.
■ Over 100, add 1 lavatory for each additional 15 persons.
■ Where exposure to poisonous, infectious or irritating materials, provide 1 lavatory for each 5 persons.
■ Where exposure to excessive heat or to skin contamination with poisonous, infectious or irritating materials, provide 1 shower for each 15 persons.
■ Provide 1 drinking fountain for first 150 persons and one for each 300 persons thereafter.
■ 24 linear inches of wash sink or 18 inches of a circular wash sink, when adequate water outlets for such space is provided, shall be considered equivalent to 1 lavatory.

[1]Urinals can replace only $1/3$ of the minimum required water closets. For example, if 3 water closets are required, 1 could be a urinal.

From the UPC® with permission of IAPMO ©1997

Figure 21-13
Minimum plumbing fixtures for places of employment (employees only):
Manufacturng, industrial, warehouses, workshops, foundries or similar establishments

female facilities. Other codes for the same type occupancy use a percentage ratio of 75 percent male and 25 percent female. Obviously, if these ratios were used rigidly, many installations would have an imbalance of toilet fixtures. The plumbing plans examiner may request a notarized letter from the owner giving the maximum number of probable male and female employees in this particular plant. Then the examiner can determine the correct number and type of plumbing fixtures from the local requirements.

Provide at least one drinking fountain for up to 150 employees, accessibly located within 50 feet of all operational processes. You can't locate the drinking fountain in a restroom or vestibule to a restroom.

You can substitute wash-up sinks for lavatories where the type of employment warrants their use (manufacturing plants, for example).

Establishments that have 10 or more offices or rooms and employ 25 persons or more must have a service sink on each floor. Manufacturing plants that could subject the employees to excessive heat, infection or irritating materials must provide a shower for bathing for each 15 persons.

Businesses that employ up to nine people and that don't cater to the public (such as storage warehouses and light manufacturing buildings) have less rigid requirements. *Some codes* would consider one water closet and one lavatory for both sexes adequate in these applications. But observe the following conditions:

▋ If the minority sex exceeds three persons, separate toilet facilities are required. For example, where four males and five females (or vice versa) are employed, separate toilet facilities must be provided.

▋ If the number of males employed exceeds five, a urinal must be provided.

Businesses that are frequented by the public must provide toilet facilities for the number of employees and the public that's reasonably anticipated, unless they have special permission to do otherwise. There are two classifications of public use which determine the number and type of plumbing fixtures required:

1) Establishments that provide countable seating capacity, such as churches, theaters, stadiums and restaurants, are in the first classification.

Standard Plumbing Code	
Type of occupancy	**Square feet net = one person**
Office and public buildings	100 sq. ft.
Retail stores	200 sq. ft.
Restaurants	40 sq. ft.
Dining rooms, clubs, lounges, etc.	40 sq. ft.
Do-it-yourself laundries	50 sq. ft.
Barber and beauty shops	50 sq. ft.
Dormitories	50 sq. ft.
Assembly areas, standing or waiting rooms	70 sq. ft.
Theaters, auditoriums, churches, etc.	70 sq. ft.

Note The occupant content and the number of required plumbing facilities for occupancies not listed above will be determined by the plumbing official.

From the Standard Plumbing Code®

Figure 21-14
Square feet per occupant based on net floor area

2) Establishments such as in retail stores, office buildings, and similar establishments that have no countable seating capacity make up the second classification.

In the *Standard Plumbing Code*, the square foot area determines the facilities required for the second classification. Figure your required fixtures by using either Figure 21-14, or the occupant load based on the egress requirements in your local code.

Public Places with Seating Capacities

Public assembly facilities such as churches, theaters, stadiums and similar establishments use the percentage ratio for determining the number and type of plumbing fixtures required. Some plumbing codes use different ratios for churches. For example, the *SPC* uses a percentage ratio of 50 percent male

Uniform Plumbing Code						
Males				**Females**		
No. of males	Water closets	Urinals	Lavatories	No. of females	Water closets	Lavatories
1 - 150	1	1	1	1 - 75	2	1
151 - 300	2	2	1	76 - 150	2	1
				151 - 225	4	2

Notes
- Lavatories: 1 per 2 water closets.
- The total number of water closets for females must equal the total number of water closets and urinals required for males.
- Drinking fountains: 1 for the first 150 persons and 1 additional fountain for each 300 persons thereafter.

From the UPC® with permission of IAPMO ©1997

Standard Plumbing Code								
Males					**Females**			
No. of males	Water closets	Urinals	Persons (total)	Lavatories	No. of females	Water closets	Persons (total)	Lavatories
1 - 50	2	1	1 - 200	1	1 - 50	2	1 - 200	1
51 - 100	2	1		1	51 - 100	3		1
101 - 200	2	2		1	101 - 200	4		1

Notes
- In male restrooms urinals are required at a rate equal to 1/2 of the required water closets.
- 2 drinking fountains are required for the first 350 persons. When over 350 persons, add 1 fountain for each 400 persons thereafter.

From the Standard Plumbing Code®

Figure 21-15
Comparison of two codes — Minimum plumbing fixtures for churches

and 50 percent female. The *UPC* uses a ratio of one-third male to two-thirds female for churches. Figure 21-15 shows the tables from the two codes.

Water closets for public use must be separated from the rest of the room and from each other by stalls made of some impervious material.

Toilet rooms connected to public rooms or passageways must have a vestibule or must otherwise be screened or arranged to insure decency and privacy. The toilet rooms of both sexes can't share a common vestibule.

Food and Drink Establishments

Use Figure 21-16 to determine the minimum toilet facilities for establishments where food and drink are served and consumed on the premises. This includes cafeterias, restaurants, private clubs and similar

establishments with countable seating capacities. Most codes follow the percentage ratio of 50 percent male to 50 percent female.

Any establishment that caters to drive-in service (like fast food restaurants) must provide adequate toilet facilities. According to the *UPC*, the number of occupants is equal to the number of parking stalls. So 40 parking stalls represent 40 people. Assume that 20 are males and 20 are females. To determine the required toilet facilities, use Figure 21-16. You'd have to provide one water closet, one urinal and one lavatory for males, and two water closets and one lavatory for females.

Public food service establishments that offer only take-out service aren't required to provide guest toilet facilities. Only toilet facilities for employees are required.

The floors and walls of public toilet rooms must have tile or other impervious materials to a height of 5 feet.

If the seating capacity is unknown, the *SPC* uses the square foot method (see Figure 21-14) to determine the number of people who'll occupy the premises. Other codes base the occupant load on the egress requirements of the building code. Check your local code for its particular requirements.

Toilet rooms must have easy and convenient access for both patrons and employees. The restrooms must be located within 100 feet along a line of travel from the nearest exit to the dining room, bar or food service area. Toilet rooms must be located on the same floor as the area they serve.

In food or drink establishments where dishes, glasses, or cutlery are reused, provide for a dishwashing machine or suitable three-compartment sink. Where food or drink are prepared or served, you must install a hand sink for employees' use. It's not enough to have lavatories in adjoining toilet rooms.

Public Places Without Seating Capacities

In computing restroom facilities for public places like shopping centers, retail stores and office buildings, you need to determine the amount of habitable floor space (area). You can deduct uninhabitable space like corridors, stairways, vertical shafts, and equipment rooms from the gross floor area. Then use the net square footage to find the occupant load factor for the business. Figure 21-14 shows net square footage per expected occupant, according to the *SPC*. Other codes may base the occupant load on the egress requirements instead of the square footage.

Roughing-In

Roughing-in the fixtures is possibly the single most critical point where plumbers display their skill (or lack of skill). Proper roughing-in of the waste and water outlets requires both knowledge and good workmanship. The plumbing fixtures will be an important part of the building throughout the life of the building. They need to be done right.

It's important that you have a thorough knowledge of roughing-in measurements for various types of plumbing fixtures. You must know the height, distance and location for waste and water outlets for

wall-hung and floor-mounted fixtures. Memorize these measurements for common fixtures or jot them down in a notebook you can keep in your tool box. For special fixtures, you can get complete roughing-in information from the manufacturer or distributor.

We'll cover roughing-in dimensions for common fixtures — bathtubs, water closets, lavatories, showers, kitchen sinks, service sinks, urinals, bidets and drinking fountains. The roughing-in measurements in Figures 21-17 through 21-21 are for American Standard fixtures, but they're similar to those of other manufacturers.

Note that the standard roughing-in measurement for all water closets is 12 inches from the finished wall. If the water closet outlet is roughed too close or too far from the finished wall, there are special water closets that you can set at 10 or 14 inches from the wall.

Uniform Plumbing Code						
Males				**Females**		
No. of males	Water closets	Urinals	Lavatories	No. of females	Water closets	Lavatories
1- 50	1	1	1	1- 50	2	1
51-150	2	1	1	51-150	3	1
151-300	3	2	2	151-300	5	2

Notes
▌ Over 300 persons, add 1 water closet for each additional 200 persons.
▌ Over 150 persons, add 1 urinal for each additional 150 persons.
▌ From 201 to 400 persons, add 1 lavatory. Over 400 persons, add 1 lavatory for each additional 400 persons.
▌ The total number of water closets for females shall equal the total number of water closets and urinals required for males.

From the UPC® with permission of IAPMO ©1997

Figure 21-16
Minimum plumbing facilities for drink and/or eat establishments

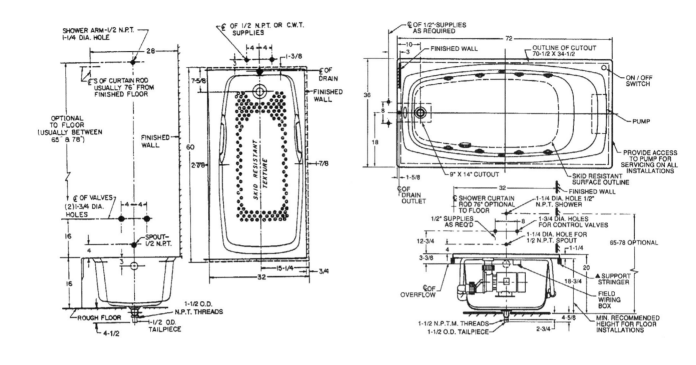

A Spectra™ bath

B Oxford™ whirlpool

C Custom-line™ sink

D Custom-line™ sink — double compartment

Courtesy: American Standard

Figure 21-17
Roughing-in measurements for tubs and sinks

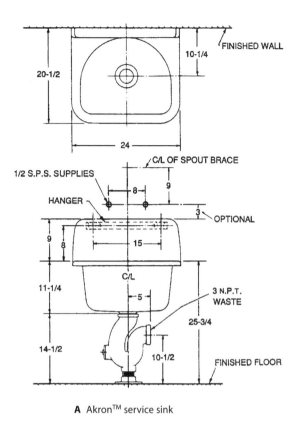

A Akron™ service sink

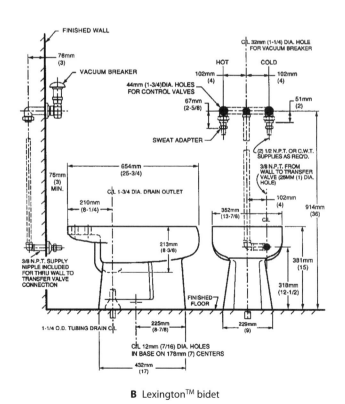

B Lexington™ bidet

Figure 21-18
Roughing-in measurements for service sink and bidet

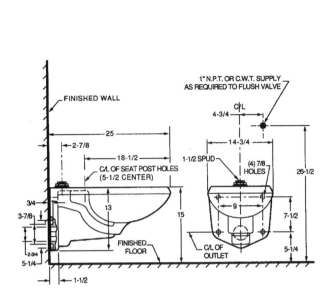

A Afwall toilet

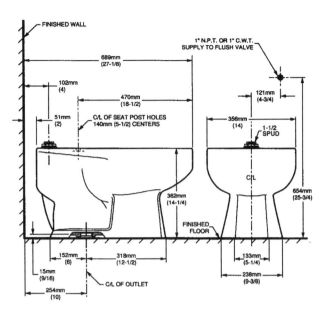

B Madera toilet

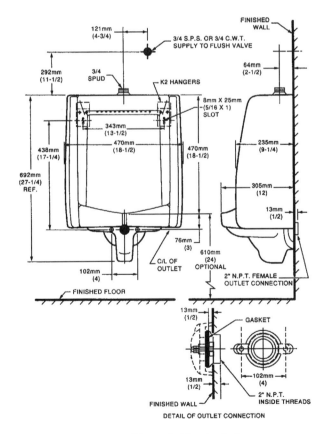

C Wall-hung Washbrook urinal

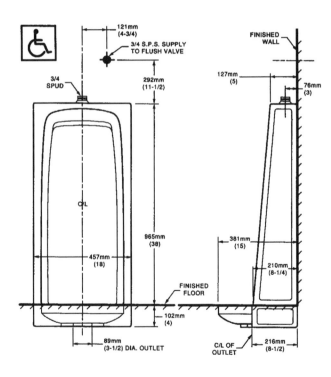

D Stallbrook™ urinal

Courtesy: American Standard

Figure 21-19
Roughing-in measurements for toilets and urinals

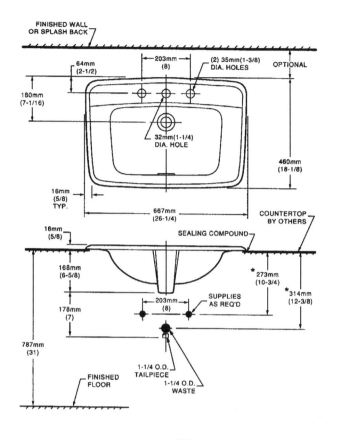

A Plaza Suite™ lavatory

B Wall-hung Penlyn™ lavatory

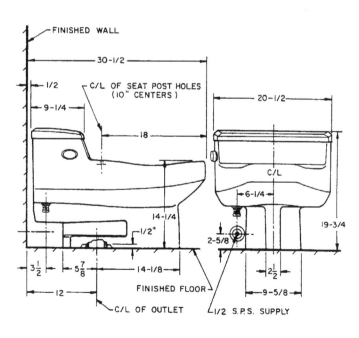

C One-piece Roma toilet

Courtesy: American Standard

Figure 21-20
Roughing-in measurements for lavatories and toilets

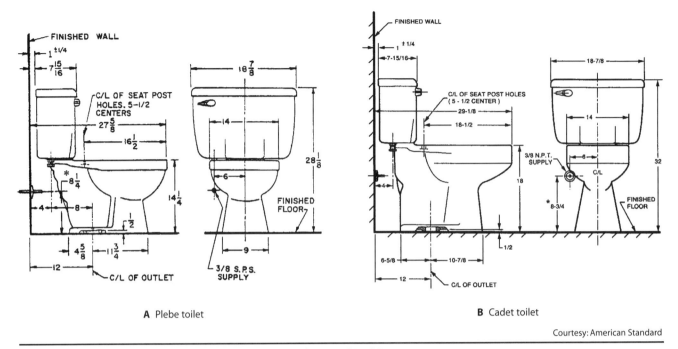

A Plebe toilet

B Cadet toilet

Courtesy: American Standard

Figure 21-21
Roughing-in measurements for toilets

Plumbing Fixture Carriers

Toilet rooms are the most susceptible of all rooms to unsanitary conditions. The bases of on-the-floor fixtures are natural areas for accumulated filth that's nearly impossible to remove. Bathroom floors made of wood tend to deteriorate next to and beneath toilet fixtures. Off-the-floor water closets can solve both of these problems. They've been gaining popularity for both commercial and residential use.

Some apprentices and journeymen seldom have an opportunity to work with off-the-floor plumbing fixtures. You can use the illustrations in Figures 21-22 through 21-25 to become familiar with their installation methods.

Figure 21-22 illustrates a flush valve on-the-floor and off-the-floor water closet commonly used in commercial buildings. There are many factors to evaluate when planning an installation, including cost, material, labor required, space available, and handling requirements. With on-the-floor water closets, you have to penetrate and sleeve the slabs at each fixture to accommodate waste piping, which often has to be

suspended below the slab. Suspended piping like that needs to be concealed in multistory installations, with drop or furred ceilings in the rooms below.

Figure 21-23 shows tank type on-the-floor and off-the-floor water closets commonly used in residential buildings. You can see some of the advantages of the off-the-floor installation. You don't have to penetrate the slab or floor at each fixture like you do with on-the-floor installations. There's a clear, unobstructed floor for cleaning. And deterioration isn't a problem.

Figure 21-24 is a typical floor plan showing a battery of water closets. This illustration is important because it shows you how to identify left and right hand closet carriers. Figure 21-25 shows three ways to vent a horizontal run of fixtures.

A single-family residence never has battery installations like commercial buildings. Residential carriers have fewer parts so they're easier to assemble. They're designed to receive the waste from a single water closet or at most two water closets (from back-to-back installations). Most residential carriers are compatible with the newer piping materials.

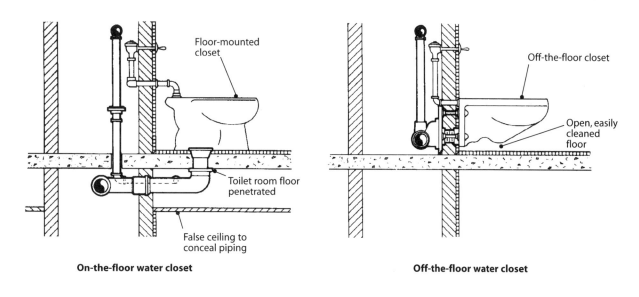

Floor-mounted closet

Off-the-floor closet

Open, easily cleaned floor

Toilet room floor penetrated

False ceiling to conceal piping

On-the-floor water closet

Off-the-floor water closet

Figure 21-22
Commercial water closet installation

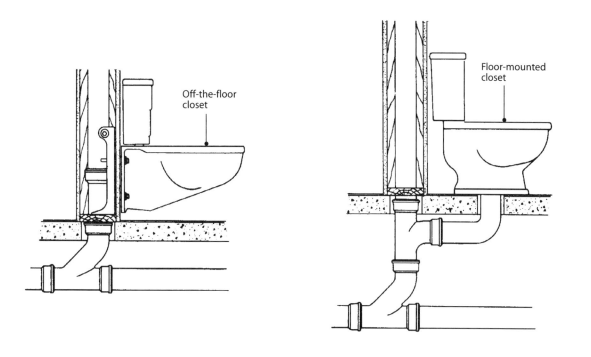

Off-the-floor closet

Floor-mounted closet

Figure 21-23
Residential water closet installation

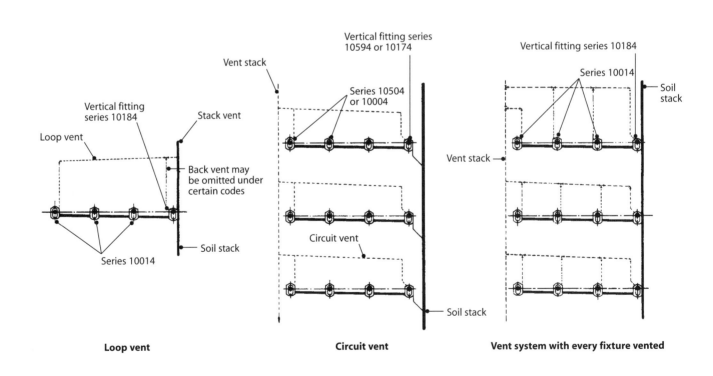

Stack

Flow from left to right →

Flow from right to left ←

Face stack from
fixture side

Left hand   Left hand   Right hand   Right hand

Notes
1. If flow in waste line is from the right of the stack, the closet fitting should be a right hand fitting.
2. If flow in waste line is from the left of the stack, the closet fitting should be a left hand fitting.

Figure 21-24
Diagram to determine left or right hand closet fittings

Vent stack

Vertical fitting series
10594 or 10174

Vertical fitting series 10184

Vertical fitting
series 10184

Series 10504
or 10004

Series 10014

Soil
stack

Loop vent

Stack vent

Back vent may
be omitted under
certain codes

Vent stack

Series 10014

Soil stack

Circuit vent

Soil stack

Loop vent

Circuit vent

Vent system with every fixture vented

Figure 21-25
Venting a horizontal run of fixtures — Josam carriers

Review Questions (answers are on page 316)

1. What is required of plumbing fixtures constructed of pervious materials such as tile or marble?

2. What must you provide in a bathroom where there's no natural ventilation available?

3. What potential danger exists when a bathroom doesn't have adequate lighting or ventilation?

4. What toilet bowl design must be installed in facilities intended for public use?

5. What seat type is required for public toilet bowls?

6. What purpose does a toilet tank refill tube serve?

7. What purpose does an overflow tube in a toilet tank serve?

8. What must a toilet have if it uses a flushometer rather than a tank?

9. What operation must a toilet flushometer accomplish after being manually activated?

10. How many toilets can a flushometer serve?

11. Why should a toilet flushometer be installed so it's readily accessible?

12. How should a wall-hung urinal be supported?

13. How should a wall-hung lavatory be finished at the wall contact point?

14. What are the two basic designs for urinals?

15. Why must stall urinals be recessed slightly below the finished floor?

16. What type of strainer is required for the waste opening of a floor-mounted stall urinal?

17. How are cabinet-mounted lavatories secured to the countertop?

18. What's the minimum outside diameter for lavatory waste outlets?

19. Where circular-type multiple wash sinks are used, how many inches of wash sink circumference represents one lavatory?

20. What's the minimum bathtub waste and overflow size?

21. Why do some codes today prohibit the use of a trip waste?

22. What's required of the joints for a bathtub that's recessed into the finished wall?

23. What type of wall materials must you use for a recessed bathtub?

24. What's the minimum size waste outlet required for a shower compartment?

25. What are the design requirements for shower strainers?

26. What's the minimum floor area required for any shower compartment?

27. What minimum weight per square foot do most codes require for lead shower pans?

28. How do you protect lead or copper shower pans from corrosion when they're installed on concrete floors?

29. How high should the sides of a shower pan extend above the finished curb?

30. At what point in a building's construction should a shower pan be prepared for inspection?

31. When may shower pans be omitted?

32. How high must the walls of a shower compartment extend above the floor?

33. What's the minimum diameter required for a laundry tub waste?

34. How large should the waste opening be on a domestic kitchen sink with a waste disposer unit?

35. Name two common fixtures provided with an overflow.

36. Why must the overflow pipe or passageway from a fixture be connected on the inlet side of the fixture trap?

37. What type of fixtures are not required to have a strainer or stopper?

38. What determines the minimum size of a fixture strainer?

39. What do some codes require in new construction when a waste disposer is installed on a two-compartment sink?

40. What fitting must you use when you install a waste disposer on an existing two-compartment sink using a single trap?

41. Through what device may a commercial food waste grinder in a restaurant *not* discharge?

42. What's the minimum size waste opening required for a commercial sink?

43. How should a commercial food waste grinder be trapped and vented?

44. When an air gap fitting is required for a dishwasher waste pipe, where does the code require you install it?

45. By most code standards, what's the maximum distance allowed between a dishwasher and the sink waste connection?

46. What must the waste pipe from the dishwasher connect to if there's a food disposal unit installed in a sink?

47. Where should the flange for a floor fixture be set?

48. What gasket materials are approved for use with plumbing fixtures having a flanged connection?

49. Into what category does the code place floor drains?

50. Why must floor drain traps have a permanent water seal?

51. In what area can drinking fountain waste *not* be discharged into a floor drain?

52. What type of equipment installed in a room disallows the installation of any kind of plumbing fixture?

53. How do you protect the water supply from backsiphonage when installing special fixtures with waste and water connections?

54. What must you consider in regard to spacing and clearances when installing any plumbing fixture?

55. What's the minimum center-to-center spacing required for water closets when set in battery installations?

56. What's the minimum required distance from the front of a urinal to any finished wall?

57. Why are center-to-center measurements not applicable to lavatories?

58. What's the minimum clearance required from the opening of a shower compartment or stall to any finished wall?

59. Why are public buildings and privately owned multistory apartment buildings now required to have toilet facilities for the physically handicapped?

60. What buildings are *not* required to provide toilet facilities for the physically handicapped?

61. In a public toilet room with six toilets, how many must be provided for use by the physically handicapped?

62. What are the minimum fixture requirements for a single-family residence?

63. What type of water facility is optional in some codes and mandatory in others?

64. How do you determine the number of toilet fixtures required in a place of employment?

65. What must be provided if toilet rooms are connected to public rooms or passageways?

66. What determines the minimum toilet facilities needed for food establishments catering to drive-in service?

67. If the seating capacity in a restaurant is unknown, what method does the *Standard Plumbing Code* use to determine the number of persons who'll occupy the premises at one time?

68. Rather than using the square-foot method, what do some codes use to determine the occupant load for a restaurant when the seating capacity is unknown?

69. How do you compute the restroom facility required for public places such as shopping centers, retail stores or large office buildings?

70. What knowledge is required for proper roughing-in of the waste and water outlets for various types of plumbing fixtures?

71. Where can you get complete roughing-in information for *special* plumbing fixtures?

72. What's the standard roughing-in measurement for all water closets?

73. Of all the rooms in a house, which room is the most susceptible to unsanitary conditions?

74. What type of bathroom plumbing fixtures do some apprentices and journeymen seldom have the opportunity to work with?

75. With what materials are most residential carriers designed to be compatible?

Answers to Review Questions

Chapter 2

1. **The two parts of a private sanitary drainage system are:**

   1) all the pipes installed within the wall line of a building on private property for the purpose of receiving liquid waste or other waste substances (whether in suspension or in solution), and 2) the pipes which convey this waste to a public sewer or a private, approved sewage disposal system. *See page 7*

2. **The aim of municipal codes in relation to drainage systems is:**

   to protect the public health. *See page 7*

3. **Most requests for clarification and resolution brought to the Boards of Rules & Appeals center on the code section dealing with:**

   the drain, waste and vent systems. *See page 7*

4. **The section of the code that most questions and isometric drawings for the journeyman and master's examinations are taken from is:**

   the section on sanitary drainage and vent systems. *See page 7*

5. **Isometric drawings serve to provide:**

   a clear means of communication between plumbing professionals. *See page 7*

6. **It's important for a plumbing contractor to know how to make and interpret isometric drawings in order to:**

   1) be able to estimate the cost of a job, and 2) show the job foreman how to rough-in a particular job. *See page 7*

7. **The three basic pipe angles you need to illustrate in an isometric drawing of a plumbing system are:**

   1) the horizontal pipe, 2) the vertical pipe, and 3) the 45-degree angle pipe. *See page 7*

8. **The lines on an isometric drawing represent:**

   pipe and fittings. *See page 9*

9. **The purpose of a horizontal twin tap sanitary tee is:**

   to permit two similar fixtures to connect to the same waste and vent stack at the same level. *See page 9*

10. **Plumbing fixtures are identified in isometric drawings and floor plans by:**

    several common abbreviations, either letters (such as L for lavatory) or shortened words (such as LAV, also used for lavatory). *See pages 9 & 12*

11. **Another term for a public sewer is:**

    municipal sewer. *See page 14*

12. **Two other terms used for a building sewer are:**

    private sewer and sanitary sewer. *See page 14*

13. **A building drain is also known as a main because:**

    it acts as the principal artery to which you can connect other drainage branches of the sanitary system. *See page 14*

14. **Plumbers often refer to a fixture drain by the terms:**

    sink arm or lavatory arm. *See page 14*

15. **The kind of waste a waste pipe carries is:**

 only liquid waste from plumbing fixtures, excluding water closets and bed pan washers. It does not convey waste that includes fecal matter. *See page 14*

16. **Another term that's used to identify a waste pipe is:**

 a wet vent. *See page 14*

17. **A soil stack is:**

 the vertical section of pipe in the plumbing system that receives the discharge of water closets, with or without the discharge from other fixtures. *See page 14*

18. **The function a branch interval serves is:**

 the same function as a soil stack. It becomes an integral part of a soil stack, except that its vertical height can never be less than 8 feet. *See page 14*

19. **A horizontal branch is:**

 the portion of a drainpipe that extends laterally from a soil or waste stack. It receives the discharge from one or more fixture drains. *See page 14*

20. **The main factor you use to determine pipe size within a drainage system is:**

 the maximum fixture unit load. *See page 15*

21. **Besides maximum fixture unit value, the three additional factors you must take into consideration when sizing drainage piping are:**

 1) the types of fixtures used, 2) the slope of the drainpipe, and 3) the vertical length of drainpipe. *See page 15*

22. **You would use the code book tables that list the various fixture load values to:**

 compute the total fixture load for any type of plumbing system. *See page 15*

23. **Special fixtures are connected to the drainage system by:**

 indirect means. *See page 15*

24. **Among the devices that are considered special fixtures are:**

 drinking fountains, bottle coolers, ice making machines, milk and soft drink dispensers, and coffee urns. (Any two answers are correct.) *See page 15*

25. **Continuous and intermittent flow devices that you can connect to a drainage system include:**

 sump ejectors, pumps, air conditioning equipment and similar devices. (Any two answers are correct.) *See page 17*

26. **For each gallon per minute of flow from continuous flow devices such as sump ejectors, the *Uniform Plumbing Code* allows:**

 2 fixture units. *See page 17*

27. **The major difference between codes is:**

 how the fixture units are assigned. *See page 17*

28. **The generally-accepted fall per foot for horizontal pipe is:**

 $^1/_4$ inch. *See page 18*

29. **Restrictions, limitations and exceptions in the code book will always supersede:**

 the established pipe sizes and fixture units in any code drainage table. *See page 20*

30. **Most codes regard fixtures with waste openings larger than the waste pipe to which they need to connect as:**

 prohibited. *See page 21*

31. **The code calls a stack that receives the discharge from a water closet:**

 a soil stack. *See page 21*

32. **The minimum size vent required by the code to serve a water closet is:**

 2 inches. *See page 22*

33. **The code-accepted minimum size for a main vent stack in a building is:**

 3 inches. *See page 22*

34. **When sizing drainage piping in a multistory building, you accumulate the fixture unit load at:**

 the base of each stack. *See page 22*

35. **The procedure you should follow in sizing vertical drainage pipes in a multistory building is:**

 to start with the top drainage fixture units and work down to the building drain, accumulating the total fixture units at the base of the stack. *See page 23*

36. **From one end to the other, the size of a vertical waste/soil stack:**

    does not vary. It must remain the same size throughout. *See page 23*

37. **A soil or waste stack can't be smaller than the largest horizontal branch pipe connected to it, except:**

    when connecting to a 3- × 4-inch water closet bend, which is not considered by some codes to be a reduction in size. See Figures 2-24, 2-25 and 2-26. *See page 23*

38. **The main considerations when sizing drain and vent pipes for future fixtures are:**

    the number and type of fixtures to be used. *See page 25*

39. **When sizing a vertical stack with an offset of 45 degrees or less, it's defined as:**

    straight, and sized as a straight vertical stack. *See page 25*

40. **The horizontal portion of a vertical stack with an offset greater than 45 degrees is sized:**

    as a building drain. *See page 26*

41. **In a vertical stack, the minimum distance for an offset above or below the horizontal branch is:**

    2 feet. *See page 26*

42. **The minimum required separation of horizontal branch drains in a multistory building is:**

    8 feet. *See page 26*

43. **Since World War II, synthetic detergents have changed the characteristics of household waste by:**

    increasing the quantity of suds being discharged. *See page 27*

44. **There are several appliances and/or fixtures that are considered suds-producing by codes, including:**

    kitchen sinks, bathtubs, clothes washing machines, dishwashers, and laundries. (Any two answers are correct.) *See page 27*

45. **When suds-producing fixtures and appliances discharge into an improperly-designed drainage and vent system, it can cause:**

    suds to bubble up into fixtures on the lower floors of the building. *See page 27*

46. **When it's impossible for waste to drain by gravity into the building drainage system, it must be discharged:**

    into an approved sump and then pumped into the building drainage system. *See page 28*

47. **Sumps and receiving tanks should be located so they are accessible for:**

    inspection, repairs and cleaning. *See page 28*

48. **Most codes require that sumps for public use be equipped with a duplex pumping system because:**

    a duplex system allows pumps to discharge waste alternately, and in case of repairs, one pump can remain in service. *See page 28*

49. **The two devices required in the discharge line between the pump and the gravity system are:**

    a check valve and a gate valve. *See page 28*

50. **For most codes, the minimum acceptable size for a sump discharge pipe is:**

    3 inches. *See page 29*

Chapter 3

1. **The two causes of fixture trap seal loss that vent systems protect against in normal fixture use are:**

    siphonage and back pressure. *See page 33*

2. **The function of a branch vent is:**

    to connect one or more individual vents to a vent stack. *See page 33*

3. **Battery venting is:**

    venting any group of two or more similar adjacent fixtures using a common horizontal drainage line. *See page 33*

4. **A common vent is used in a plumbing system to:**

    vent two fixture branches that are installed at the same level in a vertical stack. *See page 34*

5. **A continuous vent is:**

    the vertical portion that is a continuation of the drain to which it connects. *See page 34*

6. **Another term used to describe a continuous vent is:**

    a stack vent. *See page 34*

7. **The term used to describe a vent that does not receive any sewage discharge is:**

   a dry vent. *See page 34*

8. **The number of fixture traps an individual vent is installed to serve is:**

   one. *See page 35*

9. **Another term used to describe an individual vent is:**

   a back vent. *See page 35*

10. **A loop vent differs in function from a circuit vent in that:**

    it's designed to loop back and connect to the stack vent, rather than the vent stack. *See page 35*

11. **A main vent is defined in the code as:**

    the principal artery (pipe) of a venting system to which vent branches may be connected. *See page 35*

12. **The primary function of a relief vent is:**

    to provide a route for the circulation of air between the drainage and the vent systems. *See page 36*

13. **The trade name for a relief vent is:**

    a re-vent. *See page 36*

14. **The definition of a side vent is:**

    a vent connecting to a horizontal drain pipe through a fitting at an angle not greater than 45 degrees to the vertical. *See page 36*

15. **The definition of a stack vent is:**

    the dry portion of a soil or waste pipe that extends above the highest horizontal drain connected to the stack and terminates above the roof. *See page 36*

16. **The function of a vent header is to:**

    connect two or more vent pipes to the main vent at one point, or extend through the roof separately. *See page 36*

17. **The primary purpose of a vent stack is:**

    to provide circulation of air to and from all parts of a drainage system. *See page 36*

18. **The two purposes a wet vent serves are:**

    as a vent and as a means to convey waste from fixtures other than water closets. *See page 37*

19. **The purpose a yoke vent serves is to:**

    prevent pressure changes in the stack. *See page 37*

20. **Problems that may occur if vent pipes are not properly sized and arranged are:**

    1) Plumbing fixtures may drain slowly, as if a partial stoppage exists. 2) Water closets may need several flushes to remove contents from the bowl. 3) Back pressure within the drainage pipe may force sewer gases up through the liquid seals and into the building. 4) Plumbing fixtures located too far from a vent pipe may siphon the liquid trap seal when the contents are released. 5) Frost may form on the inside of the vent pipe in cold climates, restricting the free flow of air within the vent system. (Any two answers are acceptable.) *See pages 37 & 39*

21. **Another term used to describe back pressure is:**

    positive pressure. *See page 37*

22. **Negative pressure in a fixture drain can cause:**

    siphonage of the fixture trap seal. *See page 39*

23. **In frost-prone climates, the *Standard Plumbing Code* requires the vent extension through the roof to be at least:**

    3 inches in diameter. *See page 39*

24. **The *Uniform Plumbing Code* requires that a roof vent in a cold climate terminate at least:**

    10 inches above the roof. *See page 39*

25. **The free flow of air within the sanitary drainage system prevents:**

    back pressure or siphoning action from destroying fixture trap seals. *See page 39*

26. **As the maximum fixture unit load increases, the height of the vent pipe:**

    must decrease. *See page 40*

27. **The vent stack requirement for a building with a single building sewer is:**

    one vent stack (of not less than 3 or 4 inches) extending through and above the roof. *See page 40*

28. **If a water closet is located in an accessory building, the minimum size vent accepted by code is:**

    2 inches. *See page 40*

29. **The minimum size "dry" vent allowed by code when venting a water closet is:**

    2 inches. (This is also true for a wet vent. *See page 40*

30. **The smallest individual vent stack size permitted by code is:**

    1¹/₄ inches. *See page 40*

31. **The minimum size vent that can be used for a 3-inch drain pipe is:**

    1¹/₂ inches. *See page 40*

32. **Since loop or circuit vents are rarely used, when laying out a sizing plan that includes them, you should:**

    first check with your local code requirements. *See page 41*

33. **The factor that determines the maximum length of any vent pipe is:**

    its diameter. *See page 41*

34. **The distance separation required between a horizontal vent pipe and the flood level rim of the fixture served is:**

    6 inches. *See page 41*

35. **Kitchen sinks located away from walls or partitions are called:**

    island sinks. *See page 42*

36. **When a shower (or any minor fixture) is located downstream from a water closet, it must be:**

    re-vented to prevent siphonage of the minor fixture trap. *See page 44*

37. **The number of vents needed to vent a horizontal offset in a vertical stack is:**

    two, a relief vent at the top of the lower section and a vent at the base of the upper section. *See page 44*

38. **A wet vent can be used to convey waste only from fixtures:**

    with low unit ratings. *See page 44*

39. **Most codes restrict the length of horizontal wet vents to:**

    a maximum of 15 feet. *See page 44*

40. **Fixtures that can't convey waste through a 2-inch horizontal wet vent include:**

    sinks, urinals, toilets or other pressure fixtures. (Any two answers are acceptable.) *See page 45*

41. **On a 3-inch horizontal wet vent, the number of fixture units some codes will allow is:**

    up to 16 fixture units. *See page 45*

42. **The vertical drain between two fixtures connected to a stack at different levels is called:**

    a wet vent. *See page 45*

43. **The vent required by code for a sump that receives body waste is:**

    a local vent. *See page 45*

44. **A local vent is needed on a sump:**

    to prevent an air-lock from occurring in the sub-building drainage system. *See page 45*

45. **The minimum size vent required for a sump receiving body waste is:**

    1¹/₂ inches. *See page 46*

46. **According to code, sumps receiving clear water waste:**

    do not need to be vented. *See page 46*

47. **Some codes permit the use of vertical combination waste and wet vent piping in:**

    high-rise buildings. *See page 46*

48. **The fixtures prohibited from discharging into a combination waste and vent stack are:**

    water closets or urinals of all types. *See page 46*

49. **The types of establishments that horizontal combination waste and vent systems are usually installed in are:**

    restaurants, or certain other commercial establishments such as supermarkets or large warehouses with floor drainage. *See page 48*

50. **Fixtures that may connect to a horizontal combination waste and vent system are:**

    sinks, dishwashers, floor sinks, indirect waste receptors, floor drains, or similar applications. (Any three are acceptable.) *See page 48*

51. **A major requirement before you can install a horizontal combination waste and vent system is:**

    having your plans and specifications approved by your administrative authority. *See page 48*

52. **Codes prohibit the connection of appurtenances that deliver large quantities of water to a horizontal combination waste and vent system because:**

    adequate venting must be maintained, and appurtenances delivering large quantities of water may cause venting problems. *See page 48*

53. **The minimum size floor sink waste pipes required for installation underground in a horizontal combination waste and vent system is:**

    2 inches. *See page 49*

54. **A vent terminal must extend above the roof:**

    at least 6 inches. *See page 49*

55. **The minimum distance from a door that you may install a terminal for a sanitary vent system is:**

    10 feet, unless it extends 3 feet (some codes, 2 feet) above the top of the door. *See page 49*

56. **When a horizontal vent extends through a wall and turns upward, the code requires that:**

    it must be effectively screened. *See page 49*

57. **You should never terminate a vent pipe under:**

    the overhang of any building. *See page 50*

58. **The height above ground level at which a vent pipe installed outdoors must terminate is:**

    10 feet. *See page 50*

59. **The "rule of thumb" for determining the minimum aggregate cross-sectional area for the vents required for venting a building drainage system is:**

    that it can't be less than that of the largest required building sewer. *See page 50*

60. **The formula you use to find the cross-sectional area of a pipe is:**

    the square the diameter of a circle (the pipe diameter) multiplied by .7854 equals the area of the circle (or the cross-sectional area of the pipe). *See page 50*

Chapter 4

1. **The purpose of a fixture trap is:**

    to prevent obnoxious odors and sewer gases from entering the building. *See page 53*

2. **Building traps are still required today only under the following conditions:**

    where sewer gases are extremely corrosive or contain high explosive gas content. *See page 53*

3. **When connected directly to the drainage system, plumbing fixtures must be equipped with:**

    a water seal trap. *See page 53*

4. **The protection of a liquid seal must be accomplished without:**

    materially affecting the flow of sewage or other waste liquids. *See page 53*

5. **The only kind of trap that doesn't have to be self-cleaning is:**

    the interceptor trap. *See page 54*

6. **The rule that governs the size of a trap outlet compared to its connecting fixture drain is:**

    A trap outlet can never be larger than the fixture drain to which it connects. *See page 54*

7. **A trap that depends on the action of movable parts to retain its water seal:**

    Can never be used. *See page 54*

8. **Traps that most model codes prohibit are:**

    bell traps, crown-vented traps, pot traps, running traps, $^3/_4$ S traps, full S traps, drum traps and traps with slip-joint nuts and washers on the discharge side of the trap above the water seal (Any two of these are acceptable answers.) *See page 54*

9. **The minimum depth of fixture trap seals is:**

    2 inches. *See page 54*

10. **The maximum depth of fixture trap seals is:**

    4 inches. *See page 54*

11. **The trap which is exempted from the normally required depth of a trap water seal is:**

    an interceptor trap. It always requires a deeper seal. *See page 54*

12. **Cleanouts installed on fixture traps below concrete floors on fill should not be located:**

    Anywhere. Fixture traps below concrete floors on fill are prohibited from having trap cleanouts. *See page 54*

13. **When installing a fixture trap, you determine the correct level in relation to:**

    its water level. *See page 54*

14. **A water closet may not have a separate trap because:**

    the code states that fixtures with integral traps can't be separately trapped. *See page 55*

15. **Two or three lavatories adjacent to each other may use a single trap:**

    when the waste outlets do not exceed 30 inches, center to center. *See page 55*

16. **When three lavatories are connected to the same trap, the trap must be:**

    located in the center. *See page 55*

17. **According to code, a fixture can be double trapped:**

    never. *See page 55*

18. **A food waste disposal unit in a restaurant may not discharge through a pot sink trap, regardless of size. According to code:**

    Commercial food waste disposal units must be separately trapped. *See page 55*

19. **According to some codes, a food waste disposal unit may discharge through a continuous waste of a sink served by a single trap if:**

    a directional tee is used. *See page 55*

20. **Some codes may allow a domestic clothes washer to use the same trap that serves a laundry tray when:**

    it's adjacent to the laundry tray. *See page 55*

21. **The code-approved materials which may be used for concealed fixture traps are:**

    cast brass, cast iron, lead, ABS plastic or PVC plastic. (Any two of these are acceptable answers.) *See page 56*

22. **The code prohibits concealed fixture traps from having:**

    cleanouts. *See page 56*

23. **When a tubular trap is used, its minimum gauge must be:**

    17 gauge. *See page 56*

24. **The acceptable materials commonly used for accessible fixture traps are:**

    cast iron, cast brass, lead, 17 gauge tubular brass or copper. For a plastic system you must use ABS or PVC plastic. (Any two of these are acceptable answers.) *See page 56*

25. **The code-approved materials that may be used for chemical, acid or corrosive wastes are:**

    borosilicate glass, high silicon cast iron, and lead pipe with walls at least $1/8$ inch thick. (There might be others when approved by local authorities.) *See page 56*

26. **The required wall thickness of lead pipe used to convey chemical, acid or corrosive wastes is:**

    $1/8$ inch. *See page 56*

27. **The maximum vertical drop from a shower outlet to the trap water seal is:**

    24 inches. *See page 56*

28. **The maximum vertical drop of a pipe that serves floor-connected fixtures having integral traps is:**

    24 inches. *See page 56*

29. **The maximum vertical drop of a floor drain (considered a fixture) to the trap water seal is:**

    24-inch tailpiece (pipe) limit. *See page 56*

30. **One of the main reasons for code-established fixture trap sizes is:**

    to drain the fixture rapidly. *See page 57*

31. **Every fixture trap must be protected against:**

    siphonage and back pressure. *See page 58*

32. **The developed length of a fixture drain includes the measurement from the crown weir to the vent pipe and the:**

    offsets and turns. *See page 58*

33. **When, because of the fixture location, the fixture drain exceeds the limits set by code, you must install:**

    a relief vent. *See page 58*

34. **Two adverse reactions to a fixture trap when a plumbing system is improperly installed are:**

    the fixture trap may lose its seal by siphonage or back pressure. *See page 60*

Chapter 5

1. **Before cleanouts became an essential part of the drainage system, plumbers had to:**

   cut a hole in blocked drainage pipe to insert a cleaning cable (and then patch it with a cement, or cement-like, mixture). *See page 63*

2. **Three important cleanout requirements for today's model codes are:**

   location, distance between cleanouts, and size. *See page 63*

3. **The most common plumbing maintenance problem is:**

   clogged drains. *See page 63*

4. **The most common cause of clogged drains is:**

   a foreign object or other substances not designed for a drainage pipe to handle. *See page 63*

5. **Some things you'll commonly find in a kitchen that can cause a clogged drain are:**

   grease, cooking oil, butter, gravy and coffee grounds (any two are acceptable). *See page 63*

6. **The dual purpose of a cleanout (or a cleanout tee) installed where a building sewer connects to the public sewer lateral is:**

   1) to water test the building sewer, 2) to insert a sewer cable for rodding. *See page 63*

7. **The minimum size cleanout accepted by code when installing a cleanout in a 4-inch pipe at the junction of the building drain and building sewer is:**

   4 inches. *See page 64*

8. **The name of the fitting that permits upstream as well as downstream rodding is:**

   a two-way cleanout. *See page 64*

9. **When a cleanout is extended to grade in an area subject to frequent traffic, the type cleanout head that should be used is:**

   countersunk. *See page 64*

10. **According to the *Uniform Plumbing Code*, the maximum separation distance between 4-inch cleanouts is:**

    100 feet. *See page 64*

11. **Near the base of each vertical waste or soil stack, you must install:**

    a cleanout. *See page 65*

12. **When a dead end is created by a cleanout, the maximum distance it can extend outside the building wall is:**

    5 feet. *See page 65*

13. **The plumbing fixture that's sometimes considered a substitute for a cleanout is:**

    a water closet. *See page 66*

14. **P traps into which floor drains with removable strainers discharge need not have a:**

    cleanout. *See page 66*

15. **The only time a roof stack terminal in a one-story building can be used as a cleanout is:**

    when the code (or authority) permits it. *See page 66*

16. **According to code, when rain leaders connect to a horizontal storm drain, they must be equipped with:**

    a cleanout. *See page 66*

17. **The clearance required by code for a 2-inch cleanout is:**

    12 inches. *See page 67*

18. **Cleanouts should be installed:**

    in the direction of flow. *See page 67*

19. **Two prohibitions for the use of cleanout openings are:**

    1) the installation of another fixture, 2) floor drainage. *See page 67*

20. **Cleanout plugs must be equipped with:**

    a raised nut or a recessed socket for removal. *See page 67*

21. **The smallest size cleanout accepted by the *Uniform Plumbing Code* for a 6-inch building drain is:**

    4 inches. *See page 68*

22. **Most codes will permit a building sewer be installed without manholes when:**

    the sewer is less than 8 inches. *See page 67*

23. **According to the *Standard Plumbing Code*, the maximum distance between manholes on a straight run is:**

    400 feet. *See page 68*

24. **When a standard-type cleanout terminates in an area where there's vehicular traffic, the code requires:**

    installation of an approved cleanout box. *See page 68*

Chapter 6

1. **The types of waste that the code considers objectionable and harmful to the building drainage system are:**

   grease, oil, sand, plaster, lint, hair, glass, acids, flammable waste. (Any three of these are acceptable.) *See page 71*

2. **Before it's allowed to enter the drainage system, one of the following three processes must be performed on objectionable and harmful waste:**

interception, separation or neutralization. *See page 71*

3. **The primary purpose of an interceptor or separator trap is:**

to prevent objectionable waste from entering the drainage system. *See page 71*

4. **When an interceptor is used, the working blueprints must show:**

an approved detailed drawing, specifications and location of the interceptor. *See page 71*

5. **The waste that must not go through an interceptor is:**

any waste not requiring treatment. *See page 71*

6. **The following types of commercial buildings require the installation of a grease interceptor:**

restaurants, supermarkets, meat processing plants, hotel kitchens, bars, cafeterias and clubs. (Any three are acceptable.) *See page 71*

7. **Grease interceptors are not generally required in these buildings:**

Single-family residences or private living quarters, apartment buildings and establishments that sell only take-out food. (Any two are acceptable. *See page 71*

8. **The two types of grease interceptor installations usually approved by code are:**

inside and outside installations. *See page 71*

9. **Most codes don't provide this information about inside grease interceptor installations:**

an established sizing method. *See page 71*

10. **Your source for obtaining established sizing methods for grease interceptor installations is:**

either your local health department or plumbing officials. *See page 71*

11. **An inside grease interceptor is usually used in:**

a small restaurant that generates a small amount of grease. *See page 71*

12. **The maximum grease capacity permitted for an inside grease interceptor is:**

100 pounds. *See page 72*

13. **The two typical installation methods for inside grease interceptors are:**

floor-mounted or below the floor. *See page 72*

14. **An approved flow control fitting must be installed on small inside grease interceptors:**

so the flow won't exceed the rated capacity of the interceptor. *See page 72*

15. **In order to omit the fixture trap for a pot sink:**

the horizontal distance between the sink outlet and the grease interceptor must be no more than 4 feet (5 feet in the International Plumbing Code). *See page 72*

16. **The maximum vertical fixture tailpiece drop for a fixture connected to an inside grease interceptor is:**

30 inches. *See page 72*

17. **Grease interceptors must be easily accessible for:**

inspection, cleaning and removal of the intercepted grease. *See page 72*

18. **The installation of an inside grease interceptor is prohibited in the part of a building:**

where food is handled. *See page 72*

19. **Most codes prohibit the installation of a water-cooled grease interceptor because:**

the jacket could fracture or corrode creating the potential for cross-connection with the potable water supply. *See pages 72-73*

20. **Most codes prohibit a food waste disposal from discharging through a:**

grease interceptor. *See page 73*

21. **The mandated grease retention capacity of a grease interceptor is:**

2 pounds for each GPM of flow. *See page 71*

22. **The minimum retention time for a grease interceptor in a small single-service kitchen is:**

1.5 hours. *See page 74*

23. **The two major considerations when sizing a grease interceptor for a fully-equipped commercial restaurant are:**

the seating capacity and number of hours of operation. *See page 73*

24. **The grease interceptor retention time required for a fully-equipped commercial restaurant is:**

    2.5 hours. *See page 74*

25. **Most codes don't spell out sizing methods for commercial grease interceptors. Instead, they defer the sizing to:**

    the local health department or local plumbing officials. *See page 74*

26. **Identical restaurants in different geographical areas may require grease interceptors of different sizes because:**

    local codes in each location may have different formulas for sizing. *See page 74*

27. **For sizing commercial grease interceptors, the *Uniform Plumbing Code* considers 16 hours of operation to include:**

    two peak hours. *See page 75*

28. **The most commonly accepted construction material for outside grease interceptors is:**

    concrete. *See page 75*

29. **Most codes require the inlet invert in an outside grease interceptor to discharge a minimum of:**

    $2^1/_2$ inches above the liquid level line. *See page: 75*

30. **The minimum size cleanout manhole for an outside grease interceptor is:**

    20 inches in diameter. *See page 75*

31. **Outside grease interceptors must be designed and installed to avoid becoming:**

    air bound. *See page 75*

32. **Most codes require that outside grease interceptors have:**

    at least two compartments. *See page 75*

33. **An outside grease interceptor that is 21 feet long would require:**

    three manholes. *See page 75*

34. **An outside grease interceptor can't be installed closer to a building foundation than:**

    5 feet. *See page 76*

35. **An outside grease interceptor can't be installed closer to a private property line than:**

    5 feet. *See page 76*

36. **The greasy waste line system is designed and installed to function as a:**

    separate drainage system. *See page 76*

37. **A greasy waste line can connect to the building sewer if:**

    it passes through a grease interceptor first. *See page 76*

38. **The two code-approved greasy waste systems are:**

    the conventional greasy waste system and the combination waste and vent system. *See page 76*

39. **The greasy waste system that you'll most often be working on is:**

    the conventional greasy waste system. *See page 76*

40. **The guidelines for sizing pipes for a conventional greasy waste system are:**

    the same as those outlined by your local code for DWV systems. *See page 76*

41. **The difference between a conventional greasy waste system and a combination waste and vent system is:**

    the combination waste and vent system provides a horizontal wet venting system using a common waste and vent pipe. *See page 76*

42. **Some codes might permit the use of a combination waste and vent system in areas:**

    where conventional venting isn't practical, such as for extensive floor drainage areas, group showers, supermarkets, demonstration or work tables in school buildings, or in similar applications where fixtures are not located adjacent to walls or partitions. (Any three are acceptable.) *See page 76*

43. **Some codes don't recommend connecting grease-producing restaurant kitchen equipment to a combination waste and vent system because:**

    it's not self-scouring. *See page 76*

44. **Pipes in a combination waste and vent system are sized two pipe sizes larger than a conventional greasy waste system because:**

    the larger pipes provide adequate air movement within the system to maintain a flow balance. *See page 76*

45. **A fixture tailpiece should be as short as possible, never exceeding:**

    2 feet. *See page 78*

46. **You can't allow an appurtenance that delivers large quantities of waste to discharge into a combination waste and vent system because:**

    it might inhibit adequate venting. *See page 78*

47. **If a branch line in a combination waste and vent system exceeds 15 feet, you must provide:**

    separate venting. *See page 78*

48. **In a greasy combination waste and vent system, the minimum area of any vent must be:**

    one-half the size of the waste pipe it serves. *See page 78*

49. **The code requires that each vent stack in a combination waste and vent system have:**

    an accessible cleanout. *See page 78*

50. **Businesses that require lint interceptors in their drainage system are:**

    commercial or self-service laundries. *See page 78*

51. **The type of strainer usually required on a commercial or self-service lint interceptor is:**

    a nonremovable $1/2$-inch mesh screen basket or screen. *See page 79*

52. **In some codes, the horizontal drainage pipes serving commercial or self-service clothes washing machines are called:**

    indirect waste pipes. *See page 79*

53. **The advantage of an indirect waste system for a commercial or self-service laundry is that:**

    an indirect system doesn't require trapping or venting. *See page 79*

54. **In a self-service laundry, a 3-inch standpipe can accommodate:**

    two clothes washers. *See page 79*

55. **When a commercial or self-service lint interceptor connects to a building drainage system, the vent on the horizontal discharge pipe should be located:**

    as close as possible to the lint interceptor. *See page 80*

56. **Lint interceptors are not required by code in:**

    single-family houses or in apartment buildings where there's a washer in each unit. *See page 78*

57. **Most codes don't provide established sizing methods for self-service laundries, but leave it to:**

    the local plumbing officials. *See page 80*

58. **The design criteria for commercial laundries are set by:**

    the manufacturers. *See page 80*

59. **The size of a lint interceptor in a self-service laundry is determined by:**

    the number of washing machines installed, the number of cycles, waste flow rate, retention time and storage factor. *See page 80*

60. **The usual code-required retention period for a lint interceptor is:**

    2.0 hours. *See page 80*

61. **Using the local code formula in our example, the lint interceptor size for a self-service laundry with eight clothes washing machines is:**

    360-gallon liquid capacity. (Formula: $8 \times 2 \times 3 \times 7.5 = 360$ gallons). *See page 80*

62. **Using the *UPC* formula, the lint interceptor size for a self-service laundry with eight clothes washing machines is:**

    2,400-gallon liquid capacity. (Formula: 8 washers $\times$ 2 cycles per hr. $\times$ 50 gal./cycle $\times$ 2.0 hr. retention time $\times$ 1.5 storage factor = 2,400 gallons.) *See page 80*

63. **The general definition of areas where gasoline, oil and sand interceptors are required is:**

    anyplace where sand, oil, gasoline or other volatile liquids can enter the drainage system. *See page 81*

64. **The types of establishments where the code would require a gasoline, oil and sand interceptor include:**

    repair garages where motor vehicles are serviced and repaired and where floor drainage is provided; commercial motor vehicle washing facilities; gasoline stations with grease racks, grease pits or wash racks; factories that have oily and/or flammable wastes from manufacturing, storage, maintenance, repair or testing; public storage garages where floor drainage is provided; or any other place where sand, oil, gasoline or other volatile liquids can be discharged into the drainage system. (Any three answers will do.) *See page 81*

65. **The sizing and design of gasoline and oil interceptors that handle volatile liquids are determined by:**

    the amount of volatile liquids generated in the system. *See page 81*

66. **Examples of establishments that, according to the code, generate small amounts of volatile liquids and sand are:**

    commercial garages servicing or storing fewer than ten vehicles, or service stations and repair shops that service, but don't store vehicles. *See page 81*

67. **The type of floor drain usually required for an automobile repair shop is:**

    a bucket-type floor drain. (No separate sand interceptor is required.) *See page 81*

68. **The minimum liquid capacity for an oil interceptor in the floor drainage system for a service station is:**

    18 cubic feet per 20 gallons of design flow. *See page 81*

69. **In a commercial garage that services or stores fewer than ten vehicles, the inlet drain pipe should enter the oil interceptor:**

    above the liquid level line. *See page 82*

70. **If the inlet pipe to an oil interceptor is 4 inches, the minimum size for the discharge pipe is:**

    4 inches. (The discharge pipe should never be smaller than the inlet pipe.) *See page 82*

71. **You can omit the vent for the discharge pipe in an oil interceptor if:**

    a) it discharges into a catch basin,

    b) it discharges into a vented building sewer or building drain and the discharge pipe doesn't exceed a 15-foot developed length, or

    c) the interceptor is located outside the building. (Any of the three is acceptable.) *See page 82*

72. **An oil interceptor for a service station must be located:**

    outside the building. *See page 82*

73. **In addition to an oil interceptor, businesses that generate large amounts of volatile liquids must have:**

    a waste oil storage tank. *See page 82*

74. **The minimum size vent required by most codes for a waste oil storage tank for a business that generates large amounts of volatile liquids is:**

    $1^1/_2$ inches. *See page 82*

75. **Codes require that waste oil storage tanks be UL approved. The abbreviation UL stands for:**

    Underwriters Laboratories, Inc. *See page 82*

76. **The minimum height above grade for a vent serving a waste oil storage tank is:**

    12 feet. *See page 82*

77. **Before it enters an oil interceptor, floor drainage for a commercial garage building must first discharge through a:**

    sand interceptor. *See page 83*

78. **Before the liquid wastes can discharge into the building drainage system, bottling plants must first discharge their processed wastes into:**

    an interceptor. *See page 83*

79. **The drainage line of a commercial fixture used for bathing animals must contain:**

    a hair interceptor. *See page 83*

80. **The code requires interceptors or separators for all drain lines in slaughtering rooms and meat dressing rooms. The blueprints with the type, size and location of these interceptors or separators must be approved by:**

    local authorities. *See page 83*

81. **Before discharging to a legal point of disposal, most codes require that drainage pipe wastes from animal boarding businesses pass through:**

    an interceptor tank. *See page 84*

82. **The floor drainage from transformer vault rooms must discharge into:**

    an oil spill holding tank. *See page 84*

83. **The authority to size a transformer oil spill holding tank belongs to:**

    the local power company. *See page 84*

84. **Dental and orthopedic sinks must be equipped with an interceptor trap:**

    to prevent wax, plaster and other objectionable substances from discharging into the drainage system. *See page 84*

85. **The purpose of a neutralizing tank is:**

    to dilute corrosive liquids, spent acids and other chemicals that might otherwise damage a drainage system. *See page 85*

Chapter 7

1. **Wastes from fixtures, appliances and devices not regularly classed as plumbing fixtures may be drained by:**

   indirect means if they have a drip or drainage outlet. *See page 89*

2. **The purpose of the indirect drainage method for special fixtures is:**

   to prevent sewage from backing up into these fixtures in case of a stoppage. *See page 89*

3. **Plumbing fixtures and appliances that may be drained by indirect means are:**

   bar sinks, hand sinks, refrigerators, ice boxes, cooling or refrigerating coils, extractors, steam tables, egg boilers, coffee urns, stills, sterilizers, commercial dishwashers, water stations, water lifts, expansion tanks, cooling jackets, drip or overflow pans, air conditioning condensate drains, drains from overflows, relief vents from the water supply system, and similar applications. (Any three answers are acceptable.) *See page 89*

4. **Devices that may be drained by indirect means are:**

   overflow and relief pipes on the water supply system, relief pipes on expansion tanks, sprinkler systems and cooling jackets. (Any two answers are acceptable.) *See page 89*

5. **Overflow pipes on the water supply system must always be indirectly connected to the sanitary drainage system:**

   to avoid the possibility of contaminating the potable water supply through cross connection. *See page 89*

6. **When there's the possibility of a potable water supply system becoming contaminated through an unsafe source, it's called:**

   a cross connection. *See page 89*

7. **Any indirect waste piping that exceeds 5 feet (2 feet in some codes) but is less than 15 feet (25 feet in some codes) must be:**

   directly trapped. *See page 89*

8. **When a vent is required in indirect waste piping, it must be installed:**

   to extend separately to the outside air. *See page 89*

9. **The minimum size for indirect waste pipes is:**

   $^1/_2$ inch. *See page 89*

10. **The two types of indirect waste piping are:**

    air break and air gap. *See page 89*

11. **An *air break* installation requires the type of indirect waste piping arrangement in which:**

    a drain pipe from a fixture, appliance or device discharges indirectly into a fixture or receptor at a point below the flood level rim of the receptor. Refer to Figure 7-3. *See page 90*

12. **An *air gap* installation requires the type of indirect waste piping arrangement in which:**

    there's an unobstructed vertical distance through the free atmosphere between the drain pipe outlet from a fixture, appliance or device and the flood level rim of the receptor into which it discharges. Refer to Figures 7-2 and 7-4. *See page 90*

13. **For an *air gap* installation, the minimum separation between the fixture outlet and the rim of the receptor is:**

    2 inches or twice the drain pipe size. *See page 90*

14. **The main requirement for indirect waste receptor installations to allow for inspecting and cleaning is:**

    accessibility. *See page 91*

15. **The factor that determines the type of receptor to use for *air gap* indirect waste pipe is:**

    whether or not the area will have pedestrian traffic. *See page 91*

16. **The kind of strainer that the code requires for a floor sink is:**

    a beehive strainer that's at least 4 inches high. *See page 91*

17. **The type of receptor most commonly used where *air break* indirect waste pipe is installed is:**

    a floor drain. Refer to Figure 7-3. *See page 91*

18. **Most codes categorize automatic clothes washer standpipes as:**

    indirect waste receptors. *See page 91*

19. **The installation of a standpipe receptor for an automatic clothes washer must include:**

    proper traps and vents. *See page 91*

20. **The code prohibits the installation of an indirect waste receptor in:**

a toilet room, closet, cupboard or storeroom. (Any one answer is acceptable.) *See page 92*

21. **In places where gravity drainage isn't possible, indirect waste pipe may require the use of a:**

sump. *See page 92*

22. **The equipment used to lift liquids from a sump to a place of disposal is:**

a sump pump. Refer to Figure 7-7. *See page 92*

23. **Clear water wastes must empty into the building drainage system by means of:**

an air gap. Refer to Figure 7-1. *See page 92*

24. **In cases where air conditioning waste is connected to the building storm system, the type connection you should use is:**

either air gap or air break. *See page 93*

25. **The generally accepted methods of disposing of air conditioning wastes are:**

an approved receptor or other suitable fixture, a sump, a building storm or sanitary drain, the building inside rain leader, or a waste and overflow or lavatory tailpiece (not all codes accept this last method of disposing of air conditioning waste). (Any two answers are acceptable.) *See page 93*

26. **The maximum water temperature that can be discharged directly into a drainage system is:**

140 degrees F. *See page 93*

27. **Wastes from swimming pools, wading pools and spas must be connected to the building sanitary system by:**

indirect means. *See page 93*

28. **The type of waste disposal system that can *never* receive waste from a swimming pool is:**

a septic tank. *See page 94*

29. **If rainwater is not properly collected and disposed of:**

it can become a nuisance and a health hazard. *See page 94*

30. **In previous years, the type of drainage system commonly used in older cities located near lakes, rivers or the ocean was:**

a combined sewer system. *See page 94*

31. **To connect new construction into an existing combined sewer system, you need:**

advance approval. *See page 94*

32. **Roof drains must be equipped with:**

strainers. *See page 94*

33. **Most codes require installation of deck drains in the following locations:**

sun decks and parking decks. *See page 94*

34. **The maximum area most codes permit an area drain to handle is:**

100 square feet. *See page 94*

35. **The type drain you must use where there's a flat surface between a window and an outside wall is:**

a sill drain. *See page 95*

36. **When a planter drain is used, your local authority may require:**

a sand interceptor. *See page 95*

37. **The purpose of subsoil drains is:**

to intercept surface water before it reaches the building's foundation wall or footings. *See page 95*

38. **The sizing of storm water drainage pipes is usually determined by:**

a mechanical engineer. *See page 96*

39. **The two determining factors in sizing storm water drainage pipes are:**

the square footage of impervious areas, and the maximum anticipated rainfall rate in any one hour. *See page 96*

40. **As a plumber you should learn all you can about sizing commercial storm drainage and disposal systems because:**

you may be doing the installing, and because it's likely to be on your journeyman and/or master plumber's examination. *See page 96*

41. Storm drainage tables differ in local codes because:

there's a great variation in maximum anticipated rainfall in different areas. *See page 96*

42. Once you know the maximum anticipated rainfall where you work, and the square feet to be drained, you can determine the sizing of storm drainage pipes by simply using:

the tables in your local code. *See page 96*

43. When the slope is increased, it affects the sizing of horizontal storm drainage pipes as follows:

the steeper the slope, the greater the area that can be drained by each pipe size; as a result, each pipe size can be smaller. *See page 98*

Chapter 8

1. The purpose of a written plumbing code is:

to protect the public health, welfare and safety. *See page 103*

2. In general, the plumbing code covers:

the proper design, installation and maintenance of plumbing systems. *See page 103*

3. The potable water supply system ends and the sewage system begins at:

the plumbing fixtures. *See page 103*

4. It's important to design a proper drainage system:

to prevent the fowling or the depositing of solids along the walls of drainage pipes. *See page 103*

5. A drainage system must be properly vented in order to:

provide a free circulation of air. *See page 103*

6. A properly vented drainage system prevents:

the possibility of siphonage or the forcing of trap seals. *See page 103*

7. You must provide adequate cleanouts on a drainage system:

so that all portions of the system are accessible to cleaning equipment. *See page 103*

8. There are a number of fittings that are acceptable for making a change in direction in drainage pipes, including:

a short sweep, a quarter bend, an eighth bend, a 45 degree wye, a long sweep, a sixth bend, a sixteenth bend, or combination of these. (Any three answers are acceptable.) *See page 103*

9. The five fittings not acceptable for use in a drainage system are:

fittings having a hub in the direction opposite to flow, fittings having running threads, a tee branch, fittings with bands or fittings with saddles. (Any two of these answers are acceptable.) *See page 103*

10. The trenches must remain open after the installation of drainage pipes until:

the piping has been tested, inspected and accepted by the plumbing inspector. *See page 103*

11. The minimum standards for plumbing system materials are set by:

the plumbing code. *See page 103*

12. If listed or labeled materials aren't available, you may use a substitute material:

only with the approval of the building department. *See page 105*

13. There are certain organizations that approve various plumbing materials. If you see the abbreviation ASTM stamped on a cast iron soil pipe and fitting, it would indicate that the material meets the standards set by:

the American Society for Testing Materials. Refer to Figure 8-1. *See page 104*

14. It's important for you to familiarize yourself with the abbreviations of organizations that approve plumbing materials because:

some of these abbreviations may appear on your journeyman or masters examination. *See page 105*

15. The organization that approves most standards for drainage materials is:

the American Society for Testing Materials (ASTM). Refer to Figure 8-2 and/or your code book. *See page 106*

16. **It's important to keep your copy of the code updated because:**

    the code is always changing and you need to keep up with all the latest materials and installation requirements. You also need to buy the updated code supplement sheets and keep them in your code book. *See page 105*

17. **If you're using extra-heavy cast iron pipe underground within a building, most codes will require the cast iron sewer pipe be:**

    extra heavy as well. *See page 105*

18. **Some codes require Schedule 80-strength plastic sewer pipe:**

    for installations under heavy traffic areas. *See page 105*

19. **Concrete pipe isn't generally recommended for ordinary building sewers because:**

    it's highly susceptible to corrosion from acids and sewer gases. *See page 105*

20. **The use of asbestos-cement pipe in a building drainage system is limited to:**

    outside drainage systems. *See page 105*

21. **Bituminous fiber pipe can be used as a building sewer in:**

    government housing projects (HUD). *See page 105*

22. **Other than in building sewers, you can use extra strength vitrified clay pipe:**

    underground within a building. *See page 105*

23. **In a multistory building, you can only install plastic drainage piping:**

    on the first three floors. *See page 105*

24. **Where the fill is known to be deleterious, you should not use piping material:**

    that's subject to corrosion. *See page 105*

25. **According to the code, fittings used in a drainage system must conform to:**

    the material and type of piping used in the drainage system. *See page 108*

26. **All the joints in a drainage system must be:**

    gastight, watertight and root-proof. *See page 108*

27. **The minimum size required for subsoil drains is:**

    4 inches. *See page 108*

28. **There are a number of approved materials commonly used for a building subsoil drain, including:**

    clay drain tile, perforated concrete pipe, corrugated polyethylene tube, concrete drain tile, horizontally split concrete pipe, perforated or horizontally split SR plastic drain pipe, PVC sewer pipe, and vitrified clay pipe. (Any three answers are correct.) *See page 108*

29. **In addition to being compatible with the pipe used, fittings in a drainage system must:**

    not have ledges, shoulders or reductions that could restrict or obstruct flow. *See page 108*

30. **The only type of threaded fitting acceptable for use in a drainage system is:**

    the recessed type. *See page 108*

31. **Where threaded pipe is used in a vent system, the fittings may be either:**

    the drainage type or pressure type, either galvanized or black. *See page 108*

32. **The code prohibits the mixing or combining of:**

    different types of plastic materials (ABS and PVC) in the same plumbing system. *See page 108*

33. **When adding to an existing DWV system, you must use materials:**

    of like grade and quality. *See page 108*

34. **When joining different piping materials together in new work, the fitting you must use is:**

    a transition fitting. Refer to Figure 8-4. *See page 108*

35. **Galvanized steel pipe can't be used underground in a DWV system. The code states it must be kept aboveground:**

    6 inches. *See page 108*

36. **The minimum depth below grade that vitrified clay pipe must be kept is:**

    12 inches. *See page 108*

37. **The materials that are usually acceptable for use in a chemical or acid system are:**

    corrosion-resistant materials such as borosilicate glass pipe, high silicon content cast iron pipe, vitrified clay pipe, ABS and PVC Schedule 40 plastic pipe, plastic-lined pipe and lead pipe. (Any two answers are acceptable.) *See page 109*

38. The same materials are approved for use in indirect waste piping as are approved for:

potable water, sanitary drainage or storm drainage systems. *See page 109*

39. When a rain leader discharges directly into a soakage pit, you must install:

an overflow fitting. *See page 109*

40. The conventional means of protecting all exposed rainwater leaders located in areas where they may be subject to damage is:

to place a 3-inch galvanized steel pipe supported in a concrete base in front of the leader. *See page 109*

41. Joints and connections in DWV systems are pressure tested to:

ensure that they are gas and watertight. *See page 110*

42. Every lead-caulked joint in cast iron bell-and-spigot soil pipe must be firmly packed with:

oakum or hemp. *See page 110*

43. When pouring a lead joint, you must do it:

in one pour. *See page 110*

44. An alternate to lead and oakum for sealing bell-and-spigot cast iron soil pipe joints is:

a neoprene rubber gasket. Refer to Figure 8-7. *See page 110*

45. The clamp assembly used in joining hubless cast iron soil pipe and fittings for DWV systems must comply with the standards set by:

the Cast Iron Soil Pipe Institute (CISPI). Refer to Figures 8-1 and 8-8. *See page 110*

46. Joints between asbestos-cement pipe and plastic pipe should always be made with:

an approved adapter coupling with an approved rubber ring. Refer to Figure 8-9. *See page 111*

47. When it's necessary to cut asbestos-cement pipe for a new installation, the tool that you must use to ensure that the fittings will be watertight is:

a tapering tool. Refer to Figure 8-10. *See page 111*

48. The two types of joints used most often to connect plastic pipe and plastic fittings are:

solvent-cemented joints or heat joined connections. *See page 111*

49. The type sealant that should never be used to seal plastic threaded joints is:

regular pipe dope. *See page 111*

50. The fitting that you must use to join bituminous fiber pipe with other types of materials is:

a transition fitting. *See page 112*

51. The type of fitting that you use to join the plain ends of vitrified clay sewer pipe is known as:

a flexible coupling. *See page 112*

52. The only type of construction in which you can use cement mortar joints and connections for concrete sewer pipe and fittings is:

for repairs and/or connections to existing lines constructed with cement mortar joints. *See page 112*

53. The part of a regular fixture trap on which some codes prohibit the use of slip joint connectors (nuts and washers) is:

on the outlet side of the trap above water seal. *See page 112*

54. Joints between lead pipe and cast iron pipe may be made with the following fittings:

a caulking ferrule, soldering nipple or bushing. *See pages 112-113*

55. Caulked glass joints are made the same manner as caulked cast iron joints except that:

you must use an acid-proof cement, and the oakum or hemp rope must be acid-resistant. *See page 113*

56. To make a burned (welded) lead joint, you should use materials:

of the same composition as the material being joined. *See page 113*

57. Plain end ductile-iron gravity sewer pipe may be joined in the same manner as:

no-hub cast iron pipe. Refer to Figure 8-8. *See page 113*

58. In a threaded DWV system, you should always use:

recessed drainage fittings. *See page 113*

59. The types of fittings approved for copper DWV systems are:

cast brass or wrought copper fittings. *See page 113*

60. **Expansion joints may be used in vent piping or drainage stacks for:**

    the expansion and contraction of pipes. *See page 113*

61. **When piping materials have different outside diameters they may be joined together:**

    with an approved elastomeric sealing sleeve and clamping device. *See page 114*

62. **You can connect pipes and fittings of different sizes in a plumbing system using:**

    the proper size increaser, reducer or reducing fitting. *See page 114*

63. **When you need to make changes of direction in horizontal systems or in horizontal-to-vertical drainage systems, the acceptable fittings you can use to accomplish these changes are:**

    a short sweep, a one-eighth bend, a 45-degree wye, a long sweep, a quarter bend, a sixth bend, a sixteenth bend, or a combination of these. (Any two answers are acceptable.) Refer also to Figure 8-14. *See page 114*

64. **Where the direction of flow is from the horizontal to the vertical, the three fittings that are acceptable to use are:**

    a sanitary tee, a quarter bend and a one-fifth bend. Refer to Figure 8-14. *See page 114*

65. **You may use materials not covered by the standards cited in your code:**

    if you get approval from your local authority. *See page 114*

66. **The three types of cast iron soil pipe and fittings approved for building sewers are:**

    lead caulked joints, compression gasket joints for hub and spigot pipe and stainless steel shield with elastomeric gasket joints for no-hub pipe. (Any two answers are acceptable.) *See page 114*

67. **The two grades of cast iron soil pipe used today are:**

    centrifugally-spun service weight and extra heavy cast iron. *See page 114*

68. **The characteristics that make cast iron soil pipe superior for building sewers are:**

    its strength, durability and resistance to trench loads. *See page 114*

69. **When you lay pipe with hubs or couplings, you must protect it from damage in the trench by:**

    excavating for the hubs or couplings to ensure that the pipe barrel sets firmly on the soil. *See page 114*

70. **The minimum code-required depth for installing plastic pipe or one of the other fragile pipes approved for building sewers is:**

    12 inches. *See page 116*

71. **Underground or horizontal drainage, waste and vent pipe must be adequately supported in order to:**

    keep the pipe in alignment and prevent it from sagging. *See page 116*

72. **The bases of stacks must be supported by:**

    masonry or concrete. *See page 116*

73. **Drainage pipe passing through cast-in-place concrete should be protected by:**

    sleeving. *See page 116*

74. **The required clearance from the top of a drainage pipe to the bottom of the footing is:**

    2 inches. *See page 116*

75. **When drainage piping is installed in corrosive materials it must be protected with:**

    sleeves, coating, wrapping or other approved methods. *See page 117*

76. **Before installing used drainage piping in any plumbing system, you must ensure that it conforms to:**

    the standards in your code. *See page 117*

77. **The code will permit you to drill or tap a waste or vent pipe for the purpose of rodding:**

    under no circumstances. *See page 117*

78. **The code might permit a lavatory waste to connect to a water closet stub:**

    under no circumstances. The code prohibits any such connection. *See page 117*

79. **The maximum number of stories in which a plastic DWV system may be installed is:**

    three stories. *See page 117*

80. **The two types of supporting methods you must consider in drainage, waste and vent systems are:**

aboveground **horizontal** piping and aboveground **vertical** piping. *See page 117*

81. **When pipe lengths exceed 5 feet, the maximum distance you should allow between hangers for horizontal cast iron soil pipe with lead and oakum joints is:**

10 feet (10-foot intervals). *See page 118*

82. **The maximum distance you should allow between hangers for horizontal copper pipe 1$^1/_2$ inches and smaller is:**

6 feet (6-foot intervals). *See page 118*

83. **The hangers used for horizontal borosilicate glass piping require:**

padding. *See page 118*

84. **The maximum allowable distance between supports for vertical copper piping is:**

every story, but intervals must never exceed 10 feet. *See page 119*

85. **The two requirements that must be met when placing hangers for the support of horizontal and vertical piping are:**

the hangers must maintain pipe alignment and prevent sagging. *See page 117*

86. **Horizontal vent piping should be installed and sloped:**

to drain back to the soil or waste pipe by means of gravity. *See page 119*

87. **It's important not to allow moist air to be trapped in a horizontal vent pipe because:**

entrapped moist air will accelerate corrosion and reduce the life of the pipe. *See pages 119-120*

88. **When you size and install indirect waste piping you must always accommodate:**

the outlet drainage of the fixture or appliance. *See page 120*

89. **When practical, indirect waste pipe should be installed:**

below the floor. *See page 120*

90. **The indirect waste pipe installation that is made below the floor and connects through the receiving fixture above the water seal of the trap is called:**

an air break type installation. Refer to Figure 8-24. *See page 120*

91. **The outlet for above-the-floor indirect waste pipe should terminate in:**

an air gap type installation. Refer to Figure 8-25. *See page 120*

92. **The minimum size indirect waste pipe you should use for an above-floor installation is:**

$^3/_4$ inch diameter. *See page 120*

93. **The minimum indirect waste pipe size most codes require for a below-floor installation is:**

1$^1/_4$ inch diameter. *See page 120*

94. **Receiving fixtures for indirect waste piping should be located:**

where they are accessible. *See page 121*

95. **Commercial dishwashing machines must be connected to the building greasy waste drain line by:**

indirect means. *See page 121*

96. **A drinking fountain may be indirectly connected to a floor drain for the purpose of:**

resealing the trap. *See page 122*

97. **Indirect waste piping for A/C units must be installed below the bottom of the floor slab:**

at least 2 inches. Refer to Figure 8-17. *See page 122*

98. **When the waste or condensate from an A/C unit connects to the plumbing drainage system, the A/C unit is classified as:**

a plumbing fixture. *See page 123*

99. **An air conditioning unit that's over 5 tons but less than 10 tons may discharge its waste into:**

a buried pipe, filled with $^3/_4$-inch rock, that's 10 inches in diameter and 24 inches long. *See page 123*

100. **The five acceptable areas where air conditioning units 10 tons and larger may discharge their waste are:**

a drainage well, the sanitary drainage system, the storm water system, an adequate-size soakage pit and an adequate-size drainfield. (Any three answers are acceptable.) *See page 123*

Chapter 9

1. **Under present code standards, the two disposal methods no longer acceptable as a permanent means of dealing with human waste are:**

cesspools and outhouses. *See page 127*

2. **Even today, drinking water can be contaminated by untreated sewage. This can spread diseases such as:**

cholera and typhoid. *See page 127*

3. **When public sewers aren't available, the most acceptable method for sewage disposal is:**

the septic tank system. *See page 127*

4. **The governmental department that has established guidelines for septic tank safety is:**

the Department of Environmental Resource Management (DERM). *See page 127*

5. **Although septic tank and drainfield installations are important areas of plumbing work, the installation and maintenance work for these systems is usually done by:**

a licensed septic tank contractor. *See page 127*

6. **Although you probably won't do the work, it's important for you to be informed about the basic principles of septic tank and drainfield installation because:**

it's very likely that there will be questions concerning these installations on your journeyman and master's examinations. *See page 127*

7. **Septic tanks are designed to:**

separate solids from liquid wastes. *See page 127*

8. **The approximate amount of waste solids for each 100 gallons of water in a septic tank is:**

$3/4$ of a pound. *See page 127*

9. **A septic tank is sized to have a capacity equal to approximately:**

24 hours of anticipated flow. *See page 127*

10. **The solids in a septic tank are digested through:**

a biological process involving bacterial action in which sewage wastes are transformed into gases and harmless liquids. *See page 127*

11. **When the effluent enters the drainfield:**

it oxidizes and evaporates. *See page 127*

12. **After the bacterial process is completed within the septic tank, what remains are:**

small amounts of solids that settle on the bottom of the tank as sludge. *See page 127*

13. **The lighter, undigested particles that rise to the top of the liquid after the bacterial process in the septic tank is complete are called:**

scum. *See page 127*

14. **When undigested materials need to be cleaned out of a septic tank, the job must be done by:**

a certified professional with the proper equipment to perform this type of work. *See page 127*

15. **The materials not code-approved for septic tank construction are:**

blocks, bricks and wood. (Any two answers are acceptable.) *See page 128*

16. **The two most common types of septic tanks are:**

precast concrete tanks and cast-in-place tanks. *See page 128*

17. **Concrete septic tanks are protected from corrosion by:**

coating them with an approved bituminous coating. *See page 128*

18. **In order to prevent contamination by leaking sewage, the code requires that all septic tanks be:**

watertight. *See page 128*

19. **When sizing a septic tank, you must make sure that the tank will be able to accommodate:**

sludge and scum accumulations. *See page 128*

20. **Most codes require that septic tanks have:**

two compartments. *See page 128*

21. **The liquid capacity of the smallest approved septic tank is:**

750 gallons. *See page 128*

22. **The secondary compartment (outlet compartment) of the tank should have a minimum liquid capacity of:**

    250 gallons. *See page 128*

23. **On a *residential* septic tank, the type of cover slab acceptable for cleaning purposes is:**

    one that has removable sections. Refer to Figure 9-1. *See page 128*

24. **On a 500-gallon capacity commercial septic tank, the number of manholes required is:**

    two, one over the inlet tee and one over the outlet tee. *See page 128*

25. **When the first compartment of a *commercial* septic tank exceeds 12 feet in length, the number of manholes required is:**

    three, one over the inlet tee, one over the outlet tee and one over the baffle wall. *See page 128*

26. **The minimum size requirement for the vertical legs of the inlet and outlet tees of a septic tank is:**

    4 inches in diameter. *See page 128*

27. **A septic tank inlet pipe should be higher than the outlet pipe by at least:**

    2 inches. *See page 128*

28. **The air space above the liquid level in a septic tank must be:**

    a minimum of 9 inches (8 inches for some codes). *See page 128*

29. **When a septic tank is located in a parking lot, the building department requires it to have:**

    an acceptable traffic cover. *See page 128*

30. **When sizing septic tanks for single-family or *multiple* residential units, the capacity is determined by:**

    the number of bedrooms. *See page 130*

31. **The capacity requirements for septic tanks for *commercial* buildings are determined by:**

    the maximum fixture units for public use. Refer to Figure 9-4. *See page 130*

32. **You determine the capacity requirement for a septic tank for a single-family residence that has more bedrooms than there are listed in your code table by:**

    adding 150 gallons to the sizing requirement for each additional bedroom. *See page 130*

33. **You determine the capacity requirement of a septic tank for multiple *residential* units with more units than there are listed in your code table by:**

    adding 250 gallons to the size requirement for each extra unit. Refer to Figure 9-3. *See page 131*

34. **A septic tank can't be located closer to any water supply line than:**

    5 feet. *See page 132*

35. **A septic tank can't be located closer to the shoreline of an open body of water than:**

    50 feet. *See page 132*

36. **Drainfield materials commonly used to distribute the effluent from septic tanks include:**

    open-jointed drain tile, perforated drain tile, block or cradle-type drain units, or corrugated plastic perforated tubing. (Any two are acceptable.) *See page 132*

37. **The minimum inside diameter of drainfield tile is:**

    4 inches. *See page 132*

38. **The minimum slope for drainfield tile per 100 feet is:**

    3 inches. *See page 132*

39. **The minimum width for each drainfield trench is:**

    18 inches. *See page 132*

40. **The maximum length of a single tile drainfield trench is:**

    100 feet. *See page 132*

41. **The required depth of washed rock under the drain units of a reservoir-type drainfield is:**

    12 inches. *See pages 132-133*

42. **The maximum distance between centers of the distribution lines of a reservoir type drainfield is:**

    4 feet. Refer to Figure 9-9. *See page 133*

43. **The best soils for absorbing drainfield effluent are:**

    soils made up of coarse sand or gravel. *See page 134*

44. **A percolation test determines:**

    how long it takes water to be absorbed into the soil. *See page 135*

45. Other than the type of soil, the sizing of a *residential* drainfield is determined by the:

septic tank capacity in gallons. *See page 135*

Chapter 10

1. The plumbing standards and installation methods for mobile home and RV parks are regulated by:

local agencies. *See page 141*

2. Recreational vehicles are defined in the code as:

dependent trailers. *See page 141*

3. Recreational vehicles have limited plumbing facilities, usually not including:

a toilet. *See page 141*

4. A mobile home is defined by code as:

an independent trailer coach. *See page 141*

5. The code does *not* require dependent trailer parks to provide:

individual water and sewer connections. *See page 141*

6. A dependent trailer park must provide a separate building with:

toilet facilities and a waste disposal station (service building). *See page 142*

7. A dependent trailer park service building must have an individual

sewer connection. *See page 142*

8. A dependent trailer park service building must provide a minimum number of:

toilet fixtures for each sex. *See page 142*

9. The *Uniform Plumbing Code* requires one toilet for each sex in a dependent trailer park:

for the first 25 trailer sites in the park. *See page 142*

10. Each toilet room in the service building of a dependent trailer park must have:

a floor drain. *See page 142*

11. In a dependent trailer park service building, the number of lavatories that must be provided for each toilet is:

one lavatory per toilet for the first six toilets. *See page 142*

12. In the men's toilet room of a dependent trailer park service building, the portion of required toilets that may be replaced with urinals is:

one-third of the toilets. *See page 142*

13. The toilet design that must be used in the toilet rooms of a dependent trailer park is:

the elongated type with open front seat. *See page 142*

14. Each toilet installed in a dependent trailer park must have:

a separate stall with a door and latch. *See page 142*

15. A shower installed in a dependent trailer park must have a floor area of:

36 by 36 inches. *See page 142*

16. The location in a dependent trailer park's shower room that must be protected from water overflow is:

the dressing area. *See page 142*

17. According to the *Standard Plumbing Code*, the minimum fixtures required in the women's area of an independent trailer park are:

one toilet, one lavatory and one shower or bathtub. *See page 142*

18. In a trailer park, the fixture units are assigned as:

a set number of fixture units for each site drainage inlet. *See page 142*

19. The criterion used to size a park drainage system is:

the number of trailer sites. *See page 142*

20. The number of fixture units permitted on a 3-inch pipe in a trailer park drainage system is:

35 (refer to Figure 10-1). *See page 143*

21. In a trailer park, the minimum slope for a 4-inch pipe per 100 feet is:

15 inches (refer to Figure 10-2). *See page 143*

22. The installation and backfill for a trailer park drainage system is the same as:

a conventional building sewer. *See page 143*

23. In a trailer park drainage system, the first vent should be located:

no more than 5 feet downstream from the first sewer lateral. *See page 143*

24. **When a park lateral terminates with a 4-inch P-trap and a sanitary tee, a cleanout should be installed:**

    in the top of the sanitary tee. *See page 144*

25. **The minimum pressure required at each trailer site for a park's water distribution system is:**

    20 psi. *See page 144*

Chapter 11

1. **The water we obtain for public use originates from:**

    lakes, rivers and deep wells. *See page 147*

2. **Many large cities, when faced with a shortage of available water for human consumption, have had to resort to:**

    water conservation and moratoriums on new construction to extend their available water resources. *See page 147*

3. **To protect our present water sources and search for additional fresh water sources, there must be:**

    long range planning. *See page 147*

4. **The primary source of water for single-family and multifamily residential buildings is:**

    the municipal water treatment plant. *See page 147*

5. **Until recently, local codes have only considered water conservation methods that involve:**

    regulating the maximum allowable water usage for plumbing fixtures. Refer to Figure 11-1. *See pages 147-148*

6. **If the average drop of water takes only 2$^1/_2$ seconds to form, one leaky faucet wastes:**

    365 gallons of water per year. *See pages 147-148*

7. **One new method of water conservation being tried in many areas is:**

    graywater recycling. *See page 148*

8. **At this time, the only types of buildings permitted to use a graywater recycling system are:**

    single-family residences. *See page 148*

9. **Currently, the use of recycled graywater is limited to:**

    underground landscape irrigation. *See page 148*

10. **Graywater is:**

    untreated household wastewater that has not come in contact with waste from toilets, dishwashers or kitchen sinks. Refer to Figure 11-2. *See pages 148-149*

11. **A graywater recycling system cannot be connected to:**

    any potable water piping system. *See page 148*

12. **The fixtures that can connect to a graywater system are:**

    bathtubs, showers, lavatories, laundry trays and clothes washing machines. (Any three answers are acceptable.) *See page 148*

13. **The three things that determine the type of graywater recycling system that may be used are:**

    the location, the soil type, and the groundwater table. *See page 148*

14. **A graywater recycling system for a residential building must discharge the waste:**

    into underground irrigation piping or disposal fields. *See page 149*

15. **When a holding tank is installed above ground, its base must be supported by:**

    a suitably-sized 3-inch-thick concrete slab. Refer to Figure 11-4. *See pages 149-150*

16. **After you install the holding tank and piping, the code requires that you test a graywater system with:**

    a flow test by filling the tank with water to the overflow line and inspecting all the components for watertightness to the point of disposal. *See page 149*

17. **You base your estimate of the amount of graywater discharge from a single-family residence on:**

    the number of occupants and bedrooms in the home. *See page 149*

18. **In estimating graywater discharge for a single-family residence, the number of occupants considered for the first bedroom is:**

    two. *See page 149*

19. **In estimating graywater discharge for a single-family residence, the number of occupants considered for each bedroom, after bedroom number one, is:**

    one. *See page 149*

20. **To estimate the total number of gallons of graywater flow from a single-family residence, the allowance per occupant that you use is:**

     25 gpd/occupant (excluding laundry). *See page 151*

21. **To estimate the total number of gallons of graywater flow from a single-family residence, the allowance per occupant, including laundry facilities, that you use is:**

     25 gpd + 15 gpd for laundry = 40 gpd/occupant. *See page 151*

22. **The minimum number of valved zones each graywater system must have is:**

     three. Refer to Figure 11-5. *See page 151*

23. **When you excavate for a graywater subsurface irrigation/disposal field, you must not allow your excavation to extend within:**

     5 vertical feet of the highest known seasonal groundwater. *See page 151*

24. **Each graywater holding tank must have a locking gasketed access opening in order to:**

     permit inspection and cleaning. *See page 152*

25. **Each graywater holding tank must be permanently marked with:**

     its rated capacity and a sign that reads, "Graywater irrigation system, danger — unsafe water." *See page 152*

26. **The minimum capacity for a graywater holding tank is:**

     50 gallons. *See page 152*

27. **The size of a holding tank vent is determined by:**

     the total number of fixture units. *See page 152*

28. **For replacement purposes, all connecting pipes to holding tanks must be fitted with:**

     unions or other approved effective fittings. Refer to Figures 11-3 and 11-4. *See page 152*

29. **A holding tank must be designed to withstand an earth load in the amount of:**

     300 pounds psf. *See page 152*

30. **To protect against sewer backups, an underground holding tank requires:**

     a backwater valve. *See page 152*

31. **All holding tanks must be constructed of:**

     steel, with an internal and external corrosion-resistant coating. *See page 152*

32. **The minimum pipe size for use in graywater system irrigation/disposal fields is:**

     3 inches in diameter. *See page 152*

33. **Piping materials that are acceptable for graywater/disposal fields are:**

     perforated ABS, PVC pipe and perforated high-density polyethylene pipe. (Any two answers are acceptable.) *See page 152*

34. **The code requires that graywater irrigation/disposal field piping have:**

     sufficient openings for the proper distribution of the waste water into the trench area. *See page 152*

35. **The minimum size filter material for a graywater irrigation/disposal field is:**

     $3/4$ inch. *See page 152*

36. **The minimum depth of filter material required beneath the pipe in a graywater irrigation/disposal field is:**

     3 inches. Refer to Figure 11-7. *See pages 152-153*

37. **The filter material in a graywater irrigation/disposal field must extend above the pipe:**

     2 inches. Refer to Figure 11-7. *See pages 152-153*

38. **Graywater piping must be laid with a slope of:**

     3 inches per 100 feet. *See page 153*

39. **In a graywater disposal system, the maximum length of each individual line is:**

     100 feet. Refer to Figure 11-8. *See page 153*

40. **When graywater irrigation/disposal lines are installed in areas having steep grades, the code requires that:**

     the lines be stepped. Refer to Figure 11-9. *See pages 153-154*

41. **The minimum distance required from a private property line to a graywater system holding tank is:**

     5 feet. Refer to Figure 11-10. *See page 154*

42. **The minimum distance required from a graywater irrigation/disposal field to streams and lakes is:**

     50 feet. Refer to Figure 11-10. *See page 154*

43. **If it's necessary to install a graywater irrigation/disposal field parallel to a public water main, the work must be approved by:**

the local authority. Refer to Figure 11-10, footnote 7. *See page 154*

44. **The difference in the design criteria regarding typical soils for graywater irrigation/disposal fields and septic tank drainfields is that:**

graywater systems use six typical soils, while septic tanks use five typical soils. Refer to Figure 11-6. *See pages 152-153*

45. **A graywater system is designed to:**

safely manage wastewater by filtering it through the soil and returning it to the underground water table. *See page 155*

Chapter 12

1. **The first American city to build a gravity water supply system was:**

Boston, Massachusetts. *See page 159*

2. **The installation of a pressurized water system made possible:**

a safe, abundant supply of water at the turn of a handle. *See page 159*

3. **In order to make raw water safe and pleasant for drinking, it's treated for the removal of:**

unpleasant tastes, odors and impurities. (Any two answers are acceptable.) *See page 159*

4. **Potable water is:**

water that meets the health requirements for drinking, culinary and domestic purposes. *See page 159*

5. **The agency that's generally responsible for monitoring our public water supply systems is:**

the local health department. *See page 159*

6. **The purpose of a public water main is:**

to carry water for community use. *See page 159*

7. **A private service connection to a public water main must have:**

a curb stop and a connection to a water meter at the property line. Refer to Figure 12-1. *See pages 159-160*

8. **A building water service pipe begins at:**

the outlet side of the water meter, at the property line. *See page 161*

9. **The pipes that carry water from the water service pipe to the plumbing fixtures and other outlets are called:**

the water distributing pipes. *See page 161*

10. **The components of a water supply system are:**

the water service pipe, the water distributing pipe, the necessary connecting pipes, the standpipe system, fittings, control valves and all appurtenances on private property. (Any three answers are acceptable.) *See page 161*

11. **In a water distributing system, the code defines a discharge opening for water as:**

a water outlet. *See page 161*

12. **In sizing a water system, you won't ever be able to exactly predict:**

the maximum rate of flow or the demand in a building's water supply system. *See page 161*

13. **When you size water supply piping, it's important to be economical, but you must be sure to avoid:**

undersizing the water supply. *See page 161*

14. **Physical properties that can affect water in supply pipes are:**

density, viscosity, compressibility, boiling point, minimum available pressure, velocity flow and friction loss. (Any two answers are acceptable.) *See page 161*

15. **The type of water supply system that plumbers aren't expected to size is:**

a complex system (a system requiring pipe sizes larger than those given in Figures 12-4 and 12-5, or the tables in your local code). *See page 161*

16. **The psi pressure at which an ordinary faucet at ground level operates properly is:**

8 psi. *See page 164*

17. **The minimum required pressure in psi for a floor-mounted ball-cock type water closet is:**

8 psi. *See page 164*

18. **When there's a need for designing a water supply system using pipes larger than those in your plumbing code, you should:**

leave the job to a professional engineer. *See page 164*

19. **A water service pipe continues within the building to become:**

    the building water main. *See page 164*

20. **As the building water main progresses through the building, the demand on the line:**

    is likely to decrease. *See page 164*

21. **According to code, the number of fixtures that can connect to a ¹/₂-inch water supply branch is:**

    two. *See page 164*

22. **When a building exceeds four stories, or where a system's residual pressure is below what's required at the highest water outlet, you must provide:**

    an automatic control pressure pump or an adequate gravity tank. *See page 164*

23. **If a building requires a gravity tank, the design work should be done by:**

    a professional engineer. *See page 164*

24. **If each piping system is the same size, the type system on which you can install the most fixtures is:**

    a copper system. *See page 164*

25. **In most cases, as the size of the water service pipe increases:**

    the pressure loss is reduced or remains stable. *See page 164*

Chapter 13

1. **Even though it isn't required by most codes, you'll find that most buildings have:**

    a hot water system. *See page 169*

2. **When a hot water system is installed, the code very specifically requires:**

    the installation of safety devices. *See page 169*

3. **It's important to install safety equipment on hot water systems in order to:**

    protect property from damage and persons from injury due to the hazards of excessive pressure and temperature. *See page 169*

4. **Since requirements for hot water distribution systems aren't specified in the code, the design and sizing of the systems is the responsibility of:**

    professional engineers and plumbers.
    See page 169

5. **The hot water supply systems for large commercial buildings are usually designed by:**

    professional engineers. *See page 169*

6. **The hot water systems for small commercial and residential buildings are usually designed by:**

    the installing plumber. *See page 169*

7. **The two principal objectives in designing a good hot water system are:**

    (1) the system must satisfy the hot water demand for a particular type of occupancy; and (2) it must include safety features that guard against the hazards of excessive pressure and temperature. *See page 169*

8. **The normal design range for hot water temperatures in most plumbing fixtures is:**

    between 140 and 150 degrees Fahrenheit. *See page 169*

9. **Water heater thermostats are usually preset by:**

    the manufacturer. *See page 169*

10. **The three fuels used in most direct water-heating units are:**

    gas, oil and electricity. *See page 169*

11. **Electric water heaters are the most common types used in homes today because:**

    they are clean and attractive and can be installed nearly anywhere within a building. *See page 169*

12. **Gas- or oil-fueled water heaters must be located:**

    in an area where there is adequate ventilation. *See page 169*

13. **Water heaters designed to burn gas or oil must be installed with:**

    flues to carry away combustion gases. *See page 169*

14. **The type of buildings in which circulating hot water is most efficient are:**

    large buildings, such as apartment buildings, with central hot water systems. *See page 170*

15. **When installing hot water pipes in large buildings with circulating lines, an advisable conservation measure is to:**

    insulate the hot water feed lines to prevent heat loss and save energy. (This also provides substantial cost saving.) *See page 170*

16. **Though not required in a cold water system, a hot water system design must allow for:**

   the expansion and contraction of hot water lines. *See page 170*

17. **For each 100 feet of piping, you should allow an expansion variance of:**

   $1^1/_2$ inches. *See page 170*

18. **In the average home, the peak draw period for hot water is assumed to be:**

   1 hour. *See page 170*

19. **When sizing hot water storage tank capacity, the percentage of the tank's hot water supply that's assumed available during the one-hour period of peak draw is:**

   75 percent. *See page 170*

20. **For a peak draw period of one hour, a 40-gallon water heater should provide:**

   30 gallons of hot water. *See page 170*

21. **For a peak draw period of three hours, a 40-gallon water heater must provide:**

   10 gallons of hot water per hour. *See page 171*

22. **For a storage tank heater, the amount of hot water that's available during the peak demand periods is determined by:**

   the heating capacity per hour. *See page 171*

23. **The recommended storage capacity for a gas water heater installed in a two-bedroom house is:**

   30 gallons. Refer to Figure 13-3. *See page 171*

24. **The recommended storage capacity for an oil-burning water heater installed in a three-bedroom house is:**

   30 gallons. Refer to Figure 13-3. *See page 171*

25. **The recommended storage capacity for an electric water heater installed in a four-bedroom house is:**

   66 gallons. Refer to Figure 13-3. *See page 171*

26. **The determining factor in choosing the storage capacity for any water heater, regardless of fuel type, is:**

   the number of bedrooms in the house. *See page 171*

27. **It might be wise to split the hot water system and install two water heaters when:**

   the layout requires long pipe runs. *See page 171*

28. **The capacity size for a packaged high recovery rate heater that would generally be adequate for up to 12 living units is:**

   a 75-gallon storage tank. *See page 171*

29. **The code requires that all equipment used for the heating or storage of hot water be equipped with:**

   a pressure relief valve. *See page 171*

30. **The pressure relief valve on a hot water heater serves to:**

   safely relieve excess storage tank pressure. *See page 171*

31. **When you install a water heater, the following should remain visible:**

   the plates with the maximum working water pressure and other data. *See page 171*

32. **In locating a water heater, you must ensure that it's accessible for servicing or replacement without:**

   having to remove any permanent part of the building. *See page 171*

33. **The temperature relief valve that's required on all domestic hot water heaters is:**

   the reseating type, rated by its Btu capacity. *See pages 171-172*

34. **The type of relief valve commonly used today is:**

   a combination pressure and temperature relief valve. *See page 172*

35. **On a hot water heater pipe, a shut-off or check valve must never be installed:**

   between the relief valve and the hot water heater tank. *See page 172*

36. **Because of possible contamination, the code prohibits the hot water relief valve drip pipe from connecting to:**

   any plumbing drainage or vent system. *See page 172*

37. **A water heater relief valve drip pipe can never terminate above any of the following fixtures:**

   a water closet, urinal, bidet, bathtub or shower stall. (Any three answers are acceptable.) *See page 172*

38. **A water heater relief valve drip pipe should terminate:**

   to an observable point outside the building. *See page 172*

39. **The end of a water heater relief valve drip pipe may *not* be:**

    the threaded type. *See page 172*

40. **If hot water storage tanks are placed above the roof of a building, the relief valve pipe may discharge:**

    on the surface of the roof. *See page 172*

41. **A water heater relief valve drip pipe is sized by:**

    its Btu rating. *See page 172*

42. **A water heater with a 200,000 Btu rating should be served by a relief valve drip pipe with an inside diameter of:**

    $^3/_4$ inch. Refer to Figure 13-6. *See page 172*

43. **The minimum size cold water supply pipe that you can connect to any hot water heater is:**

    $^3/_4$ inch. *See page 173*

44. **A water heater drain pan is required when:**

    a water heater is located above the ground floor level of a building. *See page 173*

45. **The required depth of a water heater drain pan is:**

    $1^1/_2$ inches (2 inches, in some codes). *See page 173*

46. **The minimum size drain pan outlet is:**

    $^3/_4$ inch (1 inch, in some codes). *See page 173*

47. **The clearance between the drain pan sides and the water heater must be:**

    a minimum of 2 inches. *See page 173*

48. **In a vertical installation, the maximum number of water heater drain pans that you can connect to a 1-inch riser is:**

    three. *See page 173*

49. **On a water heater, the required location for a heat trap is:**

    on the hot water line leading away from the water heater. *See page 174*

50. **The purpose of a heat trap on a water heater is:**

    to prevent hot water from rising into the hot water line. (This also saves approximately 2 percent of the hot water generating cost.) *See page 174*

Chapter 14

1. **The three things the plumbing code regulates for water piping are:**

    the materials, the sizing and the installation methods. *See page 179*

2. **The plumbing code requires that there be an adequate supply of potable water to all fixtures so they:**

    will flush properly and remain clean and sanitary. *See page 179*

3. **Code-established safeguards protect our water supply because:**

    they prevent pollution of the water supply. *See page 179*

4. **Two reasons why you should consider the kind of water and soil in your area before you select and size the material for water supply pipes are:**

    (1) some water corrodes the interior wall of some pipes; and, (2) some types of soil can corrode the exterior of some pipes. *See page 179*

5. **The pipes that are specifically prohibited in a potable water supply system are:**

    pipes that can leave toxic substances in the water supply or were once used for other than a potable water supply system. *See page 179*

6. **All plastic pipe and fittings must display identification/acceptance from:**

    the ASTM or other recognized national standard of acceptance. *See page 179*

7. **The minimum working pressure required by code for plastic water service piping is:**

    160 psi. *See page 179*

8. **The plastic pipe and fittings first approved for a building water distribution system was:**

    CPVC. *See page 181*

9. **The highest water temperature that CPVC plastic pipe can withstand is:**

    180 degrees F. *See page 181*

10. **The unique advantage that polybutylene plastic pipe has over other approved plastic and metallic pipe and fittings is:**

    its creep resistance (expansion). *See page 181*

11. **Polybutylene (PB) pipe is undamaged by:**

    freezing temperatures. *See page 181*

12. **Three distinct advantages plastic pipe and fittings have over metallic pipe and fittings are:**

    they resist corrosion, scale, sediment buildup and aren't affected by soil conditions or electrolysis. (Any three answers are correct.) *See page 181*

13. **When installing water service pipe in the same trench as the building sewer, you must place it:**

    on a solid shelf excavated at one side of the common trench and at least 12 inches (10 inches, some codes) above the sanitary sewer line. You must also keep the joints in the water service pipe to a minimum. Refer to Figure 14-3. *See page 182*

14. **When you install metallic water service pipe on filled corrosive soil, you must protect it with:**

    one or two coats of asphaltum paint or other approved coating. *See page 182*

15. **Water service supply piping must be securely supported in order to:**

    prevent sagging, misalignment and breaking. *See page 182*

16. **Materials considered superior for outside water service supply piping include:**

    cast iron water pipe, cast iron threaded water pipe, wrought iron pipe and galvanized steel pipe. (Any two answers are acceptable.) *See page 182*

17. **Large boulders, rocks or cinder fill shouldn't be used to backfill a metallic water service supply trench because:**

    they might physically damage or encourage corrosion of the pipe. *See page 182*

18. **ABS, PVC, PE and PB plastic pipe must be treated with special care because:**

    they are considered fragile. Refer to Figure 8-16. *See page 182*

19. **The minimum separation distance required between a water service supply pipe and a sewer line when they are installed in separate open trenches is:**

    5 feet. *See page 182*

20. **In climates where pipe is subject to freezing, the trench for buried water service supply pipe should be deep enough to be:**

    below the frost line. *See page 183*

21. **Between the bottom of a building foundation or footing and the top of the water service supply pipe, you must have a clearance of:**

    2 inches. *See page 183*

22. **When connecting a lawn sprinkler system to a potable water service supply pipe, on the discharge side of each valve you must install:**

    an approved backflow preventer. *See page 183*

23. **In each building's water service supply pipe, independent of the water meter valve, you must install:**

    a separate, accessible, control valve (some codes require two). *See page 183*

24. **Water service supply piping must be electrically isolated from:**

    other pipes, conduits, soil pipe, building steel and steel reinforcing. *See page 183*

25. **The code prohibits a public water supply from interconnecting with:**

    a private water supply, such as a well. *See page 183*

26. **When you connect a swimming pool water supply to the potable water service pipe, in order to prevent cross connection you must provide:**

    a positive air gap. *See page 183*

27. **When mineral deposits in the water solidify in the distribution pipes over a period of time, it can cause:**

    a reduced flow of water and/or premature failure of the system. *See page 184*

28. **When a pipe passes through cast-in-place concrete, you should provide a clearance around its circumference of:**

    $^1\!/_2$ inch. *See page 184*

29. **In climates subject to freezing temperatures, you shouldn't install water pipe in any unheated areas unless you protect it with:**

    adequate insulation. *See page 184*

30. **You should not allow hot and cold water pipes to come into contact with each other in underground or partition installations because:**

    the heat from the hot water pipe will be transferred to the cold water pipe. *See page 184*

31. **Each water closet supply pipe is required to have:**

    an independent control valve installed above the floor. *See page 184*

32. **The number of residential plumbing fixtures that you can connect to a $1/2$-inch cold water supply pipe is:**

    two. *See page 184*

33. **You can protect water pipe installations from water hammer by providing:**

    properly located air chambers or other approved devices. *See page 184*

34. **When water pressure within a building is more than 80 psi, you must install:**

    an approved water pressure regulator with a strainer. *See page 184*

35. **You must support horizontal screwed water pipe at intervals of:**

    approximately 12 feet. *See page 185*

36. **The maximum distance allowed between supports for horizontal CPVC plastic pipe is:**

    4 feet. *See page 185*

37. **The maximum distance allowed between supports for vertical $1^1/_4$-inch copper tubing carrying cold water is:**

    4 feet. *See page 185*

38. **The code requires that potable water supply outlets terminate above the overflow rim of a fixture because:**

    this provides an air gap. *See page 185*

39. **On all water outlets equipped for hose connections (except for clothes washers) you must install:**

    a backflow preventer. *See page 185*

40. **In the water supply system, unions must have:**

    metal-to-metal joints and ground seats. *See page 185*

41. **The standards that all pipe and fitting threads must meet are:**

    those adopted by the American Standards Association. Refer to Figure 14-6. *See pages 185-186*

42. **When you use a wrench that's too large for a fitting, the result can be:**

    a bad joint or a cracked fitting. *See page 186*

43. **The wrench size recommended to tighten a fitting on a 2-inch pipe is:**

    a 24-inch pipe wrench. Refer to Figure 14-7. *See page 186*

44. **The most common type of heavy-duty steel pipe cutter is:**

    the single wheel cutter. *See page 186*

45. **When a pipe cutter wheel leaves a bur on the inside of a pipe up to 2 inches, the tool you should remove it with is:**

    a pipe reamer. Refer to Figure 14-8. *See pages 186-187*

46. **The only person that knows for certain whether an installed pipe has good threads is:**

    the plumber who installed it. *See page 186*

47. **When you're laying out and dimensioning piping arrangements, the only offsets that aren't difficult to calculate are the:**

    90-degree elbows. *See page 187*

48. **The two tools usually used to cut copper tubing are:**

    a tubing cutter and a hacksaw. *See page 187*

49. **The most common joint for copper pipe is:**

    the sweat joint. *See page 187*

50. **In a potable water supply system, the maximum allowable lead content for solders and fluxes is:**

    two-tenths (0.20) of 1 percent. *See page 188*

51. **Properly cleaning and heating a copper joint helps make a sound solder joint by:**

    allowing surface tension to spread the solder to all parts of the joined surfaces. *See page 188*

52. **When cutting plastic pipe, you can be sure of getting a square end by:**

    using a pipe cutter or a miter box and hacksaw. *See page 188*

53. **The common term for a cemented plastic pipe joint is:**

    a "welded" joint. *See page 189*

54. **To remove impurities and gloss from the surfaces of plastic pipe and fittings, use:**

    a liquid cleaner or fine sandpaper. *See page 189*

55. **You can make only one joint at a time in a plastic water system because:**

    the plastic cement sets very quickly. *See page 189*

56. After you cement the last joint, you shouldn't test a plastic water system for at least:

half an hour. *See page 189*

Chapter 15

1. The percentage of homes in the U.S. that depend on private wells as their source of water is approximately:

20 percent (one in five). *See page 193*

2. The agencies with authority to approve and inspect domestic wells are:

the local health department and the Department of Environmental Protection (DEP). (Either answer is acceptable.) *See page 193*

3. Before you can drill or drive a well, you must:

obtain a permit. *See page 193*

4. When you drive or drill a well, the bottom of the well casing must extend:

into the dry weather water table. Refer to Figure 15-1. *See pages 193-194*

5. Well water is generally classified as:

hard water. *See page 193*

6. Well water is considered hard water because:

it has a high mineral content. *See page 193*

7. Well water has a distinctly different taste from "city water" because:

it doesn't contain chlorine or other chemicals. *See page 193*

8. The minerals in untreated well water may cause:

stains in the plumbing fixtures and scale build-up in the plumbing system. *See page 193*

9. You can improve the taste and smell of well water by:

installing a water softener. Refer to Figure 15-2. *See pages 193-194*

10. The tradesmen most likely to install wells, suction lines, pumps and water pressure pumps are:

professional well drillers (Although plumbing contractors are also qualified to do this type of work, they usually subcontract it out.) *See page 194*

11. Professional well drillers are certified and licensed as:

specialty contractors. *See page 194*

12. Even though plumbing contractors seldom install domestic wells, two reasons you should learn about well systems are:

many of the people you deal with will assume you know something about wells, and plumber's examinations often include questions on well systems. *See page 194*

13. The minimum depth that local authorities usually require for a potable water supply wells is:

30 feet. *See page 194*

14. Wells that are dug or driven are classified as:

shallow wells. *See page 194*

15. The two types of driven wells are:

those with an open end casing and those with a casing equipped with a well point. *See page 194*

16. Open end well casings are commonly used:

where the ground water table is close to the ground surface and in a good rock formation. *See page 194*

17. When you're sure that you've installed a good well, before it's ready for use you must:

hook up a pump and pump water out of the well until the water is free of rocks and sand. *See page 195*

18. You would use well casings equipped with a well point:

where the water table is in loose shale or sand. *See page 195*

19. You would use a well point with a screen or fine perforations where there is:

sand. *See page 195*

20. You would use a well point with large openings where there is:

gravel or loose rock formations. *See page 195*

21. Wells that are drilled are classified as:

deep wells. *See page 195*

22. Water from deep wells is more desirable than water from shallow wells because:

there's less chance of contamination. *See page 195*

23. **Without specific approval from your local authority, you may never locate a well:**

    within a building or under the roof or projection of any building. *See page 195*

24. **The well casing in a drilled well must terminate:**

    in a suitable aquifer. *See page 195*

25. **The reason a 36-inch-wide sloping concrete collar is required around the top of the well casing is:**

    to prevent surface water from carrying pollutants down the well casing to the water reservoir. *See page 195*

26. **To provide access for inspections, measure well depth, test the static water level and allow disinfecting agents to be added to the well, you must install:**

    a tee. *See page 195*

27. **The minimum size for a suction line from the well to the pump is:**

    1 inch. *See page 195*

28. **You should install the check valve on a well suction pipe:**

    as close as practical to the well. *See page 197*

29. **The two types of check valves commonly used on suction lines are:**

    the spring-loaded type and the flapper-type. (Either answer is acceptable.) *See page 197*

30. **On the well suction line, just before the pump, you must install:**

    a union or a slip coupling. (Either answer is acceptable.) *See page 197*

31. **The two types of pressure tanks in common use are:**

    the hydropneumatic pressure tank and the diaphragm pressure tank. *See page 197*

32. **Of the two commonly-used pressure tanks, the one that's been the dominant well water supply system since the 1920s is:**

    the hydropneumatic tank. *See page 197*

33. **The psi operating range for a hydropneumatic tank is:**

    between 20 and 40 psi of water pressure. *See page 197*

34. **When there's not enough air to pressurize the tank in a hydropneumatic system:**

    the pump cycles on and off too often. *See page 197*

35. **The minimum size hydropneumatic tank needed for a single-family residence is:**

    42 gallons. *See page 199*

36. **The diaphragm tank has been approved for use in domestic well water systems since:**

    the 1950s, over 40 years. *See page 199*

37. **By adding a little extra capacity to a diaphragm tank, you can usually prevent:**

    excessive cycling (too-frequent pump starts). *See page 199*

38. **The purpose of installing a gate valve on the discharge side of the tank is:**

    to act as the house valve to control water in the building. *See page 199*

39. **The minimum size gate valve required is:**

    $3/4$ inch. *See page 199*

40. **When you locate the tank and equipment, you should take care that they are:**

    set level and placed so that they're accessible for repair or replacement. *See page 199*

Chapter 16

1. **The minimum size public water main that you can use to supply a fire standpipe system is:**

   4 inches. *See page 201*

2. **When a public water main can't provide the required quantity and pressure for a fire standpipe system, the alternative methods you can use include:**

   a fire pressure pump or a gravity tank. *See page 201*

3. **A fire standpipe system is required in buildings that are higher than:**

   50 feet. *See page 201*

4. **The special requirement that ensures a constant water supply is available for a fire standpipe system is:**

   a pressurized (wet) system. *See page 201*

5. **For buildings under construction that require standpipes, you must locate fire department connections:**

   at street level on the outside of the building and at each floor level up to the highest constructed floor. *See page 201*

6. **Fire standpipe locations must be arranged so that the standpipes are protected from:**

   mechanical and fire damage. *See page 201*

7. **The maximum distance from the nozzle end of a 100-foot fire hose that any part of a building floor can be is:**

   30 feet. *See page 201*

8. **If the stairways are enclosed, the fire standpipes should be located:**

   in the enclosed stairways. *See page 201*

9. **If the stairways aren't enclosed, the fire standpipes must be located:**

   within 10 feet of the floor landing of a stairway. *See page 201*

10. **When stairways aren't available, additional fire standpipes may be located:**

    in hallways or other accessible locations approved by the authority. *See page 201*

11. **The special fire-protection provisions required by some codes for buildings designed for theatrical performances are:**

    that there's a 2¹/₂-inch standpipe on each side of the stage, with a hose not over 75 feet long at each standpipe hose station. The Uniform Fire Code requires automatic sprinklers.
    See page 202

12. **When a fire standpipe system is required and there's no adequate public water supply, the alternate water system that the code will accept is:**

    an on-site well system. *See page 202*

13. **The maximum drawdown for an on-site fire well system when pumping at 150 percent pump capacity is:**

    4 feet. *See page 202*

14. **An on-site fire well system must be sized for a flow of:**

    500 gallons per minute at 20 psi. *See page 203*

15. **The requirement for fire department hose connections is that they must have:**

    national standard threads. *See page 203*

16. **Before the fire department having jurisdiction over an on-site well system will give you final approval, you must:**

    test the system. *See page 203*

17. **You may not have a direct connection between an on-site fire well system and:**

    the potable water supply system. *See page 203*

18. **The difference between an on-site standard fire well installation and a regular on-site well system is:**

    for an on-site fire well you don't have to install a pump, and the hose connection can be a single 4¹/₂-inch American Standard Hose connection. (Either answer is correct.) *See page 203*

19. **For fire protection in commercial, industrial and residential areas, the code usually requires:**

    fire hydrants. *See page 203*

20. **The location of all fire hydrants must be approved by:**

    the fire department. *See page 204*

21. **Underground fire line and fitting materials must be approved by:**

    the local authority having jurisdiction.
    See page 204

22. **Aboveground fire lines and fittings must be able to withstand a pressure of:**

    175 psi. *See page 204*

23. **At each change of direction in underground fire lines, you must provide:**

    concrete thrust blocks resting on undisturbed soil. *See page 204*

24. **Underground fire lines are tested to ensure that they can withstand:**

    200 psi of pressure. *See page 204*

25. **The minimum size requirement for fire standpipes in buildings up to 100 feet high is:**

    4 inches. *See page 204*

26. **The minimum size requirement for fire standpipes in buildings *over* 100 feet high is:**

    6 inches. *See page 204*

27. The maximum length allowed for a building standpipe is:

275 feet. *See page 204*

28. On buildings that are 50 feet or higher, fire standpipes are required to extend above the roof:

30 inches. *See page 205*

29. The fire department connection that you're required to install on each standpipe extension above the roof is:

a roof manifold. Refer to Figure 16-6. *See pages 205-206*

30. Standpipes located in stairway enclosures require valves for fire department hose connections that are:

$2^1/_2$ inches. Refer to Figure 16-6. *See pages 205-206*

31. The maximum distance allowed from the standpipe or hose station to a hose outlet is:

10 feet. *See page 205*

32. The pipe size generally used to connect a hose station to the fire standpipe is:

$2^1/_2$ inches. Refer to Figure 16-6. *See pages 205-206*

33. A fire hose cabinet must be located:

within 10 feet of the standpipe and where it's accessible at all times. Refer to Figure 16-6. *See pages 205-206*

34. Each fire hose must be able to withstand a working pressure of:

100 psi. *See page 206*

35. If the pressure exceeds 100 psi at the fire hose outlet, you must install:

a pressure regulating device. *See page 206*

36. A fire pump required to supply a 500 gpm flow rate must be certified by:

Underwriters Laboratory (UL listed). *See page 207*

37. The electric service required for fire pumps with a 500 gpm flow rate must be:

a separate electric service or a connection through a separate automatic transfer switch to a standby generator. *See page 207*

38. The equipment that you can use to maintain the 15 psi minimum pressure required on the roof in a fire standpipe system is:

either a jockey pump actuated by a pressure switch or a connection to a suitable domestic system through two 170 psi check valves (one with a soft seat and one with a hard seat). *See page 207*

39. Those qualified to install fire protection systems consisting of standpipes and fire hoses are:

plumbing contractors with local or state certification. *See page 207*

40. Some states require that plumbing contractors be certified by the state fire marshall before they can install fire protection systems with:

automatic sprinklers. *See page 207*

41. The minimum size for risers in a combined fire system is:

6 inches in diameter. *See page 207*

42. The water supply need for a combined fire system is determined by:

the occupancy class of the building. *See page 207*

43. The combustibility level of a Class I building is considered:

moderate. *See page 207*

44. The combustibility level of a Class II building is considered:

low. *See page 207*

45. The combustibility level of a Class III building is considered:

high. *See page 207*

46. In an installation where more than one fire standpipe riser is required, you must:

loop the risers at the lowest floor. *See page 207*

47. Each fire standpipe riser branch line must be taken off:

at the floor it serves. *See page 208*

48. The two fittings that you should install on the water supply line for automatic sprinkler systems are:

a post-indicator valve and a check valve. *See page 208*

49. **If you use two or more fire pumps in an automatic sprinkler installation, each pump must operate:**

    independently. *See page 209*

50. **You can use an engine-driven fire pump only under the condition that:**

    it's approved in advance by local jurisdictional authorities. *See page 209*

Chapter 17

1. **In order to specialize in the installation of piping and equipment for swimming pools and spas, a contractor must earn:**

    a certificate of competency. *See page 213*

2. **Even though specialists generally do swimming pool and spa work, in most parts of the country you must still be knowledgeable about it because:**

    you'll have questions about swimming pool and spa work on your journeyman and master plumber examinations. *See page 213*

3. **According to code definition, a swimming pool is:**

    any structure suitable for swimming or recreational bathing that's over 24 inches deep. *See page 213*

4. **The code defines a private swimming pool as:**

    one that's located at a single-family residence and is available only to the family and their guests. *See page 213*

5. **The code defines a public swimming pool as:**

    one that's used collectively by a number of persons for swimming or bathing, whether a fee is charged or not. *See page 213*

6. **The most common mechanical system plumbed into swimming pools today is:**

    a recirculating system. *See page 213*

7. **The basic equipment required for a recirculating-type swimming pool is:**

    a pump and a filter system. *See page 213*

8. **Other terms that you can use to identify recirculating piping are:**

    return piping and pool inlet piping. *See page 213*

9. **The chemicals you use to maintain the quality of swimming pool water are:**

    chlorine or fluorine. *See page 213*

10. **Water is lost from a swimming pool by:**

    evaporation, splashing and backwashing. (Any two answers are acceptable.) *See page 213*

11. **The purpose of having a good swimming pool filtration system is:**

    to assure that the water is clean, clear of organic matter and safe from harmful bacteria. *See page 213*

12. **In order to prevent cross-connection, a homeowner who uses a garden hose to fill his swimming pool or spa must install:**

    a vacuum breaker on the hose bibb. Refer to Figure 17-1. *See pages 213-214*

13. **To prevent cross-connection, swimming pools with a direct connection to the public water supply are required to have:**

    a fill spout with an air gap above the overflow rim of pool. *See page 213*

14. **There are several approved methods for disposing of swimming pool water, including:**

    emptying into (a) an adequately sized drainfield, (b) a sewage system (if approved by local authority), (c) a disposal well, (d) an adequately sized soakage pit or (e) an open waterway, bay or ocean (where permitted). It may also be (f) piped to a sprinkler system used for irrigation or (g) puddled on the property. In both (f) and (g), the swimming pool water must be confined to the pool owner's private property. (Any two answers are acceptable.) *See page 214*

15. **The main drain for a swimming pool should be located:**

    at the lowest point in the pool. Refer to Figure 17-6. *See page 218*

16. **The minimum number of inlets required for a swimming pool is:**

    two. *See page 216*

17. **When the main swimming pool drain is used for a return, it is considered to be:**

    an inlet. *See page 216*

18. **When the main swimming pool drain is used for a return you must size it:**

    as a suction line. *See page 216*

19. **The minimum diameter size for a swimming pool vacuum fitting is:**

    $1^1/_2$ inches. *See page 216*

20. **You connect a vacuum fitting to the swimming pool pump on:**
the suction side of the pump. Refer to Figures 17-3 and 17-4. *See pages 215-216*

21. **The filtration rate for pressure sand filters is:**
not over 5 gpm per square foot of filter area. *See page 216*

22. **The inflow and effluent lines for a swimming pool sand filter must have:**
pressure gauges. Refer to Figure 17-3. *See page 216*

23. **A swimming pool backwash line must have:**
a sight glass. Refer to Figure 17-3. *See pages 215-216*

24. **The two types of diatomite filters used for filtering swimming pool water are:**
the vacuum type and the pressure type. *See page 217*

25. **The filtration rate for a diatomite swimming pool filter is:**
2 gpm per square foot of effective filter area. *See page 217*

26. **Provisions must be made for removing caked diatomite from the swimming pool filter by:**
backwashing or disassembling the filter. *See page 217*

27. **The design and installation of a diatomite filter must permit:**
the filter elements to be removed easily. *See page 217*

28. **At the high point on each swimming pool pressure filter tank you must install:**
an air relief device. Refer to Figure 17-3. *See page 217*

29. **One surface skimming device can accommodate:**
1,000 square feet of swimming pool surface. *See page 217*

30. **The required rate of flow through a swimming pool skimming device is:**
at least 25 gpm per skimmer. *See page 217*

31. **Copper, Type K or L pipe can be used in a pool installation for:**
all lines. Refer to Figure 17-5. *See page 217*

32. **Cast iron, service weight pipe can be used in a swimming pool installation for:**
gutter lines only. Refer to Figure 17-5. *See page 217*

33. **The weight required for all fittings used with ABS or PVC plastic pipe in swimming pool installations is:**
Schedule 40. *See page 217*

34. **The type fittings that you must use in a swimming pool gutter line are:**
the drainage type. *See page 217*

35. **You may not install short radius 90-degree elbow fittings on swimming pool or spa piping:**
below grade on suction piping. *See page 217*

36. **The minimum size suction piping required for swimming pools and spas is:**
2 inches in diameter. *See page 217*

37. **If you're installing dissimilar metals in swimming pool filter piping, the kinds of fittings required are:**
dielectric fittings. *See page 217*

38. **The pressure required to water-test a swimming pool pressure piping system, including the main drain, is:**
40 psi. *See page 217*

39. **The older, flat-type main drains for swimming pools are no longer acceptable by most codes today because:**
their strong suction can hold children beneath the water and cause drowning. *See page 218*

40. **The two code-accepted main drain covers used for swimming pools and spas today are:**
the antivortex type and an 85-square-inch open grate type. *See page 218*

41. **Gas-fired swimming pool heaters and swimming pool boilers must comply with the standards set by:**
the AGA and ASME. *See page 218*

42. **A national testing agency that can approve oil-burning equipment for swimming pools is:**
Underwriters Laboratory. *See page 218*

43. **The maximum temperature acceptable for heated swimming pool water is:**
105 degrees Fahrenheit. *See page 218*

44. The code defines a residential spa as:

one used by not more than two families and their guests. *See page 218*

45. For a spa to be considered a spa and not a pool, the maximum gallon capacity cannot exceed:

3,250 gallons of water. *See page 218*

46. The maximum temperature of a spa heater is:

105 degrees Fahrenheit. *See page 219*

Chapter 18

1. Today, solar energy is used in homes for:

heating water for all domestic purposes, including swimming pools. *See page 223*

2. The professional plumber should know about installing solar energy units because:

they are becoming more and more common, the codes are changing to cover solar energy, and they require a plumbing permit for installation. Chances are you, as a plumber, will be working with solar energy installations. *See page 223*

3. The first requirement for installing, repairing or altering any solar energy system is:

to get a plumbing permit. *See page 223*

4. In most cases, the plans for a solar water heating system are prepared by:

a registered professional engineer. *See page 223*

5. The plumbing drawings for a solar water heating system must show:

the entire solar system, including structural calculations, mounting frames and anchorage detail. *See page 223*

6. When tested, a solar energy system must be able to withstand a pressure of:

125 psi for at least 15 minutes without leaking. *See page 223*

7. Using solar energy for heating domestic water isn't considered new because:

sun-rich areas such as Florida and California have been successfully using solar-heated water since the beginning of the century. *See page 223*

8. A family of four, using solar heated water for domestic purposes, could save approximately:

$200 per year. *See page 223*

9. The three major components of a solar water heating system are:

(1) the solar heat collector, (2) the circulation system and (3) the solar storage tank. *See pages 223-224*

10. If you're dealing with a pumped solar system, the fourth component you'll need is:

the control center. *See page 224*

11. The type of solar heat collector that's most practical for residential use is:

the flat plate solar heat collector. *See page 224*

12. The water temperatures produced by solar water heating systems can reach:

approximately 200 degrees Fahrenheit. *See page 224*

13. The three materials acceptable for the heat deck of a solar heat collector are:

copper, aluminum and steel. *See page 224*

14. The thermal difference between using copper, aluminum or steel for heat deck materials is:

none. Thermally, there's no difference between copper, aluminum or steel when used as heat deck materials. *See page 224*

15. The tubing and collector plate must be of the same metal because:

they need to expand and contract at the same rate. *See page 224*

16. The two purposes accomplished by ensuring that the solar heat collector box is well insulated are:

you shield the heat deck plate from the weather, and you reduce the heat loss. *See page 225*

17. The reason you should use glass with a low iron content in a solar heat collector is because:

it's transparent and admits solar radiation, but it's opaque to the long-wave energy trapped inside the collector box. This trapped heat is then transferred to the fluid in the tubing. *See page 225*

18. In order to prevent heat loss in a solar collector in a cold climate, you should:

install a solar collector box with a double layer of glass. *See page 225*

19. The quality of materials you should use for frames and braces for securing solar heat collectors to roof structures is:

exterior-quality. *See page 225*

20. **Solar collector panels that aren't an integral part of the roof must be mounted above the roof surface:**

    a minimum of 3 inches. *See page 225*

21. **At its lowest point, a solar collector panel box must have:**

    drainage holes. *See page 225*

22. **According to U.S. government figures, each of the first two people in a family will use:**

    20 gallons of hot water per day (40 gallons total for two people). *See page 225*

23. **According to U.S. government figures, each additional family member after the first two will require:**

    15 gallons of hot water per person per day. *See page 225*

24. **Each day, a 4- by 12-foot solar heat collector will provide:**

    approximately 80 gallons of hot water. *See page 225*

25. **The minimum size solar storage tank recommended for a family of four is:**

    80 gallons. *See page 225*

26. **The percentage of solar energy that strikes the glass surface of a collector and actually heats the water circulating through the tubing is:**

    only about 30 to 65 percent. *See page 226*

27. **To be the most efficient, a flat solar heat collector should face:**

    in the general direction of the sun's path across the sky, preferably south. *See page 226*

28. **The reason you should mount a solar heat collector as close as possible to the storage tank is:**

    to reduce heat losses and friction in the pipes. *See page 227*

29. **Before you install a solar heat collector as an awning or as a fixed overhang on a residence, you must have:**

    approval from your local authority. *See page 227*

30. **A natural thermosiphon solar water heating system works in the following manner:**

    the hot water (which is lighter) rises naturally to the storage tank and replaces the heavier cold water, which is then drawn into the collector. The bottom of the storage tank must be located at least 2 feet higher than the top of the collector in order for the thermosiphon to work and the water to circulate through the system on its own. *See pages 227-228*

31. **When a solar storage tank is attic-mounted, most codes require you to install:**

    an adequate-sized drain pan with a drain pipe extending to the exterior of the building. *See page 228*

32. **The minimum size piping that you can use in a thermosiphon circulation system is:**

    $3/4$ inch. *See page 228*

33. **In a pumped solar system, you can locate the hot water storage:**

    in any convenient place. *See page 228*

34. **In a pumped circulation system, the size copper tubing that's permissible to use is:**

    $1/2$ inch. *See page 228*

35. **In a closed solar energy collection system, the fluid you can use is:**

    antifreeze. *See page 228*

36. **In a closed solar energy collection system, the heat is transferred to the water in the storage tank by means of:**

    a heat exchanger. *See page 228*

37. **The *Uniform Solar Energy Code* requires that pipe and fittings used within a solar system meet the standards:**

    set by code for a potable water system. *See page 230*

38. **The two most common materials used for pipe and fittings in a solar circulation system are:**

    galvanized steel pipe and fittings and copper pipe and fittings (Type K or Type L copper pipe). *See page 230*

39. You can't use plastic pipe in a solar circulation system because:

plumbing standards forbid the use of plastic pipe where temperatures could exceed 180 degrees Fahrenheit. *See page 230*

40. You should insulate all piping that carries heated water, fluids or gases from the solar collector to storage tank in order to:

minimize heat loss in the system. *See page 230*

41. Valves up to 2 inches in diameter installed in a solar piping system must be constructed of:

brass, or other approved materials. *See page 230*

42. You must install control valves in a solar system:

where they can isolate the solar system from the potable water supply, and where they are all readily accessibly. *See page 230*

43. The required location for a combination temperature and pressure relief valve in a solar hot water storage tank is:

in the hottest water — in the top one-eighth of the tank. *See page 230*

44. If authorities require a second relief valve, besides the mandated relief valve on the storage tank, you should place it:

at the highest point of the piping system. *See page 231*

45. Automatic air discharge valves must be installed:

at the highest point of a solar piping system. *See page 231*

46. If the water usage in a household is reduced, you can prevent the water stored in the solar storage tank from becoming dangerously hot by:

installing a tempering valve in the hot water pipe leading from the tank. *See page 232*

47. The minimum working pressure for a solar storage tank is:

300 psi. *See page 232*

48. When a solar system circulating pump is running but the water in the tank isn't hot, the problem is most likely caused by:

air trapped in the collector. Refer to Figure 18-10. *See page 233*

49. The probable cause for a solar collector freezing is:

the pitch of the flat, horizontally-mounted collector doesn't permit the pipes to drain dry. Refer to Figure 18-10. *See page 233*

50. If a solar controller is malfunctioning and you've ruled out any shorts or breaks in the sensor wires, the likely cause of the malfunction is:

corrosion at the sensors' connections. *See page 232*

Chapter 19

1. The three physical states of matter are:

solid, liquid and gas. *See page 237*

2. A gas consists of:

constantly moving atoms, with neither fixed space nor volume. *See page 237*

3. Gas particles liquefy when cooled:

below their boiling point. *See page 237*

4. The first discovery of natural gas in America was in:

West Virginia in 1775. *See page 237*

5. About 70 percent of the natural gas produced in the United States is found in the states of:

Texas and Louisiana. *See page 237*

6. The warning aid added to natural gas to help curb the danger of accidental explosions is:

a chemical scent. *See page 237*

7. The three types of gases used for fuel today are:

natural gas, manufactured gas and liquefied petroleum gas. *See pages 237-238*

8. Natural gas (methane) is also known as:

dry, or sweet gas. *See page 237*

9. Although natural gas itself is not poisonous, it can cause death:

by suffocation in a closed space, or as a result of injuries in an explosion. *See page 237*

10. **The type of gas chiefly produced from coal is:**

manufactured gas. *See page 237*

11. **The poisonous substance in manufactured gas is:**

carbon monoxide. *See page 237*

12. **Other names used for liquefied petroleum gas are:**

LP or bottled gas. *See page 238*

13. **Liquefied petroleum gas consists primarily of:**

butane or propane, or a mixture of both. *See page 238*

14. **The physical change that occurs in LP gas under moderate pressure is:**

that it turns into a liquid state. *See page 238*

15. **Under normal temperature and atmospheric pressure conditions, liquefied petroleum:**

returns to its original gaseous state. *See page 238*

16. **LP gas is a convenient fuel to use in remote areas because:**

it's easily containerized and transported. *See page 238*

17. **The responsibility for sizing the gas service pipe to a building belongs to:**

the gas supplier. *See page 238*

18. **The sizing and installation methods for interior gas piping are governed by:**

the local gas code. *See page 238*

19. **Two factors you must know before sizing any gas building main or branch lines are:**

(1) the maximum gas demand at each appliance outlet, and (2) the length of piping required to reach the most remote outlet. *See page 238*

20. **The abbreviation "Btu" stands for:**

British thermal unit. *See page 238*

21. **One Btu is defined as:**

the quantity of heat required to raise the temperature of 1 pound of water 1 degree Fahrenheit. *See page 238*

22. **The number of Btu that you can assume to be in each cubic foot of natural gas is:**

1,000 cfh (cubic feet per hour). *See page 238*

23. **If you know the maximum Btu rating for an appliance, you convert it into cubic feet by:**

dividing the value in Btu by 1,000. *See page 238*

24. **When you connect a gas supply pipe to an appliance that's missing its Btu rating plate, the required pipe size is:**

the same size as, or larger than, the size of the appliance inlet pipe. *See page 238*

25. **Regardless of circumstances, the minimum size gas supply pipe outlet that you can use is:**

$1/2$ inch. *See page 238*

Chapter 20

1. **When selecting piping materials for a gas system, you must consider:**

the characteristics of your particular gas supply and its effect on pipe, especially if the gas is corrosive. *See page 243*

2. **The three piping materials that are code-accepted for both underground and above ground gas installations are:**

galvanized steel pipe, black steel pipe and galvanized wrought iron pipe. *See page 243*

3. **Piping materials that are acceptable for use in a gas system where the gas is corrosive are:**

galvanized steel pipe, black steel pipe and galvanized wrought iron pipe. (Any two answers are acceptable.) *See page 243*

4. **The percent of yellow brass pipe that must be copper if you're using it in a gas installation is:**

75 percent. *See page 243*

5. **In an underground installation, you're not allowed to install brass and copper pipe:**

under a concrete slab. *See page 243*

6. **You should not use copper piping in a gas system if:**

the gas is corrosive. *See page 243*

7. **If approved by your local code or gas supplier, the two weights of copper pipe and tubing that you can use for interior gas piping are:**

Type K or Type L. *See page 243*

8. **When joints are necessary in a copper gas piping system, the type of solder you must use is:**

   hard solder, usually a silver solder. *See page 243*

9. **You can use approved gas flare fittings in a copper gas piping system only if:**

   the joints are not concealed. *See page 243*

10. **When plastic pipe and fittings are approved for a gas system, they must conform to the specifications of:**

    the American Society for Testing and Materials (ASTM). *See page 243*

11. **You can only make connections between metallic and plastic piping:**

    outside the building and underground. *See page 244*

12. **The building locations where you should never install gas piping are:**

    air ducts, clothes chutes, elevator shafts, chimneys, vents, ventilating ducts and dumbwaiters. (Any three answers are acceptable). *See page 244*

13. **The minimum distance you must maintain between gas piping and a water pipe or a sewer line in an underground installation is:**

    8 inches. *See page 244*

14. **The minimum depth for placing underground horizontal metallic gas piping is:**

    12 inches. *See page 244*

15. **The minimum depth for placing underground horizontal plastic gas piping is:**

    18 inches. *See page 244*

16. **Ways of protecting gas piping if you're installing it in corrosive soil are:**

    with an approved wrapping, or, with one or two coats of asphaltum paint. (Either answer is correct.) *See page 244*

17. **In areas subject to freezing temperatures, the depth at which you should install gas piping is:**

    below the frost line. *See page 244*

18. **When backfilling a trench containing gas piping, the type of backfill you should use is:**

    fine material. *See page 244*

19. **You can install gas piping under a slab only if the following installation conditions are met:**

    (1) encase the pipe completely in conduit; (2) seal the termination of the conduit above the floor to prevent any gas entry; (3) seal the termination of the conduit outside the building to prevent any water entry; and (4) extend a vent above grade and secure it to the conduit to convey any leaking gas outside the building. Refer to Figure 20-2. *See page 245*

20. **The connection you use for gas equipment or appliances subject to vibration or requiring mobility is:**

    an approved flexible gas hose connector. *See page 245*

21. **You should install gas piping to serve an appliance located in the center of a room by:**

    laying the pipe in an open channel cut into the concrete floor. Refer to Figure 20-3. *See page 245*

22. **To protect gas piping in vertical masonry walls you must provide:**

    adequate chases. Refer to Figure 20-4. *See page 246*

23. **You may drill a hole in the center of a partition stud when installing:**

    a short run of horizontal gas piping that does not require additional joints. *See page 246*

24. **You shouldn't notch a partition stud deeper than one-third its total width because:**

    you'll weaken the stud. *See page 246*

25. **You should protect soft tubing in a notched partition with:**

    a metal stud guard to prevent penetration by lath nails. *See page 246*

26. **You should secure gas piping installed in metal stud partitions with:**

    tie wire. *See page 246*

27. **Bushings are permitted in a concealed gas piping system:**

    never. *See page 246*

28. **In order to prevent the loosening of a union in an existing concealed gas line:**

    you need to punch the center nut on the joint. *See page 246*

29. **You are allowed to make a new connection on an existing concealed gas piping or tubing installation:**

    at no time. *See page 246*

30. **The procedure for preparing threads for gas piping is the same as preparing threads for:**

    water piping. *See page 246*

31. **The threads for gas piping must conform to the standards of the:**

    American Standards Association (ASA). *See page 246*

32. **To catch any condensation that may form in a gas main, you must install:**

    a drip pipe. *See page 246*

33. **Gas branch pipes should connect only at the top or side of a gas feeder pipe in order to:**

    prevent condensation from entering the lines and obstructing the flow of gas. *See page 246*

34. **A shutoff valve should be installed near the gas meter in order to:**

    allow for the gas to the building to be shut off from an easily-accessible location. *See page 246*

35. **To avoid accidental or malicious tampering with the outside gas shutoff valve:**

    it's installed with a square nut head that can only be turned with a special tool. *See page 246*

36. **Each gas appliance in a building is required to have a shutoff valve that is:**

    manually-operated and accessible. *See page 246*

37. **The two types of shutoff valves manufactured for appliances are:**

    the straight pattern and the angle pattern. *See pages 246-247*

38. **The maximum distance allowed from a shutoff valve to the appliance it serves is:**

    6 feet. *See page 247*

39. **Before you can conceal a completed gas installation unit, it must be:**

    pressure tested and inspected. *See page 247*

40. **The safest way to check gas piping for leaks is:**

    to brush liquid soap around each joint to see if bubbles appear. *See page 247*

41. **The minimum height above the garage floor that you can set the combustion chamber for a gas water heater is:**

    18 inches. *See page 247*

42. **When you install a gas appliance having 100,000 Btu input or less in a separate room off the garage, you must provide:**

    two 50-square-inch ventilation openings, one 12 inches above the floor and one 12 inches below the ceiling. *See page 247*

43. **Gas water heaters should never be installed in:**

    living areas that may be closed, such as bedrooms or bathrooms. *See page 247*

44. **Gas appliances that require venting must be installed:**

    as close as possible to the vent. *See page 247*

45. **A 30-gallon gas water heater with a 4-inch draft hood requires a vent pipe:**

    no smaller than the opening of the draft hood, 4 inches in this case. *See page 247*

46. **The minimum separation required between a gas water heater with an insulated jacket and any combustible material is:**

    2 inches. *See page 248*

47. **The two acceptable types of concealed gas vent piping materials are:**

    double-wall metal pipe and fittings, and asbestos cement flue pipe. *See page 248*

48. **For gas vent pipes installed in partitions constructed of combustible material, you must provide:**

    an approved metal spacing device. *See page 248*

49. **Horizontal gas vent piping must be supported with metal straps or hangers that are:**

    at least 20 gauge sheet metal. *See page 248*

50. **All gas vent pipes extending above a roof must terminate in:**

    UL approved caps. *See page 248*

Chapter 21

1. **Fixtures constructed of pervious materials such as tile or marble must have:**

    waste outlets that can't retain water. *See page 253*

2. **In a bathroom where there's no natural ventilation available, you must provide:**

   a fan and duct. *See page 253*

3. **In a bathroom without adequate lighting or ventilation there is the potential danger of:**

   unsanitary conditions. *See page 253*

4. **The toilet bowl design that you're required to install in facilities intended for public use is:**

   the elongated type. *See page 253*

5. **The seats for public toilet bowls are required to be:**

   the elongated type with an open front, made of smooth nonabsorbent materials. *See page 253*

6. **A toilet tank refill tube serves the purpose of:**

   automatically restoring the toilet bowl water seal. *See page 253*

7. **An overflow tube in a toilet tank serves to:**

   prevent the tank from overflowing by removing excess water at the same rate that it enters the tank. *See page 254*

8. **A toilet that uses a flushometer rather than a tank must have:**

   a vacuum breaker. *See page 254*

9. **After being manually activated, a toilet flushometer must:**

   complete the normal flushing cycle automatically and deliver enough water to flush all surfaces of the bowl. *See page 254*

10. **The number of toilets a flushometer can serve is:**

    one. *See page 254*

11. **A toilet flushometer should be installed so it's readily accessible for:**

    repairs. *See page 254*

12. **A wall-hung urinal should be supported by:**

    a concealed metal carrier or other approved backing so that no strain is transmitted to the pipe connection. *See page 254*

13. **At the wall contact point, finish a wall-hung lavatory by:**

    sealing it with white cement or other suitable material. See page *254*

14. **The two basic urinal designs are:**

    the wall-hung type and the floor-mounted stall type. *See page 254*

15. **Stall urinals must be recessed slightly below the finished floor in order to:**

    provide proper drainage. *See page 254*

16. **The waste opening of a floor-mounted stall urinal is required to have:**

    a beehive-type strainer. *See page 254*

17. **Cabinet-mounted lavatories are secured to the countertop by:**

    special rim clips. *See page 255*

18. **The minimum outside diameter for lavatory waste outlets is:**

    $1^{1}/_{4}$ inches. *See page 255*

19. **Where circular-type multiple wash sinks are used, one lavatory or fixture unit is represented by:**

    each 18 inches of wash sink circumference. *See page 255*

20. **The minimum size for a bathtub waste and overflow is:**

    $1^{1}/_{2}$ inches. *See page 255*

21. **Some codes today prohibit the use of a trip waste because:**

    they are difficult to adjust to properly retain and discharge tub water. *See page 255*

22. **Bathtubs recessed into the finished wall are required to have joints that are:**

    waterproof. *See page 255*

23. **The wall materials you use for a recessed bathtub must be:**

    smooth, noncorrosive, nonabsorbent and waterproof to a height of 4 feet above the rim of the tub. *See page 255*

24. **The minimum size waste outlet required for a shower compartment is:**

    2 inches. *See page 255*

25. **Shower strainers must be designed:**

    with a minimum diameter of $3^{1}/_{2}$ inches and they must be removable so the trap can be cleaned. *See page 255*

26. **The minimum floor area required for any shower compartment is:**

    1,024 square inches. *See page 255*

27. **The minimum weight that most codes require for lead shower pans is:**

    4 pounds per square foot. *See page 255*

28. **To protect lead or copper shower pans from corrosion when they're installed on concrete floors:**

    you must paint them with asphaltum paint inside and outside. *See page 255*

29. **The sides of a shower pan should extend above the finished curb:**

    at least 2 inches. *See page 255*

30. **A shower pan should be prepared for inspection at the same point in the building's construction as:**

    the inspection for the tub and water pipe. *See page 256*

31. **Shower pans may be omitted when:**

    the shower compartment is built on a concrete slab on the ground floor. (The bottom, sides and curbs must be poured at the same time as the slab.) Refer to Figure 21-3. *See page 256*

32. **Walls of shower compartments must extend above the floor:**

    6 feet. *See page 256*

33. **The minimum diameter required for a laundry tub waste is:**

    1^1/$_2$ inches. *See page 256*

34. **The waste opening on a domestic kitchen sink with a waste disposer unit should be:**

    at least 3^1/$_2$ inches. *See page 256*

35. **Two common fixtures provided with overflows are:**

    bathtubs and lavatories. *See page 256*

36. **The overflow pipe or passageway from a fixture must be connected on the inlet side of the fixture trap in order to:**

    prevent sewer gases and odors from entering the room through the overflow. *See page 256*

37. **The type of fixtures that are not required to have a strainer or stopper are:**

    fixtures with an integral trap, such as toilets and most urinals. *See page 256*

38. **The minimum size of a fixture strainer is determined by:**

    the fixture waste outlet it serves. *See page 257*

39. **When a waste disposer is installed on a two-compartment sink in new construction, some codes require:**

    that the disposer waste discharge through a separate trap and separate waste line. *See page 257*

40. **When you install a waste disposer on an existing two-compartment sink using a single trap, you must use:**

    a special directional tee or wye fitting to flush garbage away from the other sink compartment. Refer to Figure 21-6. *See page 257*

41. **A commercial food waste grinder in a restaurant may not discharge through:**

    a grease interceptor. *See page 257*

42. **The minimum size waste opening required for a commercial sink is:**

    2 inches. *See page 257*

43. **A commercial food waste grinder should be trapped and vented:**

    individually, like any other fixture. *See page 257*

44. **When an air gap fitting is required on a dishwasher waste pipe, you should install it:**

    either deck-mounted on the sink or cabinet top, or wall-mounted. Refer to Figure 21-8. *See page 258*

45. **By most code standards, the maximum distance allowed between a dishwasher and the sink waste connection is:**

    5 feet. *See page 258*

46. **If there's a food disposal unit installed in a sink, the waste pipe from the dishwasher must connect to:**

    the tap in the body of the food disposer. Refer to Figure 21-9. *See page 258*

47. **The flange for a floor fixture should be set:**

    on top of the finished floor. Refer to Figure 21-10. *See pages 258-259*

48. **The gasket materials approved for use with plumbing fixtures having a flanged connection are:**

    graphite-impregnated asbestos and felt. *See page 258*

49. **The code places floor drains in the category of:**

    fixtures. *See page 258*

50. **Floor drain traps must have a permanent water seal:**

    to prevent evaporation from drying out the trap and allowing sewer gases into the building. *See page 258*

51. **Drinking fountain waste cannot discharge into a floor drain if the drain is:**

    in a restroom. *See page 259*

52. **No plumbing fixtures of any kind can be installed in a room containing:**

    air handling machinery. *See page 259*

53. **When installing special fixtures with waste and water connections, you protect the water supply from back-siphonage by:**

    installing an approved vacuum breaker. *See page 259*

54. **When installing any plumbing fixture, you must provide spacing and clearances:**

    that permit the fixture to be used in the manner intended, and that allow easy access for cleaning and repairs. *See page 259*

55. **The minimum center-to-center spacing required for water closets when set in battery installations is:**

    30 inches. Refer to Figure 21-12. *See page 259*

56. **The minimum required distance from the front of a urinal to any finished wall is:**

    21 inches. Refer to Figure 21-12. *See page 260*

57. **Center-to-center measurements are not applicable to lavatories because:**

    lavatories are manufactured in a variety of designs and widths. *See page 260*

58. **The minimum clearance required from the opening of a shower compartment or stall to any finished wall is:**

    24 inches. Refer to Figure 21-12. *See page 260*

59. **Both public buildings and privately owned multi-story apartment buildings are now required to have toilet facilities for the physically handicapped because:**

    Federal and state laws have mandated them. *See page 261*

60. **The only buildings that are *not* required to provide toilet facilities for the physically handicapped are:**

    single-family residences and buildings considered hazardous where handicapped people are not likely to be employed. *See page 261*

61. **In a public toilet room with six toilets, the number of toilets that must be provided for use by the physically handicapped is:**

    one. *See page 261*

62. **The minimum fixture requirements for a single-family residence are:**

    one kitchen sink, one water closet, one lavatory, one bathtub or shower unit and a provision for a clothes washing machine. *See page 261*

63. **The type of water facility that's optional in some codes and mandatory in others is:**

    hot water. *See page 261*

64. **The number of toilet fixtures required in a place of employment is determined by:**

    the number of employees and a ratio set by the local authority. *See page 262*

65. **If toilet rooms are connected to public rooms or passageways, there must be:**

    a vestibule or a screen to ensure decency and privacy. *See page 264*

66. **The minimum toilet facilities needed for food establishments catering to drive-in service is determined by:**

    the number of parking stalls in the parking lot; one parking stall equals one customer. *See page 264*

67. **If the seating capacity in a restaurant is unknown, the method used by the *Standard Plumbing Code* to determine the number of persons who'll occupy the premises at one time is:**

    the square foot method. Refer to Figure 21-14. *See pages 263 & 265*

68. **Rather than using the square-foot method, some codes determine the occupant load for a restaurant where the seating capacity is unknown by:**

    the egress requirement of the building code. *See page 265*

69. **You compute the restroom facility required for public places such as shopping centers, retail stores or large office buildings by:**

    deducting the uninhabitable spaces from the gross floor area and then using the net square footage of the remaining space to find the occupant load factor. *See page 265*

70. **Proper roughing-in of the waste and water outlets for various types of plumbing fixtures requires:**

    a thorough knowledge of roughing-in measurements. Refer to Figures 21-18 through 21-22. *See pages 265-271*

71. **You can get complete roughing-in information for *special* plumbing fixtures from:**

    the manufacturer or distributor. *See page 265*

72. **The standard roughing-in measurement for all water closets is:**

    12 inches from the finished wall. *See page 265*

73. **Of all rooms in a house, the one most susceptible to unsanitary conditions is:**

    the toilet room (bathroom). *See page 270*

74. **The type of bathroom plumbing fixtures that some apprentices and journeymen seldom have the opportunity to work with is:**

    off-the-floor plumbing fixtures. *See page 270*

75. **Most residential carriers have been designed to be compatible with:**

    the newer piping materials. *See page 270*

Examination Day

This chapter is a 200-question test which will help you evaluate your understanding of the plumbing code, and better prepare you for the journeyman's or master's examination. The multiple choice questions here are the same type, but not the same questions, that you'll find on the examination. This chapter should help you locate areas where you need additional study. Answers to the questions are at the end of the chapter.

The topics covered in this test include drainage, waste, and vent piping and fittings, private disposal systems, trailer park requirements, public and private water systems, swimming pools, fire standpipe systems, solar energy systems, gas systems, fixture requirements and mathematical problems.

You'll get the greatest benefit by putting your answers on a separate sheet of paper and completing the test before you check the answer page. When you know which answers you missed, review the section of the book covering that topic. Then review all the questions you missed until you can answer every question correctly.

If you find this type of study helpful, the author has another book, *Plumber's Exam Preparation Guide*, that is devoted entirely to preparing you for the exam. It's comprised of hundreds of multiple choice questions, with explanations included and with the wording from the plumbing code provided. It also includes study tests. A condensed version of *Plumber's Exam Preparation Guide* is also available on audiotape, so you can listen and study while you drive, or while you just sit and relax. Both the book and the audiotape are available via the order form bound into the back of this manual.

Multiple Choice Questions

1. The primary function of a relief vent is to
   (A) prevent back siphonage.
   (B) supply fresh air to a bathroom.
   (C) vent the water heater to prevent its explosion.
   (D) provide circulation of air between drainage and vent systems.

2. All piping passing under the footings of a building must have a clearance of at least ____ inches between the top of the pipe and the bottom of the footing.
   (A) 2
   (B) 3
   (C) 4
   (D) 5

3. Except when deeper seals are required for interceptors, a fixture trap must have a water seal of between
   (A) 1 and 3 inches.
   (B) 1 and 4 inches.
   (C) 2 and 4 inches.
   (D) 3 and 5 inches.

4. A waste pipe may receive the discharge from a
   (A) lavatory and urinal.
   (B) bed pan washer and lavatory.
   (C) urinal and bed pan washer.
   (D) bathtub and water closet.

5. Fixture trap inlets measured vertically from the bottom of the fixture to the top of the trap seal shall not exceed

(A) 12 inches.

(B) 15 inches.

(C) 18 inches.

(D) 24 inches.

6. Horizontal wet vents shall

(A) receive discharge from fixture drains only.

(B) not exceed 15 feet.

(C) never connect to a vertical wet vent more than 6 feet in length.

(D) be no less than 15 feet.

7. One non-metallic code-approved pipe that can't be used for building sewers is

(A) asbestos cement.

(B) Schedule 160 PVC .

(C) vitrified clay.

(D) Schedule 40 PVC .

8. The pipe that's *not* suitable for use as underground vent piping is

(A) cast iron soil pipe.

(B) lead pipe.

(C) galvanized steel pipe.

(D) brass pipe.

9. Cleanouts 3 inches and larger must be accessible, and have a clearance of

(A) 6 inches.

(B) 12 inches.

(C) 18 inches.

(D) 24 inches.

10. In a sanitary drainage system, the smallest pipe size allowed for a soil stack that carries no waste from urinals or bed pan washers is

(A) 2 inches.

(B) $2^1/_2$ inches.

(C) 3 inches.

(D) 4 inches.

11. One thing that may *not* be used in drainage pipes is

(A) 45 degree wyes.

(B) traps.

(C) supports.

(D) running threads.

12. Materials approved by the code for fire standpipes include

(A) galvanized steel pipe.

(B) cast iron pipe.

(C) lead pipe.

(D) Schedule 40 PVC.

13. Hydropneumatic water supply tanks must have an air volume control valve to

(A) remove excessive air in the tank.

(B) control water pressure.

(C) prevent rapid on-off operation of the pump.

(D) prevent air from getting into the pipes of the plumbing system.

14. The maximum area most codes will permit an area drain to handle is

(A) 50 square feet.

(B) 75 square feet.

(C) 100 square feet.

(D) 125 square feet.

15. The suction line from the water supply well to the pump (if less than 40 feet) must be at least

(A) 1 inch.

(B) $1^1/_4$ inches.

(C) $1^1/_2$ inches.

(D) 2 inches.

16. Roof drain strainers must extend at least _____ inches above the roof surface.

(A) 2

(B) 3

(C) 4

(D) 5

17. The maximum temperature at which waste water can be discharged into a building drainage system is

(A) 90 degrees.

(B) 125 degrees.

(C) 140 degrees.

(D) 180 degrees.

18. The required minimum cover over vitrified clay sewer pipe is

(A) 6 inches.

(B) 12 inches.

(C) 18 inches.

(D) 24 inches.

19. The water service pipe is the pipe from

    (A) the outlet side of the water meter to the building served.

    (B) the water main to the house valve.

    (C) the outlet side of the water meter to the first water distributing pipe.

    (D) the water meter to the building foundation.

20. A sump is a tank or pit located below the normal grade of the gravity system which must be emptied by mechanical means. Its primary function is to receive

    (A) waste from floor drains only.

    (B) clear water waste only.

    (C) waste containing chemicals in solution.

    (D) waste that requires lifting.

21. Horizontal branch drains in multistory buildings require a minimum separation of

    (A) 6 feet.

    (B) 8 feet.

    (C) 10 feet.

    (D) 12 feet.

22. Upon the completion of the entire water distribution system, it must be tested, inspected and proved tight under the ____ working pressure it's designed for.

    (A) minimum

    (B) maximum

    (C) 60-pound

    (D) 75-pound

23. To protect water pipe installations from water hammer, you must

    (A) provide for expansion.

    (B) provide for contraction.

    (C) install the system to drain dry.

    (D) install an air chamber.

24. In a multistory building a relief vent (yoke vent) is required at every

    (A) 4th floor.

    (B) 5th floor.

    (C) 6th floor.

    (D) 7th floor.

25. When a house water system is connected to both a well supply and a public water supply, the condition should be corrected by

    (A) using a by-pass line.

    (B) installing a backflow preventer on the well side.

    (C) placing a check valve on the house side.

    (D) disconnecting the well supply and capping it off.

26. Most codes require that indirect waste piping installed below a slab have a minimum diameter of

    (A) $^3/_4$ inches.

    (B) $1^1/_4$ inches.

    (C) $1^1/_2$ inches.

    (D) 2 inches.

27. Each hose station for fire protection standpipe systems in multistory buildings must be within ____ feet of the standpipe.

    (A) 4

    (B) 6

    (C) 8

    (D) 10

28. The maximum height of a fire standpipe hose station valve above the finished floor is

    (A) 60 inches.

    (B) 66 inches.

    (C) 72 inches.

    (D) 76 inches.

29. The minimum size for water supply serving $2^1/_2$ inch standpipes is

    (A) $2^1/_2$ inches.

    (B) 3 inches.

    (C) 4 inches.

    (D) 5 inches.

30. Battery venting requires a ____ to vent two or more similar adjacent fixtures which discharge into a common horizontal waste or soil branch.

    (A) branch vent

    (B) vent stack

    (C) continuous vent

    (D) loop vent

31. A vent that connects two or more individual vents with a vent stack is defined as a

    (A) relief vent.

    (B) branch vent.

    (C) wet vent.

    (D) common vent.

32. A building drainage system which cannot drain by gravity into the building sewer must discharge into

    (A) a sump.

    (B) perforated piping.

    (C) open joint piping.

    (D) a branch drain.

33. A building sewer that receives storm water is known as

    (A) an all-purpose sewer.

    (B) a special sewer.

    (C) a dual system sewer.

    (D) a combined sewer.

34. A common vent is installed to vent

    (A) 2 or more water closets.

    (B) a service sink with a 3-inch trap only.

    (C) 2 fixture drains installed at the same level in a vertical stack.

    (D) the last 2 fixtures on a horizontal drainage system.

35. According to the *UPC*, 1 fixture unit flow rate for special fixtures is considered

    (A) $7^1/_2$ gallons per minute.

    (B) $9^1/_2$ gallons per minute.

    (C) 10 per minute.

    (D) 12 gallons per minute.

36. A horizontal waste line may be connected to the vertical section of a waste stack by using

    (A) a saddle tee.

    (B) a test tee.

    (C) a single sanitary tee.

    (D) a combination.

37. Piping passing through cast-in-place concrete shall be protected by

    (A) painting the pipe with asphaltum paint.

    (B) wrapping the pipe with air conditioning tape.

    (C) sleeving to give $^1/_2$-inch annular space around the entire pipe.

    (D) wrapping the pipe with felt.

38. A 10-ton air conditioning unit centrally located below a building roof may discharge indirectly into

    (A) rain water leaders which discharge to the curb gutter.

    (B) a vent stack having a minimum size of 3 inches.

    (C) the building sanitary drainage system.

    (D) a 10-inch diameter buried pipe.

39. You may use plastic pipe and fittings for wall-hung plumbing fixtures if

    (A) the fixture doesn't weigh more than twelve pounds.

    (B) the building is a single family residence.

    (C) prior approval is obtained from the plumbing inspector.

    (D) the fixture pipe connection doesn't carry any of the load.

40. All underground soil, waste and vent piping and fittings inside a building located over deleterious fill must be

    (A) lead pipe.

    (B) centrifugally-spun service-weight cast iron pipe.

    (C) brass pipe.

    (D) Schedule 40 PVC.

41. Soil, waste and vent piping above ground inside buildings over deleterious fill may *never* be

    (A) galvanized pipe.

    (B) plastic pipe.

    (C) copper type K, L, or DWV.

    (D) asbestos cement.

42. When showers are provided in a trailer park, the minimum floor area of each shower is

    (A) 24 × 24, or 576 square inches.

    (B) 28 × 28, or 784 square inches.

    (C) 30 × 30, or 900 square inches.

    (D) 36 × 36, or 1,296 square inches.

43. A neutralizing tank is required for corrosive waste which

    (A) contains spent acids.

    (B) has a pH factor of 5.0.

    (C) has waste that needs separation.

    (D) is generated in a repair garage.

44. Fixture unit value as a load factor for special fixtures is determined by the
    (A) size of the fixture trap.
    (B) type of fixture.
    (C) manufacturer's suggested load factor.
    (D) location of fixture.

45. A building sewer, when connected to a septic tank, may be considered and sized as a building drain if the developed length does not exceed
    (A) 5 feet.
    (B) 8 feet.
    (C) 10 feet.
    (D) 12 feet.

46. A domestic kitchen sink may be installed on a waste stack that
    (A) is less than 2 inches in diameter.
    (B) is 2 inches in diameters.
    (C) vents lower fixtures.
    (D) has a diameter at least the same size as its trap.

47. Sumps and receiving tanks for liquid waste
    (A) must be constructed of pervious materials.
    (B) must be accessibly located.
    (C) need not be vented.
    (D) are for public use only.

48. Ejector pumps shall be provided with a
    (A) check valve on the discharge side of the gate valve.
    (B) gate valve only.
    (C) check valve only.
    (D) check valve located on the pump side of the gate valve.

49. According to the *Standard Plumbing Code*, not more than ____ water closet(s) may discharge into a 3-inch stack at the same point.
    (A) 1
    (B) 2
    (C) 3
    (D) 4

50. According to the *Standard Plumbing Code*, not more than ____ water closet(s) may discharge into a 3-inch stack.
    (A) 3
    (B) 4
    (C) 5
    (D) 6

51. When a clothes washing machine utilizes one side of a tap cross with a domestic kitchen sink, the waste pipe size shall be at least
    (A) 2 inches.
    (B) $2^1/_2$ inches.
    (C) 3 inches.
    (D) 4 inches.

52. Sumps receiving body waste from plumbing fixtures must have a minimum
    (A) $1^1/_2$-inch vent.
    (B) $2^1/_2$-inch vent.
    (C) 3-inch vent.
    (D) 4-inch vent.

53. Sump vents may
    (A) extend independently to above the roof.
    (B) be connected to the plumbing system.
    (C) connect to the nearest roof leader.
    (D) connect only to a stack vent.

54. The minimum size pipe used for subsoil drains when placed under the cellar or basement floor is
    (A) $2^1/_2$ inches.
    (B) 3 inches.
    (C) 4 inches.
    (D) 6 inches.

55. Air conditioning condensate drains may connect
    (A) directly to a rain water leader pipe.
    (B) by indirect means to the building drainage system.
    (C) to a properly-sized water heater drain pan pipe.
    (D) to any waste piping installed under a slab.

56. Drip pipes from walk-in refrigerator floors or store-room floors where food is stored must be installed
    (A) to drain into a $1^1/_2$-inch pipe.
    (B) to drain into a sump.
    (C) as a direct waste.
    (D) as an indirect waste.

57. Walk-in refrigerator floors or store-room floors where food is stored must be ____ above the overflow point of receiving fixtures.
    (A) 1 inch
    (B) 2 inches
    (C) 3 inches
    (D) 4 inches

58. Air conditioning condensate drains for units with not more than 5 tons capacity may discharge

   (A) onto a pervious area.

   (B) onto an impervious area.

   (C) directly into a building drainage system.

   (D) indirectly into a building storm drain.

59. Air conditioning drains of PVC shall be a minimum of ____ below the bottom of the slab.

   (A) 2 inches

   (B) 4 inches

   (C) 6 inches

   (D) 12 inches

60. For concrete sewer pipe, approximately ____ percent of the joint at the base of the socket shall be filled with jute or hemp.

   (A) 10

   (B) 15

   (C) 25

   (D) 30

61. Mortar for cement joints shall be composed of

   (A) 2 parts cement, 1 part sand.

   (B) 1 part cement, 2 parts sand.

   (C) 2 parts cement, 2 parts sand.

   (D) 3 parts cement, 1 part sand.

62. A trap depending on movable parts to retain its seal may be used

   (A) for fixtures with clear water waste only.

   (B) for fixtures having integral traps only.

   (C) for swimming pool installations.

   (D) never.

63. The waste from a commercial dishwasher must discharge

   (A) indirectly into the building greasy waste drain.

   (B) into a floor drain.

   (C) directly into the building drainage system.

   (D) into a special sump.

64. Accessible cleanouts must be located so that all building drains are within reach of a ____ -foot cable.

   (A) 25

   (B) 50

   (C) 75

   (D) 100

65. According to the *Standard Plumbing Code*, cleanouts must be the same nominal size as the pipe in which they're installed, up to ____ inches.

   (A) 4

   (B) 6

   (C) 8

   (D) 10

66. Vertical Schedule 40 plastic pipe must be supported at intervals of

   (A) 4 feet.

   (B) 6 feet.

   (C) each story.

   (D) every 2 stories.

67. Suspended horizontal lead joint soil pipe, in 10-foot lengths, must be supported at ____ intervals.

   (A) 4-foot

   (B) 6-foot

   (C) 8-foot

   (D) 10-foot

68. All extensions of soil, waste and vent stacks must terminate at least ____ inches above the roof.

   (A) 4

   (B) 6

   (C) 8

   (D) 12

69. Most codes state that where roofs are used as sun decks, all vents must extend at least ____ above the deck.

   (A) 24 inches

   (B) 36 inches

   (C) 60 inches

   (D) 84 inches

70. Vent pipes are graded to

   (A) prevent air lock.

   (B) help circulate the air.

   (C) drain to the soil or waste pipe.

   (D) prevent backflow.

71. The fixture which may *not* discharge into a horizontal 3-inch wet vent is

   (A) a bidet.

   (B) a water closet.

   (C) a shower.

   (D) a bathtub.

72. The type of vent that's prohibited is a

    (A) yoke vent.

    (B) crown vent.

    (C) common vent.

    (D) back vent.

73. The pipe or dry section of a circuit vent may have a diameter ____ pipe size(s) less than the diameter of the horizontal soil drain it serves.

    (A) $^1/_2$

    (B) 1

    (C) $1^1/_2$

    (D) 2

74. A continuous vent is also known as a

    (A) stack vent.

    (B) vent stack.

    (C) main stack.

    (D) relief vent.

75. The diameter of an individual vent can't be less than ____ inch(es).

    (A) 1

    (B) $1^1/_4$

    (C) $1^1/_2$

    (D) 2

76. According to the *SPC*, a 4-inch combination waste and vent stack no longer than 100 feet may receive the discharge from

    (A) 2 water closets.

    (B) 4 urinals.

    (C) 6 water closets.

    (D) 10 kitchen sinks.

77. According to the *SPC*, a 2-inch combination waste and vent stack that's shorter than 30 feet may *not* receive the discharge from

    (A) 4 lavatories.

    (B) 2 bathtubs.

    (C) 1 kitchen sink.

    (D) 2 showers.

78. The minimum size vent stack allowable when a water closet is installed in an accessory building is

    (A) 2 inches.

    (B) $2^1/_2$ inches.

    (C) 3 inches.

    (D) 4 inches.

79. The vent terminal of a sanitary plumbing system can't be less than ____ feet from the point of any mechanical air intake opening.

    (A) 6

    (B) 8

    (C) 10

    (D) 12

80. If a vent terminal is within ten feet of any door, window, or ventilating opening, it must extend at least ____ foot/feet above such opening.

    (A) 1

    (B) 3

    (C) 4

    (D) 5

81. Establishments that don't require a grease interceptor are

    (A) bars.

    (B) clubs.

    (C) supermarkets.

    (D) take-out food establishments.

82. Fixture trap cleanouts are prohibited on

    (A) lavatory traps.

    (B) barber shop sinks.

    (C) concealed traps.

    (D) kitchen sinks.

83. Equip all sundeck drains with a

    (A) plastic strainer to prevent corrosion.

    (B) flat surface strainer.

    (C) minimum 2-inch waste outlet.

    (D) brass strainer only.

84. The more roof drains you install,

    (A) the fewer leader pipes required.

    (B) the more expensive it will be.

    (C) the fewer puddles you'll have.

    (D) the larger the leader pipe will have to be.

85. When a planter drain is used, your local authority may require that

    (A) the waste water discharge through the storm water system.

    (B) the excess waste water pass through a sand interceptor.

    (C) a flapper valve be installed within 2 feet of the drain.

    (D) a backwater valve be installed on the waste line.

86. When you have a vertical wall that drains onto a flat roof area that you're sizing, be sure to add ____ of the vertical wall area to your horizontal projection.

    (A) 25 percent

    (B) 33$^1/_3$ percent

    (C) 50 percent

    (D) 75 percent

87. The maximum distance between hangers for 1$^1/_2$-inch horizontal copper pipe is

    (A) 4 feet.

    (B) 6 feet.

    (C) 8 feet.

    (D) 10 feet.

88. The material that shall *not* be used for above-ground storm drainage within a building is

    (A) asbestos-cement pipe.

    (B) galvanized steel pipe.

    (C) plastic pipe.

    (D) lead pipe.

89. When a rainwater leader discharges directly into a soakage pit, it requires

    (A) a backwater valve.

    (B) a cleanout at its base.

    (C) protection from vehicle traffic.

    (D) an overflow fitting at its base.

90. You can use asbestos cement pipe only for

    (A) storm sewers.

    (B) acid waste.

    (C) rain leaders.

    (D) greasy waste substances.

91. Small grease interceptors require a

    (A) water-cooled jacket to speed coagulation.

    (B) vent on the discharge side.

    (C) flow control fitting.

    (D) grease retention capacity of 50 pounds.

92. The minimum capacity for graywater holding tanks is ____ gallons.

    (A) 30

    (B) 40

    (C) 50

    (D) 60

93. An excavation for graywater disposal fields must never come within

    (A) 5 vertical feet of the highest known seasonal groundwater.

    (B) 5 feet of the building structure.

    (C) 40 feet of any stream.

    (D) 8 feet of a public water main.

94. When a restaurant sink produces so much grease that stoppages occur, you install

    (A) a filter.

    (B) a flow control device.

    (C) a vent.

    (D) an interceptor.

95. Interceptors for a commercial laundry must be maintained in efficient operating condition by periodic

    (A) removal of accumulated contents.

    (B) replacement of "limited use" parts.

    (C) flushing with chemical mixtures.

    (D) checking the outlet pipe.

96. The overflow from a fixture must be connected to

    (A) the crown vent.

    (B) the fixture branch.

    (C) a drip pan.

    (D) the inlet side of the fixture trap.

97. A lavatory must be ____ inches from any finished or stall compartment wall.

    (A) 10

    (B) 8

    (C) 6

    (D) 4

98. Where bucket type floor drains are required, most codes mandate the minimum diameter of its outlets to be

    (A) 2 inches.

    (B) 4 inches.

    (C) 6 inches.

    (D) 8 inches.

99. When a water closet is installed next to a bath tub, the minimum distance from the center of the bowl to the edge of a tub is

    (A) 10 inches.

    (B) 12 inches.

    (C) 14 inches.

    (D) 16 inches.

100. Plumbing fixtures must be constructed of approved materials. The type of material that is *not* acceptable is

    (A) china.

    (B) cultured marble.

    (C) a material that is impervious.

    (D) a material that is pervious.

101. Water closets installed for public use must have a

    (A) regular closed-front seat with or without a cover.

    (B) regular open-front seat with or without a cover.

    (C) elongated closed-front seat with or without a cover.

    (D) elongated open-front seat with or without a cover.

102. When sheet lead is used for a shower pan, it must weigh at least ____ pounds per square foot.

    (A) 2

    (B) 4

    (C) 6

    (D) 8

103. When sheet copper is used for a shower pan, it must weigh at least ____ ounces per square foot.

    (A) 4

    (B) 6

    (C) 8

    (D) 12

104. Lead and copper shower pans must be protected against the corrosive effects of concrete by

    (A) coating the inside with asphaltum paint.

    (B) coating the outside with asphaltum paint.

    (C) coating the inside and outside with asphaltum paint.

    (D) coating the inside and outside with a water seal paint.

105. Built-in tubs with overhead showers must have

    (A) an 8-inch shower arm.

    (B) a waterproof joint between the tub and wall.

    (C) a minimum of 5-foot tile walls.

    (D) either a shower door or curtain.

106. No floor drain or other plumbing fixture shall be installed in a room

    (A) containing air handling machinery.

    (B) used for sleeping.

    (C) used for the storage of food.

    (D) used for recreation purposes.

107. Floor drains serving indirect waste pipes from food or drink storage rooms shall *not* be installed in any

    (A) toilet room.

    (B) unventilated room.

    (C) storeroom.

    (D) cupboard.

108. According to the *UPC*, floor drains sized 4 inches may not require a vent if installed within ____ feet of a vented sewer line when measured horizontally.

    (A) 10

    (B) 12

    (C) 15

    (D) 20

109. The code defines graywater as

    (A) partially treated household waste water, excluding toilet waste.

    (B) untreated household waste water, excluding toilet waste.

    (C) clothes washing machine waste water only.

    (D) kitchen sink waste water only.

110. The sizing and design of gasoline and oil interceptors which handle volatile liquids is governed by

    (A) the amount of volatile liquids in a system.

    (B) the number of floor drains for public storage garages.

    (C) the type of volatile liquids that enter a drainage system.

    (D) the number of washing facilities which cater to commercial motor vehicles.

111. Unless the plumbing inspector rules otherwise, every floor drain trap directly connected to the drainage system must have a permanent water seal which can be fed from

    (A) a drinking fountain.

    (B) an ice maker.

    (C) an automatic priming device.

    (D) an A.C. condensate drain.

112. Previously-used piping material may be reused in a potable water supply system when

    (A) it's been galvanized inside and outside.

    (B) its previous use was for a potable water supply system.

    (C) it's approved by a plumbing inspector.

    (D) the water is not too hard.

113. Pressure-rated plastic service piping must have a minimum working pressure of

    (A) street level pressure.

    (B) 75 pounds per square inch.

    (C) 100 pounds per square inch.

    (D) 160 pounds per square inch.

114. Codes require that lawn sprinkler systems using potable water have an approved

    (A) gate valve.

    (B) back-flow preventer.

    (C) check valve.

    (D) ground joint union.

115. The hose-connected faucet that's *not* required to have a back-flow preventer installed is

    (A) a three-compartment commercial sink.

    (B) a service sink.

    (C) an automatic clothes washing machine.

    (D) an outside hose faucet.

116. Plastic water service piping requires a minimum cover of

    (A) 6 inches.

    (B) 8 inches.

    (C) 10 inches.

    (D) 12 inches.

117. The minimum liquid capacity of a septic tank serving a three-bedroom, three-bath residence is

    (A) 750 gallons.

    (B) 1,000 gallons.

    (C) 1,200 gallons.

    (D) 1,500 gallons.

118. Cast-in-place septic tanks located in a parking lot must be designed to support the anticipated load, but not less than ___ psf.

    (A) 100

    (B) 200

    (C) 250

    (D) 300

119. The minimum distance from a drainfield to a basement wall is

    (A) 5 feet.

    (B) 8 feet.

    (C) 10 feet.

    (D) 12 feet.

120. When a seepage pit is installed in the vicinity of a septic tank drainfield, the minimum separation distance is

    (A) not important.

    (B) 5 feet.

    (C) 8 feet.

    (D) 10 feet.

121. Where reservoir-type drainfields are used, the maximum distance between centers of distribution lines is

    (A) 1 foot.

    (B) 2 feet.

    (C) 3 feet.

    (D) 4 feet.

122. The suction line from a potable water-supply well serving a single family residence shall have a union installed before

    (A) the pump.

    (B) the hair strainer.

    (C) the check valve.

    (D) the well casing.

123. Even when the underground water table is quite shallow, your local authority may require a well depth of

    (A) 25 feet.

    (B) 30 feet.

    (C) 35 feet.

    (D) 40 feet.

124. The minimum sized hydropneumatic tank for a single family residence shall be

    (A) 20 gallons.

    (B) 30 gallons.

    (C) 42 gallons.

    (D) 50 gallons.

125. Waste from an air conditioning unit connected to the building sanitary drainage system must allow ____ fixture unit(s) per gallon per minute.

    (A) 1

    (B) 2

    (C) 3

    (D) 4

126. A 4-inch concrete pad must be poured around the well casing and extend ____ inches on all sides.

    (A) 10

    (B) 18

    (C) 20

    (D) 36

127. The maximum length of a single tile drainfield lateral is ____ feet.

    (A) 40

    (B) 60

    (C) 80

    (D) 100

128. Horizontal screwed gas piping 1¼ inches and larger shall be supported every ____ feet.

    (A) 4

    (B) 6

    (C) 10

    (D) 12

129. The location that's *not* acceptable for installation of gas piping within a building is

    (A) under a concrete floor.

    (B) a solid partition.

    (C) a hollow partition.

    (D) an elevator shaft.

130. Branch outlet pipes for a gas distribution system must never be taken from the ____ of horizontal lines.

    (A) top

    (B) side

    (C) bottom

    (D) end

131. Plastic pipe, tubing and fittings used for the installation of gas piping shall *not* be joined by

    (A) heat-fusion method.

    (B) solvent cement method.

    (C) adhesive method.

    (D) compression couplings or flanges.

132. Equipment that requires mobility during operation inside a building may use indoor gas hose connectors, providing the length doesn't exceed ____ feet.

    (A) 6

    (B) 8

    (C) 10

    (D) 12

133. For most fixtures, excluding direct flush valves, minimum service pressure at the point of discharge is ____ psi.

    (A) 4

    (B) 6

    (C) 8

    (D) 10

134. Combination pressure and temperature valves must be installed so that the temperature sensing element is within ____ of the top of the tank.

    (A) 2 inches

    (B) 4 inches

    (C) 6 inches

    (D) 8 inches

135. When the pool pressure piping system is ready for inspection, it must be water-tested at

    (A) 10 psi.

    (B) 20 psi.

    (C) 30 psi.

    (D) 40 psi.

136. A pressure discharge line from a domestic water heater may discharge into a

    (A) service sink.

    (B) pot sink.

    (C) shower.

    (D) floor drain.

332   *Plumber's Handbook*

137. The minimum size for a domestic water heater relief valve discharge is

    (A) $^3/_8$ inch I.D.

    (B) $^3/_8$ inch O.D.

    (C) $^1/_4$ inch I.D.

    (D) $^1/_4$ inch O.D.

138. Urinal traps and floor drains installed downstream from a water closet(s) in a circuit vent group must be ____ inches.

    (A) $1^1/_2$

    (B) 2

    (C) 3

    (D) 4

139. All standpipes in buildings 50 feet or higher must extend full size above the roof at least ____ inches.

    (A) 10

    (B) 20

    (C) 30

    (D) 40

140. The minimum number of fire standpipes which must extend through the roof of a building having four standpipes is

    (A) 1.

    (B) 2.

    (C) 3.

    (D) 4.

141. Liquid waste, as defined by code, does *not* include the discharge from

    (A) urinals.

    (B) rain leaders.

    (C) dental chairs.

    (D) kitchen sinks.

142. The term "sewage" as defined by the code would *not* include

    (A) liquid waste containing animal matter in suspension.

    (B) liquid waste containing minerals in solution.

    (C) rain water.

    (D) liquids containing chemicals in solution.

143. Standpipes and fittings above ground within exterior building walls must be strong enough to withstand ____ pounds per square inch water pressure at the topmost outlet.

    (A) 55

    (B) 65

    (C) 100

    (D) 150

144. Outside rain leaders, when exposed to contact with vehicles, must have a cast iron pipe which extends ____ feet above grade.

    (A) 2

    (B) 3

    (C) 4

    (D) 5

145. Any indirect waste pipe installation must have a ____ -inch minimum clearance above the floor.

    (A) 1

    (B) 2

    (C) 3

    (D) 4

146. Any cleanout plug made of heavy brass or plastic must be ____ inch thick.

    (A) $^1/_{16}$

    (B) $^1/_8$

    (C) $^3/_{16}$

    (D) $^1/_4$

147. Horizontal copper tubing 2 inches and larger must be supported at approximately ____ intervals.

    (A) 4 foot

    (B) 6 foot

    (C) 8 foot

    (D) 10 foot

148. Residential septic tanks

    (A) must have a minimum capacity of 500 gallons.

    (B) may not receive storm water.

    (C) may be constructed of masonry block with a coat of $^1/_2$-inch portland cement grout.

    (D) shall be rectangular in shape.

149. A vapor vent for a small oil interceptor must extend at least ____ feet above grade.

    (A) 6

    (B) 8

    (C) 10

    (D) 12

150. The minimum size water service pipe from the meter to the house is

    (A) $1/2$ inch.

    (B) $3/4$ inch.

    (C) 1 inch.

    (D) determined by meter size.

Using Illustration 1, identify each section of piping as specified in questions 151 through 154.

151. Section "A" is an

    (A) island vent.

    (B) loop vent.

    (C) relief vent.

    (D) circuit vent.

152. Section "B" is a

    (A) waste pipe.

    (B) fixture drain.

    (C) building drain.

    (D) fixture branch.

153. Section "C" is a

    (A) fixture branch.

    (B) wet vent.

    (C) combination waste and vent pipe.

    (D) dirty waste pipe.

154. Section "D" is a

    (A) soil pipe.

    (B) waste pipe.

    (C) building sewer.

    (D) building drain.

Using the information from the *UPC*, give the fixture units and size of pipes shown in Ilustration 2 for questions 155 through 157.

155. The minimum size of urinal trap "A" is

    (A) $1^1/2$ inches.

    (B) 2 inches.

    (C) $2^1/2$ inches.

    (D) 3 inches.

156. Fitting "B" is known in the trade as a

    (A) double combination wye and $1/8$ bend.

    (B) double combination wye and $1/5$ bend.

    (C) double combination wye and $1/16$ bend.

    (D) double drainage wye.

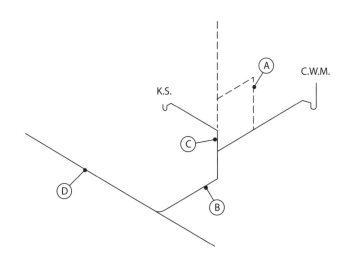

Illustration 1

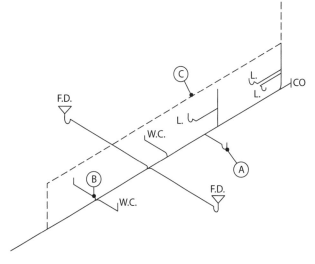

Illustration 2

157. The minimum size of horizontal vent pipe "C" is
    (A) 2 inches.
    (B) 2¹/₂ inches.
    (C) 3 inches.
    (D) 4 inches.

158. According to the *UPC*, a 3-inch building drain installed at a maximum fall (pitch) of ¹/₈ inch per foot may receive the discharge from fixtures having a total fixture unit load of
    (A) 18 F.U.
    (B) 22 F.U.
    (C) 25 F.U.
    (D) 28 F.U.

159. Using the *UPC*, a 2-inch vent pipe, 20 feet long, is adequate to serve
    (A) 14 bathtubs.
    (B) 7 automatic clothes washers.
    (C) 26 lavatories.
    (D) 13 showers.

160. Using the *UPC*, the piece of equipment that may *not* be drained by indirect means is a
    (A) drinking fountain.
    (B) hand sink.
    (C) three-compartment glass sink.
    (D) beverage cooler.

161. According to the *UPC*, a clothes washing machine standpipe must have a minimum length of
    (A) 10 inches.
    (B) 14 inches.
    (C) 18 inches.
    (D) 28 inches.

Use Illustration 3 to answer questions 162 through 164.

162. The pipe section "A" between the two lavatories and the sink serves as a
    (A) waste pipe for the kitchen sink.
    (B) waste pipe for the lavatories.
    (C) vent pipe for the kitchen sink.
    (D) common vent for both fixtures.

163. The combined fixture units at pipe section "B" is
    (A) 1 unit.
    (B) 2 units.
    (C) 3 units.
    (D) 4 units.

164. The combined number of fixture units at pipe section "C" is
    (A) 1 unit.
    (B) 2 units.
    (C) 3 units.
    (D) 4 units.

165. Using Illustration 4, the maximum length required by the *UPC* for the 4-inch pipe "A", which serves a floor drain from a vented building drain, is
    (A) 3 feet.
    (B) 10 feet.
    (C) 15 feet.
    (D) 20 feet.

166. Lead pipe is permitted to be fully wiped and joined to
    (A) plastic.
    (B) brass.
    (C) stainless steel.
    (D) wrought iron.

Illustration 3

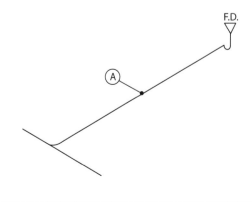

Illustration 4

167. A flaring tool is used to make flare joints in

(A) Type K copper pipe.

(B) Type L copper pipe.

(C) Type L copper tubing.

(D) Type M copper tubing.

168. The material that's considered semi-rigid is

(A) Type K copper pipe.

(B) Type L copper pipe.

(C) Type L copper tubing.

(D) Type M copper tubing.

169. The piping which may be used to convey corrosive gases is

(A) brass.

(B) galvanized steel.

(C) Type K copper pipe.

(D) stainless steel extra heavy.

170. Vent piping serving a gas water heater that penetrates the roof near a 2-foot parapet wall should be a minimum of

(A) 1 foot high.

(B) 2 feet high.

(C) 3 feet high.

(D) 4 feet high.

171. A water closet using a flushometer valve must have a vacuum breaker located at least ____ inch(es) above the rim of the bowl.

(A) 1

(B) 3

(C) 6

(D) 8

172. A single flushometer valve may be used to serve

(A) 1 water closet.

(B) 2 water closets.

(C) 1 water closet and 1 urinal.

(D) 1 water closet and 2 urinals.

173. In a water piping system, the code prohibits the use of

(A) CPVC plastic.

(B) brass.

(C) type M copper pipe.

(D) aluminum pipe.

174. Steel fittings used within a building in a water piping system must

(A) be beaded.

(B) have standard pipe threads.

(C) be plain.

(D) be galvanized.

175. Fire protection standpipe systems for multistory buildings exceeding a 275-foot height must be

(A) sized by hydraulic calculcations.

(B) zoned for multiple standpipes.

(C) designed with minimum 4-inch piping.

(D) designed by the local authority.

176. The size of a gas supply piping outlet serving any gas appliance must not be less than

(A) $3/8$ inch.

(B) $1/4$ inch.

(C) $3/4$ inch.

(D) the appliance inlet pipe.

177. The combustion chamber of a gas-fired water heater installed in a private garage must be at least ____ above the floor level.

(A) 12 inches

(B) 18 inches

(C) 24 inches

(D) 28 inches

178. The *UPC* requires that new or replacement residential lavatory faucets comply with water conservation requirements by having a maximum flow rate of

(A) 1.2 gpm.

(B) 1.8 gpm.

(C) 2.2 gpm.

(D) 2.8 gpm.

179. A sanitary tee may

(A) be used to change direction from horizontal to horizontal.

(B) be used to change direction from vertical to horizontal.

(C) be used to change direction from horizontal to vertical.

(D) *not* be used to change direction in drainage pipes.

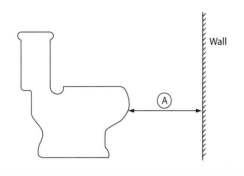

Illustration 5

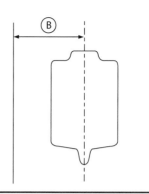

Illustration 6

180. Most codes require a minimum air space of ____ inches in a septic tank.

    (A) 4

    (B) 9

    (C) 10

    (D) 12

181. Each drain unit for a reservoir-type drainfield serving a 900 gallon septic tank should equal

    (A) 2 square feet.

    (B) 2 linear feet.

    (C) 4 square feet.

    (D) 4 linear feet.

182. For each trailer park site drainage inlet, the *UPC* assigns ____ fixture units.

    (A) 10

    (B) 12

    (C) 14

    (D) 16

183. According to the *UPC*, a 4-inch sewer installed at a slope at 15 inches per 100 feet may serve ____ trailers.

    (A) 21

    (B) 23

    (C) 25

    (D) 28

184. Vent pipes penetrating a roof must be made watertight by using

    (A) approved flashing material.

    (B) approved roofing cement.

    (C) 30 lb. felt sealed with hot tar.

    (D) approved caps.

185. The fittings for galvanized drainage, waste and vent pipes must be

    (A) galvanized inside only.

    (B) galvanized outside only.

    (C) the straight type.

    (D) the recessed type.

186. In Illustration 5, distance "A" between a water closet and a partition must be at least

    (A) 12 inches.

    (B) 15 inches.

    (C) 21 inches.

    (D) 30 inches.

187. In Illustration 6, distance "B" for a wall-hung urinal must be at least

    (A) 10 inches.

    (B) 12 inches.

    (C) 15 inches.

    (D) 16 inches.

188. Using Illustration 6 once again, distance "B" for a stall urinal must be at least

    (A) 10 inches.

    (B) 12 inches.

    (C) 15 inches.

    (D) 16 inches.

189. An insulated gas-fired water heater must be at least ____ inch(es) from any door or wall constructed of combustible materials.

    (A) 1

    (B) 2

    (C) 3

    (D) 4

190. A thermosiphon solar system requires

    (A) a circulating pump.

    (B) a thermostat sensor.

    (C) a minimum $1/2$-inch pipe in the collector and circulation system.

    (D) a minimum $3/4$-inch pipe in the collector and circulation system.

191. The material *not* recommended for a solar heat collector deck is

    (A) plastic pipe.

    (B) aluminum pipe.

    (C) copper pipe.

    (D) steel pipe.

192. A gas appliance consumption of 220,000 Btu per hour is equal to

    (A)   22 cubic feet.

    (B)   220 cubic feet.

    (C) 2,200 cubic feet.

    (D) 22,000 cubic feet.

193. The *UPC* requires a ____ -inch cleanout fitting for a 6-inch building sewer.

    (A) 3

    (B) $3^1/2$

    (C) 4

    (D) 6

194. A commercial food waste disposer unit must be connected directly into

    (A) the sanitary system.

    (B) the greasy waste system.

    (C) a bucket type trap.

    (D) a grease trap.

195. According to the *UPC*, the drainage system vent receiving waste from plumbing fixtures (less water closets) having a total of 7 fixture units may be

    (A) $1^1/4$ inches.

    (B) $1^1/2$ inches.

    (C) 2 inches.

    (D) 3 inches.

196. A sump check valve must be located

    (A) in the basement drain.

    (B) on the sump side of the gate valve.

    (C) on the building drain side of the gate valve.

    (D) 2 feet from the building drain.

197. Fixtures connected directly to the sanitary drainage system must be equipped with

    (A) a water seal trap.

    (B) a vent.

    (C) a backwater valve.

    (D) a waste pipe.

198. A 2-inch pipe has an aggregate cross-sectional area of

    (A) 1.7671.

    (B) 3.1416.

    (C) 4.9087.

    (D) 7.0686.

199. The minimum floor area of a shower stall must be ____ square inches.

    (A)   950

    (B) 1,024

    (C) 1,110

    (D) 1,200

200. The material used in a gas system that's *not* permitted outside a building, underground, is

    (A) plastic pipe.

    (B) aluminum pipe.

    (C) copper pipe.

    (D) brass pipe.

Multiple Choice Answers

1. D (p. 36)
2. A (p. 116, Fig. 8-17)
3. C (p. 54)
4. A (p. 14)
5. D (p. 56, Fig. 4-11)
6. B (pp. 44-45, Fig. 3-34)
7. B (p.105, Fig.14-2)
8. C (p. 108)
9. C (p. 67)
10. C (p. 21)
11. D (p. 103)
12. A (p. 204)
13. B (p. 197)
14. C (p. 94)
15. A (p. 195)
16. C (p. 94)
17. C (p. 93)
18. B (p. 116, Fig. 8-16)
19. C (p. 161)
20. D (p. 28)
21. B (p. 26)
22. B (p. 185)
23. D (p. 184)
24. B (p. 7)
25. D (p. 183)
26. B (p. 120)
27. D (p. 205)
28. C (p. 205)
29. C (p. 201)
30. A (p. 33)
31. B (p. 33)
32. A (p. 28)
33. D (p. 94)
34. C (p. 34, Fig. 3-3)
35. A (p. 17, Fig. 2-13)
36. C (pp. 114-115)
37. C (p. 116, Fig. 8-15)
38. C (p. 123)
39. D (pp. 253-255)
40. D (p. 105)
41. D (p. 108)
42. D (p. 142)
43. A (p. 85, Fig. 6-22)

44. A (p. 15, Fig. 2-12)
45. C (p. 21)
46. B (p. 21)
47. B (p. 28)
48. D (p. 29, Fig. 2-35)
49. B (p. 17, Fig. 2-15)
50. D (p. 17, Fig. 2-15)
51. B (p. 22)
52. A (p. 46)
53. A (p. 46)
54. C (p. 108)
55. B (p. 93)
56. D (p. 122, Fig. 8-28)
57. B (p. 122, Fig. 8-28)
58. A (p. 123)
59. A (p. 122)
60. C (p. 112)
61. B (p. 112)
62. D (p. 54)
63. C (p. 121)
64. C (p. 64)
65. A (pp. 67-68, Fig. 5-16)
66. C (p. 119)
67. D (pp. 118-119, Fig. 8-22)
68. B (p. 49)
69. D (p. 50)
70. C (p. 119)
71. B (p. 45)
72. B (p. 54, Fig. 4-2)
73. B (p. 41)
74. A (p. 34)
75. B (p. 40, Fig. 3-24)
76. D (p. 46, Fig. 3-36)
77. C (p. 46, Fig. 3-36)
78. A (p. 40)
79. C (p. 49)
80. B (p. 49)
81. D (p. 71)
82. C (p. 54)
83. B (p. 94)
84. C (p. 96)
85. B (p. 95)

86. C (p. 97)
87. B (p. 118, Fig. 8-22)
88. A (p. 107)
89. D (p. 109, Fig. 8-5)
90. A (p. 109, Fig. 8-3)
91. C (p. 72)
92. C (p. 152)
93. A (p. 151, Fig. 11-10)
94. D (p. 72)
95. A (p. 79)
96. D (p. 256)
97. D (p. 260)
98. B (p. 81)
99. B (p. 259)
100. D (p. 253)
101. D (p. 253)
102. B (p. 255)
103. D (p. 255)
104. C (p. 255)
105. B (p. 255)
106. A (p. 123)
107. A (p. 92, Fig. 8-30)
108. A (p. 58, Fig. 4-18)
109. B (p. 148)
110. A (p. 81)
111. C (p. 258)
112. B (p. 179)
113. D (p. 179)
114. B (p. 183)
115. C (p. 185)
116. D (p. 116, Fig. 8-16)
117. B (p. 128, Fig. 9-2)
118. D (p. 128)
119. B (p. 137)
120. B (p. 137)
121. D (p. 133, Fig. 9-8)
122. A (p. 197)
123. B (p. 194)
124. C (p. 199, Fig. 15-5)
125. B (p. 123)

126. B (p. 195, Fig. 15-4)
127. D (p. 132)
128. D (p. 185, Fig. 8-22)
129. D (p. 244)
130. C (p. 247, Fig. 20-6)
131. C (p. 244)
132. A (p. 245)
133. C (p. 164, Fig. 12-3)
134. C (p. 172, Fig. 13-5)
135. D (p. 217)
136. D (p. 172)
137. A (p. 172, Fig. 13-6)
138. C (p. 41, Fig. 3-25)
139. C (p. 205)
140. D (p. 205)
141. B (p. 14)
142. C (p. 14)
143. B (pp. 204 & 206)
144. D (p. 109)
145. C (p. 120)
146. B (p. 67)
147. D (p. 118)
148. D (pp. 128-129)
149. D (p. 82)
150. B (p. 164, Fig. 12-4 & 12-5)
151. C (p. 58, Fig. 4-17)
152. A (p. 14)
153. B (p. 14)
154. D (p. 14)
155. B (p. 254)
156. A (p. 10, Fig. 2-3)
157. C (p. 41)
158. D (p. 18, Fig. 2-16)
159. B (p. 18, Fig. 2-16)
160. B (p. 89)
161. C (p. 91)
162. A (p. 14)
163. D (pp. 15 & 16, Fig. 2-9 & 2-11)

164. D (pp. 15 & 16, Fig. 2-9 & 2-11)
165. B (p. 59, Fig. 4-18)
166. B (pp. 112-113)
167. C (p. 188)
168. C (p. 188)
169. B (p. 243)
170. C (p. 249, Fig. 20-8 (D))
171. C (p. 254)
172. A (p. 254)
173. D (p. 224)
174. D (p. 179)
175. B (p. 204)
176. D (p. 238)
177. B (p. 247)
178. C (pp. 147-148, Fig. 11-1)
179. C (p. 115, Fig. 8-14)
180. B (p. 128)
181. C (p. 133)
182. B (p. 142)
183. A (pp. 142-143, Fig. 10-1)
184. A (p. 49)
185. D (p. 113)
186. C (pp. 259-260, Fig. 21-12)
187. C (p. 260, Fig. 21-12)
188. C (p. 260, Fig. 21-12)
189. B (p. 248)
190. D (p. 228)
191. A (p. 230)
192. B (p. 238)
193. B (p. 68, Fig. 5-16)
194. A (p. 257)
195. B (p. 18, Fig 2-16)
196. B (p. 29, Fig. 2-35)
197. A (p. 53)
198. B (p. 50, Fig. 3-40)
199. B (p. 255)
200. B (p. 244, Fig. 20-1)

Appendix: Definitions and Abbreviations

The terms included here are found in most plumbing codes. Some words in the code have become so descriptive and specialized that their meaning is different from what a dictionary might give.

Two words that appear repeatedly in every code are *shall* and *may*. To be able to comply with the code, you must clearly understand the specialized meanings of these words.

- *Shall* is mandatory. It requires compliance without deviation. For example, part of the requirements for waste disposal is that "sewage and liquid waste shall be treated and disposed of as hereinafter provided."

- *May* is a permissive term. When used in the code, it means allowable or optional, but not required. For example, "Drinking fountains may be installed with indirect waste only for the purpose of resealing required traps of floor drains."

"Building drain" and "building sewer" are two terms often used improperly. Many professionals assume that both terms apply to the same part of the drainage system. However, note the code definition for each term:

- A *building drain* is the main, horizontal collection system within the walls of a building that extends to 2 feet beyond the building line. (This distance may vary in some codes.)

- A *building sewer* is defined as that part of the horizontal drainage system outside the building line that connects to the building drain and conveys the liquid waste to a legal point of disposal.

Effective and constructive code interpretation is possible only when the words and terms used in the code are understood. Many definitions can be illustrated through isometric drawings. It's possible to have some variation and still remain within the intent of the code definitions. The alert professional will discover that most isometric drawings fit into these definitions. Isometric illustrations provide a better understanding of code definitions.

Definitions

Absorption The process of being absorbed in a drainfield absorption area.

Air gap (in a water supply system) The unobstructed vertical distance through the free atmosphere between the lowest opening from any pipe or faucet supplying water to a tank, plumbing fixture, or other device, and the flood level rim of the receptacle.

Air lock A stoppage or slowing of the passage of water or liquid waste through a piping system caused by air that has become trapped in the pipe or pipe connection. For example, air trapped in the flexible hose connection from a mobile home or trailer to the park sewer connection, caused by a sag in the hose, can slow or stop the flow of waste or sewage.

Anaerobic Living without free oxygen. For example, anaerobic bacteria function in septic tanks to digest organic matter.

Approved Having the approval of the plumbing official or other authority given jurisdiction by the code.

Area drain A receptacle designed to collect surface or rain water from an open area.

Back siphonage The flow of water or other liquids, mixtures or substances into the distributing pipe of a potable supply of water, or into any other fixture, device, or appliance from any source other than its intended course, due to a negative pressure in such pipe.

Backfill That portion of the trench excavation up to the original earth line that is required to be refilled after sewer or other piping has been laid.

Backflow The flow of water or other liquids, mixtures, or substances into the distributing pipe of a potable supply of water, or into any other fixture or appliance from any source other than its intended course.

Backflow connection Any arrangement under which backflow can occur. Also known as *backflow condition*.

Backflow preventer A device or means to prevent the flow of water or other liquids, mixtures, or substances into the distributing pipe of a potable supply of water, or into any other fixture or appliance from any source other than its intended course.

Backwash piping Piping, including the piping connected to the backwash outlet of the filter, that conveys waste water to an approved point of disposal.

Base The lowest point of any vertical pipe.

Battery of fixtures Any group of two or more similar adjacent plumbing fixtures that discharge into a common horizontal waste or soil branch.

Boiler blow-off An outlet on a boiler that permits the emptying or discharging of water or sediment from the boiler.

Branch Any part of the piping system other than a main, a riser or a stack.

Branch drainage pipe That portion of the drainage system extending from a trailer park sanitary drainage system to the trailer site, including the terminal end that connects to the trailer drain hose.

Branch interval A length of soil or waste stack (vertical pipe), generally one story in height (approximately 9 feet, but not less than 8 feet), which connects the horizontal branches from one floor or story of a building to the stack.

Branch vent A vent that connects one or more individual vents with a vent stack or a stack vent.

Branch water pipe That portion of the water distribution system extending from a trailer park service main to a trailer site, including the terminal end which connects to the trailer water supply pipe.

Building drain The main horizontal sanitary collection system, inside the wall line of the building, that conveys sewage to the building sewer which begins 2 feet (more in some codes) outside the building wall. The building drain excludes the waste and vent stacks that receive the discharge from soil, waste and other drainage pipes, including storm water.

Building sewer That part of the horizontal piping of a drainage system that connects to the end of the building drain and conveys the contents to a public sewer, private sewer, or individual sewage disposal system.

Building storm drain A drain used to receive and convey rain water, surface water, ground water, subsurface water and other clear water waste, and discharge these waste products into a building storm sewer or a combined building sewer beginning 2 feet outside the building wall.

Building storm sewer The pipe that connects to the end of the building storm drain to receive and convey its contents to a public storm sewer, combined sewer, or other approved disposal point.

Building subdrain Any portion of a drainage system which cannot drain by gravity into the building sewer.

Caulking Any approved method for rendering a joint water and gas tight. For cast iron pipe and fittings with hub joints, the term refers to caulking the joint with lead and oakum.

Circuit vent A vent that serves two or more fixture traps. It connects to a horizontal branch drain from the downstream side of the highest (last) fixture connection and then to the vent stack.

Code Regulations and their subsequent amendments, or any emergency rule or regulation, lawfully adopted by the administrative authority having jurisdiction to control plumbing work.

Combined building sewer A building sewer which receives storm water, sewage and other liquid waste.

Common vent The vertical vent portion which serves to vent two fixture drains that are installed at the same level in a vertical stack.

Conductor See *Leader*.

Continuous waste A drain connecting two or more fixtures or a single fixture with more than one compartment to a common trap.

Cross connection Any physical connection or arrangement between two separate piping systems, one containing potable water and the other containing water of unknown or questionable safety. The code prohibits cross connections.

Dead end A branch leading from a soil, waste or vent pipe, building drain or building sewer that is terminated by a plug or other closed fitting at a developed distance of 2 feet or more. A dead end is also classified as an extension for future connection, or as an extension of a cleanout for accessibility.

Dependent travel trailer Any motorized vehicle used as a temporary dwelling unit for travel, vacation and recreation. Usually has limited built-in sanitary facilities but not a plumbing system suitable for connection to a park sewage and water supply system.

Dependent travel trailer sanitary service station A trailer park location equipped for emptying intermediate waste holding tanks.

Developed length The length of a pipe measured along the center line of the pipe and fittings.

Diameter The nominal diameter (distance straight through the middle) of a pipe or fitting as designated commercially, unless specifically stated otherwise.

Downspout See *Leader*.

Drain Any pipe which carries liquid, waste water or other water-borne wastes in a building drainage system to an approved point of disposal.

Drain hose connection An approved flexible hose that is easily detachable and is used to connect a trailer drain to a park's sewer inlet connection.

Drainage system All of the piping within public or private premises that conveys sewage, rain water, or other liquid wastes to a legal point of disposal.

Drainage well Any drilled, driven or natural cavity which taps the underground water table and into which surface waters or treated waste water, industrial waste or sewage is disposed. Always requires approval from local authority.

Dry vent That portion of a vent system that receives no sewage or waste discharge.

Durham system An all-threaded pipe system of rigid construction, using recessed drainage fittings to correspond to the types of piping being used in drain, waste and vent systems.

Effective opening The minimum cross-sectional area of the diameter of a pipe at the point of water supply discharge.

Effluent The liquid waste discharged from a septic tank into the drainfield.

Filter piping All the piping, fittings, and valves necessary to connect the filter system of a swimming pool together as a unit.

Fire lines The complete wet standpipe system of a building, including the water service, standpipe, roof manifold, Siamese connections and pumps.

Fixture branch The drain from the trap of a fixture to the junction of that drain with a vent. Some codes refer to a fixture branch as a *fixture drain*.

Fixture drain The drain from the fixture branch to the junction of any other drain pipe. Some codes refer to a fixture drain as a *fixture branch*.

Fixture unit A design factor used to determine the load-producing effects of different types of plumbing fixtures on the plumbing system. For instance, most codes accept that 1 fixture unit equals up to 7.5 gallons per minute of flow.

Flood level rim The top edge of a plumbing fixture or other receptacle from which water or other liquids can overflow, such as in the rim of a bathtub.

Floor drain An opening or receptacle located at approximately floor level which is connected to a trap, designed to receive the discharge from indirect waste and floor drainage.

Floor sink An opening or receptacle, usually made of enameled cast iron, located at approximately floor level which is connected to a trap, designed to receive the discharge from indirect waste and floor drainage. A floor sink is more sanitary and easier to clean than a regular floor drain, and is usually used for restaurant and hospital installations.

Flushometer valve A device actuated by direct water pressure which discharges a predetermined quantity of water to fixtures for flushing purposes.

Grade The slope or pitch of drainage piping, known as the *fall*, usually expressed as a fraction of an inch per foot.

Horizontal branch A drain pipe extending laterally from a soil or waste stack or building drain. It may or may not have vertical sections or branches.

Horizontal pipe Any pipe or fitting that makes an angle of more than 45 degrees with the vertical.

Independent trailer coach Any trailer coach designed for permanent occupancy, equipped with kitchen and bathroom facilities and a plumbing system suitable for connection to a sewage, water and gas supply system in a trailer park.

Indirect waste A waste pipe that conveys liquid wastes (other than body wastes) by discharging them into an open plumbing fixture or receptacle such as a floor drain or floor sink. The overflow point of such a fixture or receptacle is at a lower elevation than the item drained.

Inground pool Any pool where the sides rest partially on, or have full contact with, the surrounding earth.

Inlet coupling The terminal end of a trailer park's water system at each trailer site. The water service connection from the trailer coach is made by a swivel fitting or threaded pipe end.

Insanitary Contrary to sanitary principles; unclean enough to endanger health.

Interceptor A device designed and installed to separate and retain deleterious, hazardous, or undesirable matter from normal wastes, and permit normal sewage or liquid wastes to discharge by gravity into the disposal terminal or sewer.

Intermediate waste holding tank An enclosed tank mounted on a travel trailer for temporary retention of waterborne waste.

Leader The vertical water conductor from the roof to the building storm drain, combined building sewer, or other approved means of disposal. Also called a *downspout*.

Liquid waste The liquid discharge from any fixture, appliance or appurtenance that connects to a plumbing system that does not receive body waste.

Load factor The percentage of the total connected fixture unit flow rate that is likely to occur at any point with the probability factor of simultaneous use. It varies with the type of occupancy, which, in turn, determines the total flow unit.

Loop or circuit waste and vent A combination of plumbing fixtures on the same floor level, in the same or in adjacent rooms, that are connected to a common horizontal branch soil or waste pipe.

Main The principal artery of any system of continuous piping, to which branches may be connected.

Main vent The principal artery of the venting system, to which vent branches may be connected.

May A permissive term used in the code which means allowable or optional, but not required.

Mezzanine An intermediate floor placed on any story or in a room. When the total area of a mezzanine floor exceeds $33\frac{1}{3}$ percent of the total floor area in that room or on that story, it is then considered an additional story rather than a mezzanine.

Mobile home, left side The side farthest from the curb when a mobile home is being towed or is in transit.

Mobile home lot A space in a mobile home park designed for the accommodation of one mobile home.

Mobile home or travel trailer park A parcel of land designated and improved to accommodate one or more trailers. Such trailers may be used for temporary or permanent living quarters.

Mobile park sanitary drainage system The entire drainage piping system in the mobile park used to convey waterborne waste to a legal point of disposal.

Mobile park water service main That portion of the park's water-distributing system that extends from park's water supply source to each branch service line.

Non-permanent pool Any pool constructed to be disassembled and re-assembled to its original integrity.

Non-swimming area Any portion of a pool which is too shallow or which has underwater ledges or walls that prevent normal swimming activity.

Onground pool Any pool where the bottom and sides rest fully above the surrounding earth.

Park sanitary drainage system The entire drainage piping system used to convey sewage or other liquid waste from all the trailer drain connections to a public sewer or private sewage disposal system.

Park water main The portion of the water supply piping that extends from the public water supply or other source of supply to the branch service lines.

Permanent pool Any pool constructed in the ground, on the ground, or in a building, that cannot be disassembled for storage.

Pitch Grade or slope of the piping. See also *Grade.*

Plumbing (includes any or all of the following) (1) The materials, including pipe, fittings, valves, fixtures and appliances, which are attached to and are a part of a system for the purpose of creating and maintaining sanitary conditions in a building, camp or swimming pool on private property where people live, work, play, assemble or travel. (2) That part of a water supply and sewage and drainage system extending from either the public water supply mains or private water supply to the public sanitary, storm or combined sanitary and storm sewers, or to a private sewage disposal plant, septic tank, disposal field, pit, box filter bed or any other receptacle, or into any natural or artificial body of water or water course upon public or private property. (3) The design, installation or contracting for installation, removal and replacement, repair or remodeling, of all or any part of the materials, appurtenances or devices attached to and forming a part of a plumbing system, including the installation of any fixture, appurtenance or devices used for cooking, washing, drinking, cleaning, fire fighting, mechanical or manufacturing purposes.

Plumbing fixtures Receptacles, devices, or appliances that are supplied with water or that receive or discharge liquids or liquid borne waste, with or without discharge, into the drainage system to which they may be directly or indirectly connected.

Plumbing official inspector The chief administrative officer charged with the administration, enforcement and application of the plumbing code and all amendments thereto.

Plumbing system The drainage system, water supply, water supply distribution pipes, plumbing fixtures, traps, soil pipes, waste pipes, vent pipes, building drains, building sewers, building storm drain, building storm sewer, liquid waste piping, water treating equipment, water using equipment, sewerage treatment, sewerage treatment equipment, fire standpipes, fire sprinklers, and relative appliances and appurtenances, including their respective connections and devices, within the private property lines of a premise.

Potable water Water which is satisfactory for drinking, culinary and domestic purposes and that meets the requirements of the health authority having jurisdiction. Potable water is considered *purified*, having been treated by one or several processes as required by its original untreated condition.

Private property For the purposes of the code, private property includes all property except streets or roads dedicated to the public, and public easements (excluding easements between private parties).

Private or private use fixtures The plumbing fixtures in residences and apartments, and in private bathrooms of hotels and similar installations, where the fixtures are intended for the use of a family or an individual.

Private sewer A sewer privately owned and not directly controlled by a public authority.

Private swimming pool A swimming pool located at a single-family residence and available for use only by the family of the household and their guests.

Public or public use fixtures Plumbing fixtures in commercial and industrial establishments, restaurants, bars, public buildings, comfort stations, schools, gymnasiums, railroad stations or places to which the public is invited or which are frequented by the public without special permission or special invitation, and other installations (whether paid or free) where a number of fixtures are installed so that their unrestricted use is available to the public.

Public sewer A common sewer directly controlled by public authority.

Public swimming pool A pool, together with its buildings and appurtenances, where the public is allowed to bathe or which is open to the public for bathing purposes by consent of the owner. It may be operated by an owner, lessee, operator, licensee, or concessionaire whether a fee is charged or not. A public pool may be one of four types:

1) *Competition pool* — a pool used for competitive swimming events.

2) *Public pool* — a pool intended for public recreational use.

3) *Semi-public pool* — a pool operated in conjunction with buildings such as hotels, motels and apartments.

4) *Special purpose pool* — a pool operated for water therapy treatments rather than for recreational purposes.

Recirculating piping (also called *return piping* or *pool inlet piping*) The piping connected to the discharge side of the pool pump. It returns water to the pool after filtering.

Refuse All solid wastes, including garbage, rubbish and ashes, but excluding body waste.

Relief vent A vent, the primary function of which is to provide air circulation between drainage and vent systems.

Rim An unobstructed open edge at the overflow point of a fixture.

Rock drainfield Washed rock from ³/₄ to 2¹/₂ inches used in septic tank drainfield absorption areas.

Roof drain An outlet installed to receive water that collects on the surface of a roof and that discharges that water into a leader or downspout.

Roughing-in The installation of all the parts of a plumbing system that can be completed prior to the installation of the plumbing fixtures; includes drainage, water supply, and vent piping, and the necessary fixture supports.

Sanitary sewer A pipe which carries sewage, excluding storm, surface and ground water.

Second hand A term applied to material or plumbing equipment which has been installed and used, or removed after use.

Septic tank A watertight receptacle which receives the discharge of a drainage system or part thereof, so designed and constructed as to separate solids from liquid, digest organic matter through a period of detention, and allow the liquids to discharge into the soil outside the tank through a subsurface system of open-joint or perforated piping, or other approved methods.

Service building A building in a trailer park with laundry facilities as well as toilet and bathing facilities for men and women.

Service connection The portion of the water distribution system that extends from the mobile home park branch service line to the inlet fitting at the trailer.

Sewage Any liquid waste containing animal, mineral or vegetable matter in suspension or solution. May also include liquids containing chemicals in solution.

Shall A mandatory term. When used in the code, it requires compliance without deviation.

Side vent A vent which connects to a horizontal drain pipe through a fitting at an angle no greater than 45 degrees to the vertical.

Slope See *Grade*.

Soil pipe Any pipe which conveys the discharge of water closets or fixtures having similar functions, with or without the discharge from other fixtures, to the building drain or building sewer.

Stack The vertical pipe of a soil, water or vent piping system.

Stack vent The extension of a soil or waste stack above the highest horizontal drain connected to the stack. Also called a *waste vent* or *soil vent*.

Stack venting A method of venting fixtures through the soil or waste stack.

Standpipe system A system of piping installed for fire protection purposes which has a primary water supply constantly or automatically available at each hose outlet.

Storm sewer A sewer used for conveying rainwater and/or surface water.

Subsurface drain A drain that receives only subsurface or seepage water and conveys it to a place of disposal.

Sump A tank or pit, located below the normal grade of the gravity system, which receives sewage or liquid waste and which must be emptied by mechanical means, such as a pump.

Supply well Any artificial opening in the ground designed to conduct water from a source bed through the surface when water from such well is used for public, semi-public or private use.

Supports Devices for supporting and securing pipe and fixtures to walls, ceilings, floors or structural members. Also known as *hangers* or *anchors*.

Swimming pool Any structure suitable for swimming or recreational bathing that's over 24 inches deep.

Trailer A mobile home, truck coach, travel trailer or recreation vehicle that can be used as a dwelling.

Trailer coach Any vehicle which can be licensed for use on public streets but is designed for permanent occupancy as a dwelling for one or more persons.

Trailer coach space A site or lot within a trailer park designated for use by one trailer coach.

Trap A fitting or device designed and constructed to provide a liquid seal that prevents the back passage of air and sewer gases without materially affecting the flow of sewage or water.

Trap seal The maximum vertical depth of liquid that a trap will retain, measured between the crown weir and the top of the dip of the trap.

Vacuum fitting A device which connects the pool sweeper hose to the side of the pool for cleaning purposes. Every pool must have a vacuum fitting, located a maximum of 10 inches below the water surface inside the pool, in an accessible location.

Vent stack A vertical vent pipe installed primarily for the purpose of providing circulation of air to and from any part of the drainage system.

Vent system A pipe or pipes installed to provide a flow of air to or from a drainage system, or to provide a circulation of air within such system which prevents back pressure and siphonage from breaking water trap seals serving the fixtures on the system.

Vertical pipe Any pipe or fitting installed in a vertical position or that makes an angle of not more than 45 degrees with the vertical.

Waste pipe Any pipe which receives the discharge of any fixture, except water closets or fixtures having similar functions, and conveys it to the building drain or to the soil or waste stack.

Water-distributing pipe A pipe within a building which conveys water from the water service pipe to the plumbing fixtures, appliances and other water outlets.

Water main A water supply pipe for public or community use. The public water distribution system is located in the street, alley or a dedicated easement adjacent to individual parcels of land.

Water service pipe The pipe from the water main or other source of water supply to the building served.

Water supply system A piping system consisting of the water service pipe, the water-distributing pipes, the standpipe system and the necessary connecting pipes, fittings, control valves and all appurtenances related to the water supply in or on private property.

Wet vent A waste pipe designed to vent and convey waste from fixtures other than water closets.

Yoke vent A pipe connecting upward from a soil or waste stack for the purpose of preventing pressure changes in the stacks.

Abbreviations

The abbreviations here are often found on blueprints (building plans) and in plumbing reference books (including the code) to identify plumbing fixtures, pipes, valves and nationally-recognized associations.

A	area
A.D.	area drain
AGA	American Gas Association
AISI	American Iron and Steel Institute
ASA	American Standard Association
ASCE	American Society of Civil Engineering
ASHRAE	American Society of Heating, Refrigeration and Air Conditioning Engineers
ASME	American Society of Mechanical Engineers
ASSE	American Society of Sanitary Engineering
ASTM	American Society for Testing and Materials

AWWA	American Water Works Association		I.W.	indirect waste
B.S.	bar sink		I.P.S.	iron pipe size
B.	bidet		K.S.	kitchen sink
B.T.	bathtub		L. or LAV.	lavatory
Btu	British thermal unit		L.T.	laundry tray
C to C	center to center		L	length
CI	cast iron		lb.	pound
CISPI	Cast Iron Soil Pipe Institute		Max.	maximum
C	condensate line		Mfr.	manufacturer
CO	cleanout		Min.	minimum
C.W.	cold water		M.H.	manhole
cu. ft.	cubic feet		NAPHCC	National Association of Plumbing Heating and Cooling Contractors
cu. in.	cubic inches		NBFU	National Board of Fire Underwriters
C.W.M.	clothes washing machine		NBS	National Board of Standards
C.V.	check valve		NFPA	National Fire Protection Association
D.F.	drinking fountain		NPS	nominal pipe size
D.W.	dish washer		O	oxygen
E to C	end to center		O.D.	outside diameter
E.W.C.	electric water cooler		oz.	ounce
°F	degrees Fahrenheit		P.D.	planter drain
F	Fahrenheit		P.P.	pool piping
F.B.	foot bath		psi	pounds per square inch
F.F.	finish floor		Rad.	radius
F.CO	floor cleanout		R.D.	roof drain
F.D.	floor drain		red.	reducer
F.D.C.	fire department connection		R.L.	roof leader
F.E.C.	fire extinguisher cabinet		San.	sanitary
F.G.	finish grade		Sh.	shower
F.H.C.	fire hose cabinet		Spec.	specification
F.L.	fire line		sq.	square
F.P.	fire plug		S.B.	sitz bath
F.S.P.	fire standpipe		sq. ft.	square feet
F.U.	fixture unit		S.P.	swimming pool
gal.	gallons		S.S.	service sink
gpd	gallons per day		Std.	standard
gpm	gallons per minute		SV	service
Galv.	galvanized		SW	service weight
G.S.	glass sink		S & W	soil and waste
G.V.	gate valve		T	temperature
H.B.	hose bibb		U or Ur.	urinal
Hd or H.D.	head		V	volume
H.W.	hot water		VTR	vent through roof
H.W.R.	hot water return		W	waste
H.W.T.	hot water tank		W.C.	water closet
in.	inch		W.H.	water heater
I.D.	inside diameter		XH	extra heavy

Index

Practical References for Builders

CD Estimator

If your computer has *Windows*™ and a CD-ROM drive, *CD Estimator* puts at your fingertips 85,000 construction costs for new construction, remodeling, renovation & insurance repair, electrical, plumbing, HVAC and painting. You'll also have the *National Estimator* program — a stand-alone estimating program for *Windows*™ that *Remodeling* magazine called a "computer wiz." Quarterly cost updates are available at no charge on the Internet. To help you create professional-looking estimates, the disk includes over 40 construction estimating and bidding forms in a format that's perfect for nearly any word processing or spreadsheet program for *Windows*™. And to top it off, a 70-minute interactive video teaches you how to use this CD-ROM to estimate construction costs. **CD Estimator is $68.50**

How to Succeed With Your Own Construction Business

Everything you need to start your own construction business: setting up the paperwork, finding the work, advertising, using contracts, dealing with lenders, estimating, scheduling, finding and keeping good employees, keeping the books, and coping with success. If you're considering starting your own construction business, all the knowledge, tips, and blank forms you need are here. **336 pages, 8¹/₂ x 11, $24.25**

National Plumbing & HVAC Estimator

Manhours, labor and material costs for all common plumbing and HVAC work in residential, commercial, and industrial buildings. You can quickly work up a reliable estimate based on the pipe, fittings and equipment required. Every plumbing and HVAC estimator can use the cost estimates in this practical manual. Sample estimating and bidding forms and contracts also included. Explains how to handle change orders, letters of intent, and warranties. Describes the right way to process submittals, deal with suppliers and subcontract specialty work. Includes a CD-ROM with an electronic version of the book with *National Estimator*, a stand-alone *Windows*™ estimating program, plus an interactive multimedia video that shows how to use the disk to compile construction cost estimates. **352 pages, 8¹/₂ x 11, $48.25. Revised annually**

Construction Forms & Contracts

125 forms you can copy and use — or load into your computer (from the FREE disk enclosed). Then you can customize the forms to fit your company, fill them out, and print. Loads into *Word for Windows, Lotus 1-2-3, WordPerfect, Works,* or *Excel* programs. You'll find forms covering accounting, estimating, fieldwork, contracts, and general office. Each form comes with complete instructions on when to use it and how to fill it out. These forms were designed, tested and used by contractors, and will help keep your business organized, profitable and out of legal, accounting and collection troubles. Includes a CD-ROM for *Windows*™ or Mac. **400 pages, 8¹/₂ x 11, $39.75**

Home Inspection Handbook

Every area you need to check in a home inspection — especially in older homes. Twenty complete inspection checklists: building site, foundation and basement, structural, bathrooms, chimneys and flues, ceilings, interior & exterior finishes, electrical, plumbing, HVAC, insects, vermin and decay, and more. Also includes information on starting and running your own home inspection business. **324 pages, 5¹/₂ x 8¹/₂, $24.95**

Basic Plumbing with Illustrations, Revised

This completely-revised edition brings this comprehensive manual fully up-to-date with all the latest plumbing codes. It is the journeyman's and apprentice's guide to installing plumbing, piping, and fixtures in residential and light commercial buildings: how to select the right materials, lay out the job and do professional-quality plumbing work, use essential tools and materials, make repairs, maintain plumbing systems, install fixtures, and add to existing systems. Includes extensive study questions at the end of each chapter, and a section with all the correct answers. **384 pages, 8¹/₂ x 11, $33.00**

Roofing Construction & Estimating

Installation, repair and estimating for nearly every type of roof covering available today in residential and commercial structures: asphalt shingles, roll roofing, wood shingles and shakes, clay tile, slate, metal, built-up, and elastomeric. Covers sheathing and underlayment techniques, as well as secrets for installing leakproof valleys. Many estimating tips help you minimize waste, as well as insure a profit on every job. Troubleshooting techniques help you identify the true source of most leaks. Over 300 large, clear illustrations help you find the answer to just about all your roofing questions. **432 pages, 8¹/₂ x 11, $35.00**

National Renovation & Insurance Repair Estimator

Current prices in dollars and cents for hard-to-find items needed on most insurance, repair, remodeling, and renovation jobs. All price items include labor, material, and equipment breakouts, plus special charts that tell you exactly how these costs are calculated. Includes a CD-ROM with an electronic version of the book with *National Estimator*, a stand-alone *Windows*™ estimating program, plus an interactive multimedia video that shows how to use the disk to compile construction cost estimates. **560 pages, 8¹/₂ x 11, $49.50. Revised annually**

Construction Estimating Reference Data

Provides the 300 most useful manhour tables for practically every item of construction. Labor requirements are listed for sitework, concrete work, masonry, steel, carpentry, thermal and moisture protection, door and windows, finishes, mechanical and electrical. Each section details the work being estimated and gives appropriate crew size and equipment needed. Includes an electronic version of the book on computer disk with a stand-alone *Windows*™ estimating program FREE on a CD-ROM **432 pages, 11 x 8¹/₂, $39.50**

Audiotapes: Plumber's Exam Prep Guide

These tapes are made to order for the busy plumber looking for a better-paying career as a licensed apprentice, journeyman or master plumber. Howard Massey, who developed the tapes, has written many of the questions used on plumber's exams, and has monitored and graded the exam. He knows what you need to pass. This two-audiotape set asks you over 100 often-used exam questions in an easy-to-remember format. This is the easiest way to study for the exam. **Two 60-minute audiotapes, $19.95**

National Construction Estimator

Current building costs for residential, commercial, and industrial construction. Estimated prices for every common building material. Provides manhours, recommended crew, and gives the labor cost for installation. Includes a CD-ROM with an electronic version of the book with *National Estimator*, a stand-alone *Windows*™ estimating program, plus an interactive multimedia video that shows how to use the disk to compile construction cost estimates. **560 pages, 8¹/₂ x 11, $47.50. Revised annually**

Contractor's Guide to the Building Code Revised

This new edition was written in collaboration with the International Conference of Building Officials, writers of the code. It explains in plain English exactly what the latest edition of the *Uniform Building Code* requires. Based on the 1997 code, it explains the changes and what they mean for the builder. Also covers the *Uniform Mechanical Code* and the *Uniform Plumbing Code*. Shows how to design and construct residential and light commercial buildings that'll pass inspection the first time. Suggests how to work with an inspector to minimize construction costs, what common building shortcuts are likely to be cited, and where exceptions may be granted. **368 pages, 8¹/₂ x 11, $39.00**

Plumbing & HVAC Manhour Estimates

Hundreds of tested and proven manhours for installing just about any plumbing and HVAC component you're likely to use in residential, commercial, and industrial work. You'll find manhours for installing piping systems, specialties, fixtures and accessories, ducting systems, and HVAC equipment. If you estimate the price of plumbing, you shouldn't be without the reliable, proven manhours in this unique book. **224 pages, 5¹/₂ x 8¹/₂, $28.25**

Planning Drain, Waste & Vent Systems

How to design plumbing systems in residential, commercial, and industrial buildings. Covers designing systems that meet code requirements for homes, commercial buildings, private sewage disposal systems, and even mobile home parks. Includes relevant code sections and many illustrations to guide you through what the code requires in designing drainage, waste, and vent systems. **192 pages, 8½ x 11, $19.25**

Plumber's Exam Preparation Guide

Hundreds of questions and answers to help you pass the apprentice, journeyman, or master plumber's exam. Questions are in the style of the actual exam. Gives answers for both the Standard and Uniform plumbing codes. Includes tips on studying for the exam and the best way to prepare yourself for examination day. **320 pages, 8½ x 11, $29.00**

Rough Framing Carpentry

If you'd like to make good money working outdoors as a framer, this is the book for you. Here you'll find shortcuts to laying out studs; speed cutting blocks, trimmers and plates by eye; quickly building and blocking rake walls; installing ceiling backing, ceiling joists, and truss joists; cutting and assembling hip trusses and California fills; arches and drop ceilings —all with production line procedures that save you time and help you make more money. Over 100 on-the-job photos of how to do it right and what can go wrong. **304 pages, 8½ x 11, $26.50**

National Repair & Remodeling Estimator

The complete pricing guide for dwelling reconstruction costs. Reliable, specific data you can apply on every repair and remodeling job. Up-to-date material costs and labor figures based on thousands of jobs across the country. Provides recommended crew sizes; average production rates; exact material, equipment, and labor costs; a total unit cost and a total price including overhead and profit. Separate listings for high- and low-volume builders, so prices shown are specific for any size business. Estimating tips specific to repair and remodeling work to make your bids complete, realistic, and profitable. Includes a CD-ROM with an electronic version of the book with *National Estimator*, a stand-alone *Windows*™ estimating program, plus an interactive multi-media video that shows how to use the disk to compile construction cost estimates.
304 pages, 8½ x 11, $48.50. Revised annually

Pipe & Excavation Contracting

Shows how to read plans and compute quantities for both trench and surface excavation, figure crew and equipment productivity rates, estimate unit costs, bid the work, and get the bonds you need. Explains what equipment will deliver maximum productivity for a job, how to lay all types of water and sewer pipe, and how to switch your business to excavation work when you don't have pipe contracts. Covers asphalt and rock removal, working on steep slopes or in high groundwater, and how to avoid the pitfalls that can wipe out your profits on any job.
400 pages, 5½ x 8½, $29.00

Contractor's Index to the 1997 *Uniform Building Code*

Finally, there's a common-sense index that helps you quickly and easily find the section you're looking for in the *UBC*. It lists topics under the names builders actually use in construction. Best of all, it gives the full section number and the actual page in the *UBC* where you'll find it. If you need to know the requirements for windows in exit access corridor walls, just look under Windows. You'll find the requirements you need are in Section 1004.3.4.3.2.2 in the *UBC* — on page 115. This practical index was written by a former builder and building inspector who knows the *UBC* from both perspectives. If you hate to spend valuable time hunting through pages of fine print for the information you need, this is the book for you.
192 pages, 8½ x 11, $26.00. Loose-leaf edition, $29.00.

Basic Lumber Engineering for Builders

Beam and lumber requirements for many jobs aren't always clear, especially with changing building codes and lumber products. Most of the time you rely on your own "rules of thumb" when figuring spans or lumber engineering. This book can help you fill the gap between what you can find in the building code span tables and what you need to pay a certified engineer to do. With its large, clear illustrations and examples, this book shows you how to figure stresses for pre-engineered wood or wood structural members, how to calculate loads, and how to design your own girders, joists and beams. Included FREE with the book — an easy-to-use version of NorthBridge Software's *Wood Beam Sizing* program.
272 pages, 8½ x 11, $38.00

Craftsman Book Company
6058 Corte del Cedro
P.O. Box 6500
Carlsbad, CA 92018

☎ 24 hour order line
1-800-829-8123
Fax (760) 438-0398

Name _____

Company _____

Address _____

City/State/Zip _____

○ This is a residence

Total enclosed_____(In California add 7.25% tax)

We pay shipping when your check covers your order in full.

In A Hurry?
We accept phone orders charged to your
○ Visa, ○ MasterCard, ○ Discover or ○ American Express

Card#_____

Exp. date_____Initials_____

Tax Deductible: Treasury regulations make these references tax deductible when used in your work. Save the canceled check or charge card statement as your receipt.

Order online
http://www.craftsman-book.com

10-Day Money Back Guarantee

○ 19.95 Audio: Plumber's Exam
○ 38.00 Basic Lumber Engineering for Builders
○ 33.00 Basic Plumbing with Illustrations
○ 68.50 CD Estimator
○ 39.50 Construction Estimating Reference Data with FREE stand-alone *Windows*™ estimating program on a CD-ROM.
○ 39.75 Construction Forms & Contracts with a CD-ROM.
○ 39.00 Contractor's Guide to Building Code Revised
○ 26.00 Contractor's Index to the 1997 *UBC* — *Paperback*
○ 29.00 Contractor's Index to the 1997 *UBC* — *Looseleaf*
○ 24.95 Home Inspection Handbook
○ 24.25 How to Succeed w/Your Own Construction Business
○ 47.50 National Construction Estimator with FREE *National Estimator* on a CD-ROM.
○ 48.25 National Plumbing & HVAC Estimator with FREE *National Estimator* on a CD-ROM.
○ 49.50 National Renovation & Insurance Repair Estimator with FREE *National Estimator* on a CD-ROM.
○ 48.50 National Repair & Remodeling Estimator with FREE *National Estimator* on a CD-ROM.
○ 29.00 Pipe & Excavation Contracting
○ 19.25 Planning Drain, Waste & Vent Systems
○ 29.00 Plumber's Exam Preparation Guide
○ 28.25 Plumbing & HVAC Manhour Estimates
○ 35.00 Roofing Construction & Estimating
○ 26.50 Rough Framing Carpentry
○ 32.00 Plumber's Handbook Revised
○ FREE Full Color Catalog

Prices subject to change without notice

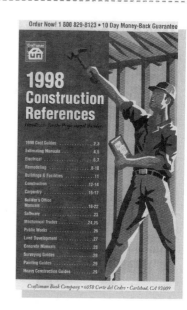